Stem Cell Biology and Regene

Volume 72

Series Editor

Kursad Turksen, Ottawa Hospital Research Institute, Ottawa, ON, Canada

Our understanding of stem cells has grown rapidly over the last decade. While the apparently tremendous therapeutic potential of stem cells has not yet been realized, their routine use in regeneration and restoration of tissue and organ function is greatly anticipated. To this end, many investigators continue to push the boundaries in areas such as the reprogramming, the stem cell niche, nanotechnology, biomimetics and 3D bioprinting, to name just a few. The objective of the volumes in the Stem Cell Biology and Regenerative Medicine series is to capture and consolidate these developments in a timely way. Each volume is thought-provoking in identifying problems, offering solutions, and providing ideas to excite further innovation in the stem cell and regenerative medicine fields.

Series Editor
Kursad Turksen, Ottawa Hospital Research Institute, Canada

Editorial Board
Pura Muñoz Canoves, Pompeu Fabra University, Spain
Lutolf Matthias, Swiss Federal Institute of Technology, Switzerland
Amy L Ryan, University of Southern California, USA
Zhenguo Wu, Hong Kong University of Science & Technology, Hong Kong
Ophir Klein, University of California SF, USA
Mark Kotter, University of Cambridge, UK
Anthony Atala, Wake Forest Institute for Regenerative Medicine, USA
Tamer Önder, Koç University, Turkey
Jacob H Hanna, Weizmann Institute of Science, Israel
Elvira Mass, University of Bonn, Germany

More information about this series at https://link.springer.com/bookseries/7896

Francisco Jimenez · Claire Higgins
Editors

Hair Follicle Regeneration

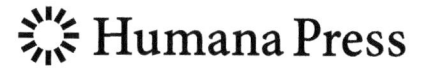

Humana Press

Editors
Francisco Jimenez
Mediteknia Dermatology and Hair
Transplant Clinic
Universidad Fernando Pessoa Canarias
Las Palmas de Gran Canaria, Spain

Claire Higgins
Department of Bioengineering
Imperial College London
London, UK

ISSN 2196-8985 ISSN 2196-8993 (electronic)
Stem Cell Biology and Regenerative Medicine
ISBN 978-3-030-98333-8 ISBN 978-3-030-98331-4 (eBook)
https://doi.org/10.1007/978-3-030-98331-4

© The Editor(s) (if applicable) and The Author(s), under exclusive license to Springer Nature Switzerland AG 2022

This work is subject to copyright. All rights are solely and exclusively licensed by the Publisher, whether the whole or part of the material is concerned, specifically the rights of translation, reprinting, reuse of illustrations, recitation, broadcasting, reproduction on microfilms or in any other physical way, and transmission or information storage and retrieval, electronic adaptation, computer software, or by similar or dissimilar methodology now known or hereafter developed.

The use of general descriptive names, registered names, trademarks, service marks, etc. in this publication does not imply, even in the absence of a specific statement, that such names are exempt from the relevant protective laws and regulations and therefore free for general use.

The publisher, the authors, and the editors are safe to assume that the advice and information in this book are believed to be true and accurate at the date of publication. Neither the publisher nor the authors or the editors give a warranty, expressed or implied, with respect to the material contained herein or for any errors or omissions that may have been made. The publisher remains neutral with regard to jurisdictional claims in published maps and institutional affiliations.

This Humana imprint is published by the registered company Springer Nature Switzerland AG
The registered company address is: Gewerbestrasse 11, 6330 Cham, Switzerland

Francisco Jimenez.- To Dow Stough, Ralf Paus, and Enrique Poblet, the three most influential physicians in my career as hair clinician and surgeon.

Claire Higgins.- To Colin Jahoda, who inspired me to pursue a PhD and career in this fascinating field

Preface

How to create a hair follicle de novo capable of producing thick hairs in a reproducible and efficient manner, or how to transform the existing miniaturized hair follicles present in a balding scalp into terminal follicles capable of cycling and producing thick hair shafts, is one of the unmet challenges in hair biology that, if resolved, would have a tremendous impact in clinical practice. In this book, we conduct an in-depth review of the topic of hair follicle regeneration that ranges from consideration of its most fundamental biological bases to the current approaches undertaken to translate the information and knowledge acquired through research into clinical practice.

This book is divided into four parts. In the first part, *Historical Context and Overview for Hair Follicle Regeneration*, Dr. Higgins and Prof. Jahoda discuss the very relevant early experiments that led to the notion that the hair dermis can direct hair development and instruct hair growth in mature skin (chapter "The Historical Studies Underpinning the Concept of Hair Follicle Neogenesis"). In the chapter "Hair Regeneration and Rejuvenation: Pipeline of Medical and Technical Strategies", Drs. Limbu and Kemp review the medical and technical strategies that have been used over the years to treat androgenetic alopecia, with special emphasis on the research conducted in the early 2000s by the companies Intercytex and Aderans.

The second part of the book addresses the *Cells and Structures Involved in Hair Follicle Regeneration*. In the first chapter of this section, Drs. Tsai and Garza provide the reader with a general overview of the cells and structures involved in hair cycling and regeneration, including hair follicle stem cells, dermal papillae and the dermal sheath, adipocytes, immune cells, lymphatics, blood vessels, and nerves. The concept of exosomes is also introduced, along with their role in intercellular communications. Drs. Pantelireis, Goh, and Clavel then describe in the chapter "The Dermal Papilla and Hair Follicle Regeneration: Engineering Strategies to Improve Dermal Papilla Inductivity" the unique characteristics of the human DP, its hair inductive capacity, how to maintain DP inductivity using specific extracellular factors and 3D culture, and which bioengineered methods could be used to harness the inductive potential of DP to provide a cell-based therapy for hair loss. The chapter "Dermal Sheath Cells and Hair Follicle Regeneration" by Drs. Yoshida, Tsuboi, and Kishimoto points to the DS as a key element for hair follicle regeneration. The authors discuss the potential use of DS cells injected into the scalp of androgenetic alopecia patients. We felt that this section would be incomplete if we were not to include two other structures which are currently the object of intense research in hair biology: the adipose cells and the lymphatic vessels. The chapter "Epithelial-Mesenchymal Interactions Between Hair Follicles and Dermal Adipose Tissue" by Dr. Ramos and Prof. Plikus describes the physiology of the dermal white adipose tissue (dWAT), highlighting the paracrine communication between hair follicles (HFs) and dWAT as well as the dWAT implications in mammal skin physiology and regenerative mechanisms. The chapter "Lymphatic Vasculature and Hair Follicle Regeneration" by Drs. Cazzola and Perez-Moreno introduces the reader to the LV organization and function, the association of LV to the hair follicle stem cell (HFSC) niche, the HFSC-LV crosstalk in the modulation of HF growth, and their potential implications in HF regeneration.

The third part of this book is titled *Therapeutic Strategies for Hair Follicle Augmentation.* This section focuses entirely on the translational application of cell-based therapies and bioengineering technologies for hair regeneration. As an introduction, Drs. Bertolini, Piccini, and McElwee describe "In Vitro and Ex Vivo Hair Follicle Models to Explore Therapeutic Options for Hair Regeneration" including 2D and 3D cultures, 3D HF organoid models, induced pluripotent stem cell (iPSC) models, cultured skin equivalents, ex vivo organ HF culture, and skin explant organ culture. The chapter "Extracellular Vesicles Including Exosomes for Hair Follicle Regeneration" by Drs. Edith Aberdam, Le Riche, Bordes, Closs, Park, and Daniel Aberdam reviews the current state of knowledge on the use of EVs for improving hair follicle growth. Detailed information about the different cellular sources of the exosomes is given, and the clinical data and potential use of exosomes in androgenetic alopecia are discussed. The chapter "Stem Cell-Based Therapies for Hair Loss: What is the Evidence from a Clinical Perspective?" by Drs. Park and Choi discusses the potential clinical application and outcomes of the studies performed with a variety of cell-based therapies including platelet rich plasma, stromal vascular fraction, stem cells, and stem cell derived conditioned medium for the treatment of hair loss disorders, mainly androgenetic alopecia. The chapter "Induced Pluripotent Stem Cell Approach to Hair Follicle Regeneration" by Drs. Pinto and Terskikh describes the generation of hair follicles from human pluripotent cells. They explain how human DP cells can be generated from iPSCs via a mesenchymal route or a neural crest intermediate. They also describe the steps for the spontaneous formation of HFs within 3D skin organoids and reflect on the clinical translation of this form of therapy. This section ends with a chapter on "Biofabrication Technologies in Hair Neoformation" written by Drs. Abreu, Gasperini, and Marques. This chapter reviews the cellular and biomaterial elements commonly considered when bioengineering the HF and describes biofabrication strategies to promote HF neoformation, from simpler cell-based approaches to those assisted by more advanced biotechnologies.

The fourth and last section of the book is titled *Hair Follicle Regeneration and Injury.* In the chapter "Wound Healing Induced Hair Follicle Regeneration", Drs. Jiang and Myung describe the cells and signals involved in this rare regenerative event observed in mammals in which de novo HFs are formed at the wound site.

They also discuss how uncovering the mechanisms behind wound-induced follicular neogenesis can inform clinical studies for human applications. Finally, the chapter "Hair Follicles in Wound Healing and Skin Remodelling" by Drs. Plotczyk and Jimenez explores the role that human HFs play in wound healing and the clinical and experimental evidence that support this connection. They then summarize the latest translational clinical work using follicular hair transplantation as a therapeutic tool to stimulate the healing response.

As a final reflection, we would like to say that given that anecdotal improvements of hair loss can occur with almost any form of therapy, including placebos, we have opted to keep to a minimum discussion about alternative therapies that we consider are not yet supported by sufficient scientific evidence. Our intention in this respect is to avoid conveying incorrect judgments and/or jumping to erroneous conclusions which could potentially lead to confusion and move us away from true scientific knowledge.

We thank all the authors who have contributed to this book and hope you enjoy reading it.

Las Palmas de Gran Canaria, Spain Francisco Jimenez, MD
London, UK Claire Higgins, Ph.D.

Contents

Part I Historical Context and Overview for Hair Follicle Regeneration

1 **The Historical Studies Underpinning the Concept of Hair Follicle Neogenesis** 3
Claire A. Higgins and Colin A. B. Jahoda

2 **Hair Regeneration and Rejuvenation: Pipeline of Medical and Technical Strategies** 25
Summik Limbu and Paul Kemp

Part II Cells and Structures Involved in Hair Follicle Regeneration

3 **Cells and Structures Involved in Hair Follicle Regeneration: An Introduction** 39
Jerry Tsai and Luis A. Garza

4 **The Dermal Papilla and Hair Follicle Regeneration: Engineering Strategies to Improve Dermal Papilla Inductivity** 59
Nikolaos Pantelireis, Gracia Goh, and Carlos Clavel

5 **Dermal Sheath Cells and Hair Follicle Regeneration** 91
Yuzo Yoshida, Ryoji Tsuboi, and Jiro Kishimoto

6 **Epithelial-Mesenchymal Interactions Between Hair Follicles and Dermal Adipose Tissue** 107
Raul Ramos and Maksim V. Plikus

7 **Lymphatic Vasculature and Hair Follicle Regeneration** 135
Anna Cazzola and Mirna Perez-Moreno

Part III Therapeutic Strategies for Hair Follicle Augmentation

8   In Vitro and Ex Vivo Hair Follicle Models to Explore
    Therapeutic Options for Hair Regeneration 155
    Marta Bertolini, Ilaria Piccini, and Kevin J. McElwee

9   Extracellular Vesicles Including Exosomes for Hair Follicle
    Regeneration 205
    Edith Aberdam, Alizée Le Riche, Sylvie Bordes, Brigitte Closs,
    Byung-Soon Park, and Daniel Aberdam

10 Stem Cell-Based Therapies for Hair Loss: What is
    the Evidence from a Clinical Perspective? 219
    Byung-Soon Park and Hye-In Choi

11 Induced Pluripotent Stem Cell Approach to Hair Follicle
    Regeneration 237
    Antonella Pinto and Alexey V. Terskikh

12 Biofabrication Technologies in Hair Neoformation 255
    Carla M. Abreu, Luca Gasperini, and Alexandra P. Marques

Part IV Hair Follicle Regeneration and Injury

13 Wound Healing Induced Hair Follicle Regeneration 277
    Yiqun Jiang and Peggy Myung

14 Hair Follicles in Wound Healing and Skin Remodelling 291
    Magdalena Plotczyk and Francisco Jimenez

Index 305

Part I
Historical Context and Overview for Hair Follicle Regeneration

Chapter 1
The Historical Studies Underpinning the Concept of Hair Follicle Neogenesis

Claire A. Higgins and Colin A. B. Jahoda

Abstract *Introduction* In an age of intense interest in cellular reprogramming the mature hair follicle mesenchyme or dermis is a fascinating paradigm. This is mainly due to its capacity to instruct and direct a switch of differentiation in surrounding and adjacent adult epithelia. *Methods* To collate the historical studies that contributed to the concept that the hair follicle dermis can instruct or induce development, we reviewed literature on Pubmed using keywords such as 'hair follicle', 'development', 'induction', 'appendage development', 'recombination experiments' and 'neogenesis'. *Results* Several historical experiments led to the notion that the hair dermis can direct hair development and instruct hair growth in mature skin. While many processes in development are universal and can be applied to human systems, there are others that are more specific, and restricted by taxonomic class, or appendage subtypes. Indeed in humans, even the cellular mechanisms that underpin androgenetic alopecia (male pattern baldness) are poorly understood. *Discussion* In this chapter, we will address some of the concepts related to hair development that remain elusive or contentious, including the cellular mechanisms controlling condensation and pattern formation. We also examine how the follicle dermis acquires the capacity to instruct growth and retain inductive capabilities. We will review the changes that occur in human hair follicles affected by androgenetic alopecia and briefly discuss translational strategies to reverse miniaturisation seen in affected hair follicles.

Keywords Hair morphogenesis · Hair development · Condensation · Induction

---

C. A. Higgins (✉)
Department of Bioengineering, Imperial College London, London, UK
e-mail: c.higgins@imperial.ac.uk

C. A. B. Jahoda (✉)
School of Biological and Biomedical Sciences, Durham University, Durham, UK
e-mail: colin.jahoda@durham.ac.uk

© The Author(s), under exclusive license to Springer Nature Switzerland AG 2022
F. Jimenez and C. Higgins (eds.), *Hair Follicle Regeneration*, Stem Cell Biology and Regenerative Medicine 72, https://doi.org/10.1007/978-3-030-98331-4_1

Key Points

- Interactions between mesenchymal and epithelial tissues during embryonic development are required to facilitate hair follicle morphogenesis.
- In adult hair follicles, the hair follicle dermis is absolutely necessary for both hair growth and hair cycling.
- Hair follicle miniaturisation, seen in androgenetic alopecia, is characterised by a decrease in the number of cells and amount of extracellular matrix within the dermal papilla.

Introducing the Foundations for Hair Growth

The hair follicle is comprised of two tissue types; connective tissue and epithelial tissue. Interaction between these two tissues fuels both hair follicle development and hair cycling and growth in the adult. While later book chapters within this volume will discuss other cell types both within the hair follicle and the surrounding macroenvironment, this 1st chapter will introduce the hair follicle dermis as a conductor of hair follicle development.

The principal connective/mesenchymal components of adult hair follicles are the dermal papilla and dermal sheath with the latter sometimes referred to as the connective tissue sheath. In mouse body skin, the follicle papilla, sheath and arrector pili muscle, as well as the papillary fibroblasts of the upper dermis, originate from Blimp1 + dermal progenitors in the embryo [1]. Both papilla and sheath cells arise from the dermal condensate, an aggregation of mesenchymal cells that becomes segregated from the surrounding dermis at the start of hair follicle development [2, 3]. Following condensate interaction with an epidermal thickening termed the placode, the epidermal structure buds downwards, initially with the dermis still surrounding the epidermis. With continued epidermal downgrowth, dermal cells at the base of the developing follicle are largely engulfed by the epidermis and, at this second key stage, become the specialised fibroblasts of the dermal papilla (Fig. 1.1). Those not incorporated into the new "bulb" structure form the dermal sheath, connected to the dermal papilla through the very base of the follicle, where the epithelial advance arrests before completing its engulfment. Interaction between the papilla cells and the adjacent epithelium induces the latter to proliferate, differentiate and grow into a hair fibre. Being an instructive component of the follicle is not the only defining trait of a papilla. The size and morphology of the papilla, how it changes shape during the hair follicle cycle, and the way it interacts with the epidermis and macroenvironment, all contribute to making a dermal papilla a functional unit.

In mature hair follicles, the dermal papilla functions to maintain hair growth, while interaction between the mesenchymal and epithelial components of the follicle also drives progression through stages of the hair cycle [4]. Follicles go through cycles of growth (anagen), in which they actively produce a hair fibre, regression (catagen), and rest (telogen). The club fibre, produced by the follicle during catagen and telogen is eventually shed from the follicle in a process termed exogen [5]. There is very strong

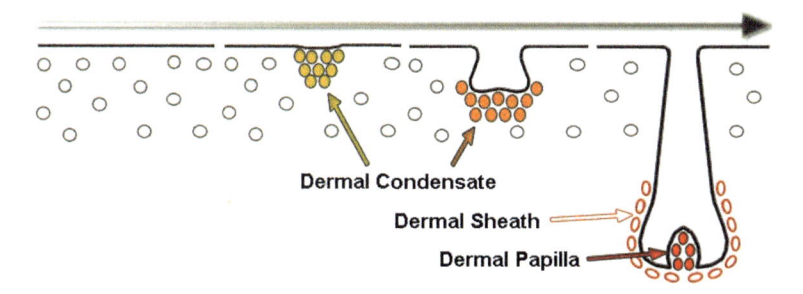

Fig. 1.1 Hair follicle dermal development. Condensation formation occurs beneath follicular placodes. As development progresses, condensates become engulfed within the downward growing follicle epithelium. As condensate cells specialise, they become dermal papilla and dermal sheath of the adult follicle

evidence that in the mouse pelage, the Wnt, BMP, TGF and FGF families are involved in regulating these transitions through the hair cycle [6]. Such signals not only arise within the follicle microenvironment, but a growing body of evidence now points to the contribution of the other hair follicle mesenchyme such as the dermal sheath, the dermal macroenvironment, and even signals emitted from the dermal papilla of surrounding follicles, in regulation of the hair cycle in rodents [7–9]. The dynamic cyclical nature of fibre production and the heterogeneity of epithelial stem cells in the follicle [10] means that the follicle dermis, as a relatively stable population, remains as a key choreographer of dynamic morphogenetic changes and signalling switches [11]. The requirements of adult cycling require the follicle dermal cells to have retained characteristics associated with their developmental precursors, so logically the basis of their inductive properties must lie in developing skin.

Rules of Appendage Development

Many embryological events are regulated by interactions between mesenchymal and epithelial tissues, enabling organ morphogenesis (Fig. 1.2). More particularly, these interactions are crucial in the development of ectodermal appendages, seen in several species, and including teeth, feathers, hair, and scales [12]. Sengel and Dhouailly in France were prominent in performing embryonic skin recombination experiments, splitting epithelial and mesenchymal elements, recombining them with a foreign partner and growing the associations further to observe developmental outcomes [13–15]. These studies established the developmental "rules" of interaction demonstrating, for example, that the appendage size and pattern was dictated by the dermis [16]. Crucially they also showed that development was essentially a two-phase process, the first involving signals that are common across different types of appendage, even those from a different vertebrate class, followed by appendage specific signals [16]. Thus, in recombinations between hair forming dermis and

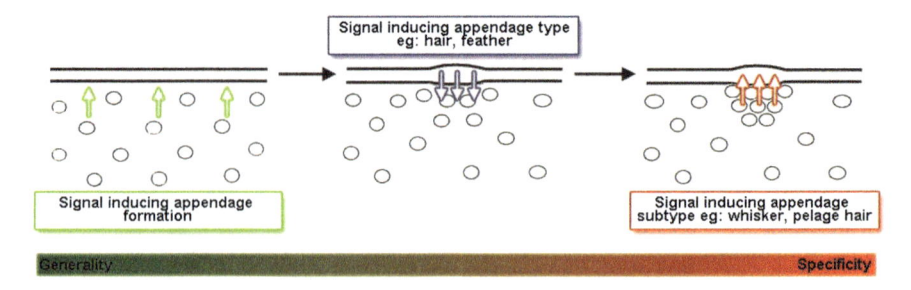

Fig. 1.2 Developmental rules of interaction. As development progresses the specificity of the appendage forming signals increases. We start with a broad appendage induction signal from the dermis, while later signalling from the dermal condensate specifies appendage subtype

scale forming epidermis, the epidermis responds to dermal signals to initiate follicle formation, but the process is arrested after an early structure has been formed [17], before creation of a hair follicle papilla.

Despite common 'rules' of development there are many differences between early chick feather and mouse backskin follicle primordial which should be highlighted for the reader to consider throughout this chapter. For example, feathers are initiated in a wave from a central mid dorsal line, while hair follicles form more synchronously. There are also important differences in follicle identity. Vibrissa (whisker) follicles which develop two days earlier than pelage follicles in mice have very distinct patterning in rows [18, 19]. Vibrissa start off as dermal ridges [20] where retinoic acid may be involved in the control of primordial size and timing [21, 22]. Condensations in vibrissae are larger and may be generated differently from those of backskin follicles. When it comes to human hair follicle development very little is known about the comparative early events [23].

While Sengel and Dhouailly focused on embryonic tissues, Lillie and Wang used transplantation to demonstrate the importance of mesenchymal-epithelial interactions in mature feather follicles [24]. These pioneering studies paved the way for a key advance in the ectodermal appendage field; the recognition that adult as well as embryological activities are regulated by interactions between the mesenchyme and epithelium.

The hair follicle is a very elegant and accessible model for studying interactions between the mesenchyme and epithelium. Following Lillie and Wang's work on feather follicles, Oliver demonstrated that adult rat whisker follicle papillae could be isolated, and transplanted into adult skin where, when placed against epithelium, they are capable of reinitiating the mesenchymal-epithelial interacting cascade that is observed during development to make new follicles [25, 26]. Historically, therefore, well before reprogramming of adult cells became a widely recognised or fashionable concept, mature follicle dermal cells were shown experimentally to be capable of directing transdifferentiation in non-follicular epithelial cells [27]. Subsequently, a multitude of experiments have demonstrated that dermal papillae and cultured dermal papilla cells isolated from rodents [28–31], sheep [32], dogs [33], rabbits [34], human teeth [35] and human hair [36], can induce follicle formation and hair growth not only in skin, but several other types of epithelia [37–39]. Ex vivo follicle dermal sheath tissue can also induce fibre formation in follicle epithelium [40], and induce new follicles in human skin [41], however they are limited in their inductive capacity once cultured. These experiments were the foundational studies that underpinned later clinical studies utilising this inductive capacity of hair follicle mesenchyme to instruct new hair growth. These studies, and the challenges associated with maintaining the inductiveness of the dermal papilla will be discussed in later chapters within this book.

What Mechanisms Underlie Patterning and Condensate Formation?

Given that most follicles develop within a particular spatial arrangement, understanding the morphogenesis of individual structures is interlinked with research into appendage patterning [42, 43]. However, when considering these mechanisms globally, there are caveats in that different appendages have very different spatial organisation, and the possibility must be considered that some elements of early follicle morphogenesis may be independent from patterning. In mouse and chick skin the mesenchyme in appendage forming areas initially presents as a homogeneous dermis, with relatively equally spaced fibroblasts. Prior to follicle development there is a "densification" of the upper layer of dermal cells nearest the epidermis [15, 44, 45]. Wnt ligands in the epidermis are thought to activate mesenchymal Wnt signalling, leading to fibroblast proliferation [46]. When the dermis reaches a critical cellular density threshold, proliferation stops and condensations are organised [47, 48]. These condensations first appear as clusters, with a cellular density approximately twice of that observed in surrounding fibroblasts [45, 49], and positioned directly beneath epidermal follicle placodes. There is little evidence for cell division within feather or hair follicle condensates [49] so with this ruled out both mechanical and biochemical stimuli have been proposed as alternative models of aggregation (Fig. 1.3). Early studies showed that collagen fibrils were hexagonally arranged in sheets of embryonic feather follicle dermis maintained in vitro [50]. Originally it was thought that this collagen lattice served to align dermal cells, and it was proposed to form a cellular highway along which cells can drag themselves and form the condensate [50]. This cell traction model starts with cells pulling on their fibrous extracellular substrate resulting in local cells coming into contact. It is based on Oster-Murray-Harris principles and has the virtue of explaining both patterning and individual condensate morphogenesis in a single process [51]. However, experiments on feather forming skin in particular do not conform to this mechanism and work has since shown that the lattice appears as a consequence of the dermal condensate, rather than the cause [52].

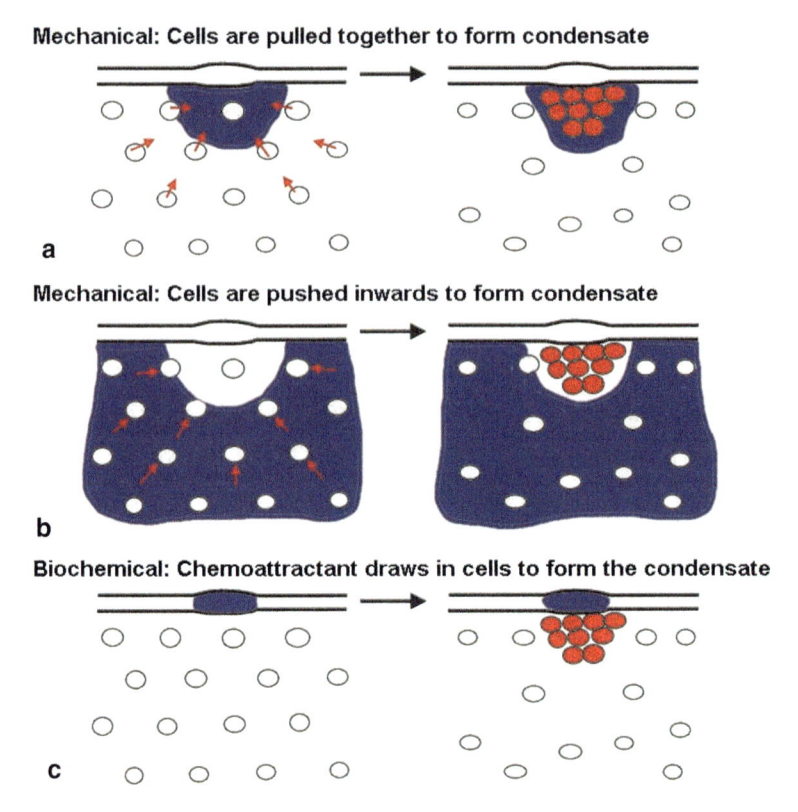

Fig. 1.3 Mechanical and biochemical stimuli for condensation. **a** An increase in cell adhesion molecules such as NCAM within the condensate are believed to draw cells together, leading to formation of the condensate. **b** A loss of bulky extracellular matrix within the condensate is thought to lead to mechanical stimulus, essentially pushing cells together to fill the void. **c** A chemoattractant model proposes that cells are drawn toward the stimuli from the follicle placode, coming together to form the condensate

Other than cell traction, a range of mechanisms can lead to a local increase in cell density, one of which is via cell adhesion [53, 54]. Factors present within the epithelium, including TGFβ2 [55], have been proposed to increase expression of adhesion molecules such as NCAM and tenascin within the dermis. These factors may have direct or indirect targets, since after expression in the epithelial placode some start to be expressed within the condensate itself [48, 55]. Mesenchymal cell density increases in a regularly spaced pattern at sites of NCAM expression, thought to be as a result of increased cell adhesion between cells, while application of antibodies against NCAM can result in formation of unevenly sized condensations in feather follicles [56].

The disappearance of interstitial matrix is another mechanical model that has been proposed as a mechanism for condensation [57]. Skin dermis contains large amounts of hyaluronic acid, a large matrix molecule that separates cells. Within the condensate there is a notable exclusion of hyaluronic acid, and expression of the hyaluronan receptor CD44, a molecule that is involved in hyaluronan degradation [58]. Differential interstitial stresses between the surrounding dermis and the presumptive condensate are thought to influence cell migration. As cells like to follow the path of least resistance [59] the model proposes that fibroblasts migrate inwards to fill the void created by the loss of interstitial matrix, hence forming the condensate [57].

As an alternative to the mechanical models, the biochemical model is based on the Turing reaction–diffusion model coupled with chemotaxis, proposed to create spatial patterns of increased cell density with increased chemical concentration [60]. In this, a chemical (chemoattractant) causes cells within the dermis to migrate up the gradient towards a focal stimulus. In feather follicles FGF2, FGF4, and BMP7 are expressed in the placode region, and can stimulate migration, and thus condensation in underlying mesenchyme [48, 61, 62]. More recently, FGF20 was identified in mouse pelage hair follicle placodes, and has also been proposed as a stimulus for condensation [63]. Mice and chickens lacking FGF20 fail to make condensations and don't develop normal numbers of follicles [64, 65]. In line with the Turing model, an inhibitor is also required to restrict condensate size, so condensates do not continue to enlarge and achieve regular spacing [66]. For developing feather follicles BMP2 is the proposed candidate [48].

For both the cell adhesion and reaction–diffusion models above the assumption is that placodal signals form the foci for dermal condensation mechanisms. There are, however, other questions worth exploring, one of which is whether there is a mechanism of autonomous patterning within the dermis that precedes placode formation. Apparently militating against this, is the absence of molecular heterogeneity in the key locations. For example, β-catenin activity is seen in the nuclei of cells of the upper dermis in mouse backskin well before placode formation, so there is active early wnt signalling in this compartment [46]. However this expression is uniformly distributed, and so the observation doesn't challenge the view that condensation signals are engendered by the epidermis [67]. Indeed, Linsenmayer showed that when embryonic chick epithelium which had already formed placodes was combined with younger embryonic dermis, then condensations formed beneath the existing placodes [68]. However, contrary to this, experiments performed by Novel showed that when placodal stage skin was split, and the epidermis rotated 90° so the placode and condensate were no longer aligned, follicle structures first disappeared over 30 h. When new ones reformed they followed the pattern of the dermis, however, their orientation was determined by the epidermis [69, 70].

A landmark paper by the Headon group in Edinburgh working on embryonic mouse skin provides an explanation that, to a large extent, reconciles these lines of argument [71]. These authors combined live cell imaging and tracking of GFP labelled cells showing active Wnt/ β-catenin expression, with global transcriptomic analysis, pathway manipulation experiments, and mathematical modelling. Given the rapidity of the dermal condensation event, the researchers focused their investigation on searching for genes with short mRNA half-lives belonging to key developmental pathways in pre-condensation skin. They demonstrated an axis of BMP,

FGF, and WNT signalling that, when dissected, provided evidence consistent with there being two different patterning mechanisms. One, a variation of the reaction–diffusion model operating largely in the epidermis, was based on high FGF low BMP levels of signalling localised in the follicle primordium, that, directed mesenchymal aggregation. However, intriguingly the authors also showed that if they created the molecular environment of the placode uniformly over the whole skin, the dermal mesenchyme underwent self-organisation and distinct patterning in the absence of histologically distinct epidermal placodes or key placode molecular markers. Thus, without an overlying and pre-established reaction diffusion system, they revealed a second, intrinsic but latent chemotactic mechanism for dermal aggregation and patterning within the mouse mesenchyme.

Intriguingly a paper published around the same time investigating patterning of feather follicles in chick skin, also concluded that mesenchymal condensation was not simply due to epidermal signalling [72]. The mechanistic interpretation of these authors, however, was linked to the idea of mechanical forces and tissue architecture, with the argument that mesenchymal cells within the dermis were primed to aggregate but this phenomenon was hindered by the stiffness of the tissue. Their work was largely based around skin explant experiments, in particular those demonstrating that mesenchymal cells behaved differently according to the mechanical status of the embryonic tissues. Specifically, they showed that skin cultured on substrates of medium stiffness permitted dermal cell aggregation while softer or more rigid substrates did not.

In concluding this discussion therefore, a number of caveats and uncertainties need to be considered, including aforementioned differences in appendage development between species, and intraspecies follicle types. One cannot automatically transpose data and mechanisms from one model to the other and indeed there may be very different cellular as well as molecular mechanisms at play. Clearly timing is important, and this was another difference observed between the above mouse and chick papers, both of which recognize the role of Wnt pathway signalling in eliciting hair follicle primordium associated gene activity. In the mouse hair follicles, it was found this Wnt activity was detected before the onset of visible morphogenetic follicle indicators [71], whereas in their paper on feather patterning Shyer et al. observed that Wnt pathway expression was coincidental with morphological events signifying the start of follicle formation [72]. In relation to the cellular processes that create condensations, while the balance of interest has switched to chemotaxis from other mechanisms, it is entirely possible that in primary mouse hair follicles, a combination of different events are ultimately involved in cell clustering. For example, once cells have moved to create a focal point then extracellular matrix loss and local changes in adhesion could cement the process.

How and When Is Dermal Condensate Fate Determined?

Shortly after a condensation is formed, the closely packed cells within start to express genes essential for appendage development [73, 74], demonstrating a change in identity. The boundary between condensates and interfollicular dermis is also well defined by differential expression patterns, including extracellular matrix constituents [58]. Whether, in relation to cell status, condensate specification is autonomous (either due to density dependence or predetermined), or specification is conditional, and arises due to external influences such as the epithelium is the subject of much ongoing research (Fig. 1.4).

Experimental evidence from feather bearing skin is that early commitment at least is reversible and that cells within condensates are interchangeable with unincorporated ones. In a landmark experiment, feather dermal condensates were labelled with vital dye, dissociated enzymatically and then together with unlabelled non-condensate cells incorporated into reconstituted dermis. When recombined with intact epidermis the epithelial placodes stimulated new condensate formation, in which all cells had the same potential to become part of a condensate, regardless of whether or not they had originally been in the condensate [75]. This demonstrated a

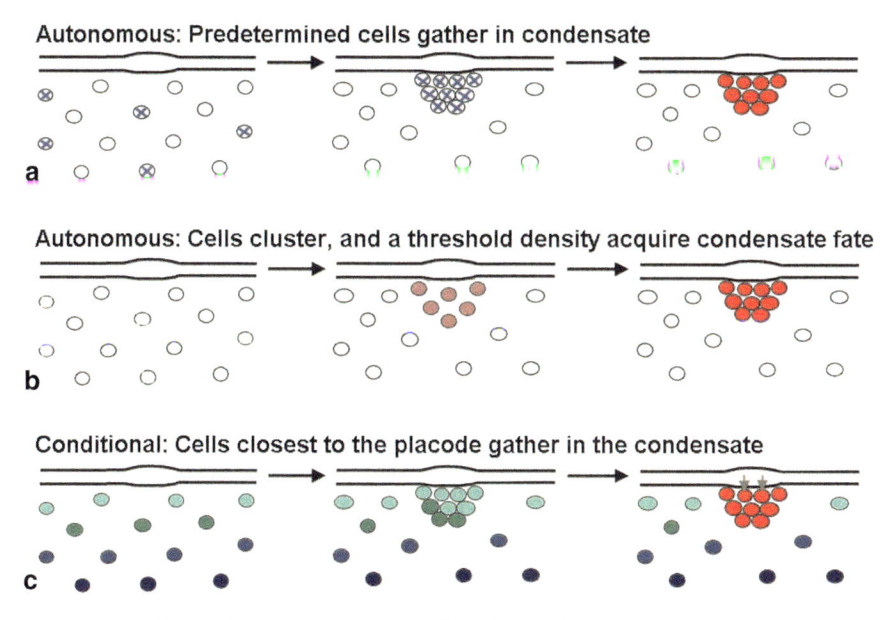

Fig. 1.4 Is condensate fate autonomous or conditional? **a** Cells may be predetermined to become part of the condensate. In response to a cue for condensation, only a subset of cells are capable of responding and migrating towards the placode. **b** Cells arrive in the condensate and as a result of cell autonomous signaling due to a threshold density they change identity **c** Inclusion into the condensate may be conditional upon distance from the placode. In response to a cue leading to condensation, the cells closest to the placode move together. Upon arrival in the condensate, their fate is subsequently defined by non-cell autonomous signalling

self-organising capacity in support of a non-autonomous patterning model. It further showed that the cells of the dermis can be reset by dissociation, however, this does not preclude that in normal development cells could be preselected to become part of the condensate. In ovine follicle development it has been demonstrated that Delta+ cells are present as a minority population in the pre-follicle mesenchyme but are in the majority in condensations. One interpretation is that these cells are being selectively incorporated into condensations, and this would indicate that their fate is determined prior to follicle initiation, and support autonomous pattern formation [76]. Alternatively, it could also be that Delta+ expression is upregulated in cells as a consequence of condensate formation. As before, caveats regarding developmental differences between species and follicle types should be borne in mind. Most recently in support of a non-cell autonomous mechanism of condensate commitment is work from Rendl in New York, who used single cell RNA-seq to identify a subset of pre-dermal condensate (pre-DC) cells that upregulated expression of condensate genes and exited from the cell cycle prior to condensate formation in mouse skin [77]. These condensate precursors were derived from fibroblasts and were dependent on FGF20 signalling from the overlying epithelium. While this is convincing evidence that condensate identity is dependent on signalling from the epithelium, it raises many more questions, including whether all fibroblasts are equally capable of responding to epithelial signals that direct condensate formation? Here again the patterning paper from the Headon group brings new information to the debate. Analysis of the time lapse images shows clearly that the cells entering the condensation are those in close proximity. There is no evidence of longer-range travel in response to chemotactic stimuli [71], nor of collective migration, and this strongly militates against the predetermination model (Fig. 1.A).

One condition that may shed light on this issue is Terminal Hypertrichosis, where patient follicles are characterised by an enlarged dermal papilla and hair follicle. This inherited disorder is often caused by position effects, and candidate genes thought to be causative include *Trps1* and *Sox9*, found in both the epidermal and dermal components of the follicle [78, 79]. In Terminal Hypertrichosis hair follicle types all over the body are affected, but there is also an increase in the overall number of hair follicles. In other follicle types, such as the feather, when a condensation becomes too large it leads to a depletion of cells around, and a glabrous area around one large condensation is observed [48]. In hypertrichosis no such glabrous areas is observed, so we speculate that a large pool of dermal cells is present during development that are capable of becoming condensations, and can generate an increased number of follicles. Whether a signal from the epidermis to establish condensations is more concentrated, the dermal cells have a pronounced cellular response, or there is lack of an inhibitor regulating spacing of condensations and follicles is unknown.

Lastly, over expression of β-catenin in the epidermis of adult mouse skin results in the epidermis reprogramming the dermis to a more plastic state that is capable of contributing to follicles and forming ectopic follicles with associated de novo dermal papillae [80]. These ectopic follicles contain papillae that are polyclonal in origin [81], however, it is not known when the selection, or differentiation of these non-follicular dermal cells into condensate/papilla cells takes place. Interestingly, ectopic follicles induced in the adult period never form Sox2 positive hairs that are seen in the backskin of mice [82]. In considering this issue it should be recalled that mouse pelage follicles develop in three phases with the primary or guard hair follicles induced first, followed by second and tertiary waves initiating other follicle types later on [83]. While the e14.5 and 16.5 condensations express Sox2, the e18.5 condensations do not [82]. On the other hand, cells of the upper interfollicular dermis continue to express markers of the hair follicle dermis such as Trps1, from e17.5 up until birth [84]. When Wnt is activated to induce new follicles in adult skin [85] analysis shows that the dermal cells are reprogrammed into a neonatal-like dermis [80]. Therefore, since these cells are not reverted to the dermal phenotype from which the earliest sets of condensates are made, in terms of developmental precedent it follows that they will not make Sox2 positive condensates and thus guard or awl/auchene type follicles.

Several points emerge from these observations. First, the greater attention should be given to how condensations form in later developing follicles, looking at whether the mechanisms as well as the cell types involved are different. Even more importantly the Wnt activation experiments in mice underscore the crucial importance of the epidermis in modulating dermal differentiation status, as well as the necessity of proximity since one of the key observations of this work is that only the upper dermal cells adjacent to the epidermis are reprogrammed. In a more practical sense, understanding the timing, and cues of lineage commitment within a dermal condensate is necessary if we want recapitulate these events in vitro, perhaps for formation of condensate cells or dermal papilla from iPSCs [86, 87]. Moreover, safe modulation of Wnt signalling has a potential for follicle neogenesis in human skin [88, 89].

Are Dermal Papilla Necessary for Hair Growth?

The presence of a dermal condensate is fundamental for follicle development, as its early dispersal following irradiation results in an arrest of follicle development [90]. Likewise, genetic and molecular modifications that result in condensations not being formed or being ablated block follicle morphogenesis at an early stage [63, 64, 91, 92]. In adults, as in development, a close anatomical relationship between the dermal papilla and the epithelium is critical for hair follicle growth and cycling. Removal of the dermal papilla has long been shown to stop epithelial growth and survival [93] and this physical displacement of the papilla from the adjacent epithelium causes an immediate arrest of growth [94]. Similarly, in the hairless (hr/hr) mutant mouse, it is separation of the papilla from the follicle epithelium during the follicle cycle that arrests growth and prevents a new phase of anagen [95]. Partial ablation of anagen dermal papilla in murine backskin follicles does not result in an arrest of hair growth, but rather growth of smaller follicle types (zigzag hairs instead of guard/awl/auchene) [96]. After transition through the follicle cycle, both the type of follicle, and size of the dermal papilla can be recovered. While the source of cells contributing to this recovery has not been traced [96] one possibility is that the recently described dermal stem cells, located in the dermal sheath may be key in this recovery [97]. Adding to this, removal of the entire dermal papilla of growing follicles (anagen) stops hair growth but not permanently, as cells from the dermal sheath are capable of replacing and reconstituting the dermal papilla in rodent and human follicles [19, 97, 98]. However, when both dermal papilla and dermal sheath cells are ablated by directly targeting the combined papilla and sheath during telogen, the resting phase of the follicle cycle, then follicle growth is arrested [99]. Follicles remain in telogen, and without a dermal presence lose their capacity to reactivate the follicle epithelium and stimulate progression back into follicle downgrowth and fibre production [99]. So, while a condensate is essential for follicle development, the presence of a dermal papilla in adult follicles is required both for maintenance of fibre production, and for the follicle to cycle (Fig. 1.5).

These data again underline the importance of dermal-epidermal contacts. They also emphasize the plastic relationship between the dermal papilla and the dermal sheath which arise from the same condensate precursors. Not only can the sheath replace the papilla [40, 97], in those experiments in which adult papillae induce new follicle formation, the sheath cells of the new structures derive from the papilla.

So to what extent is the collective association of papilla cells, or signals produced within this structure, or both, required for hair growth? Several secreted factors present within the dermal papilla have been shown to promote hair growth, including FGF7. Beads coated in FGF7, implanted near the papilla of resting hair follicles can induce follicles to re-enter anagen by activating germ cells in the follicle epithelium [100]. In these experiments FGF7 was able to recapitulate the dermal papilla signal required for growth initiation, however, the papilla was still physically present. It

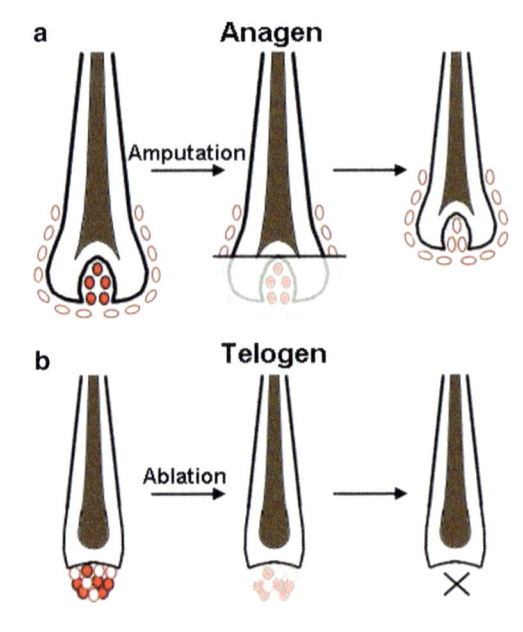

Fig. 1.5 Regeneration of the hair follicle dermis. **a** Amputation of the lower follicle during anagen results in complete removal of the dermal papilla and partial removal of the dermal sheath. As a consequence regeneration of the follicle can occur. **b** Ablation of the follicle mesenchyme during telogen results in removal of both the dermal papilla and sheath, and subsequently regeneration of the follicle cannot occur

remains to be seen whether introducing secreted factors to replicate those emitted by the dermal papilla could maintain follicle growth or reactivate resting hair follicles if there were no papillae physically present. The smallest follicles have papillae with very few cells, and a structure only one of two cells in diameter, therefore it would be interesting to test directly whether an aggregation of papilla cells is required to provide the correct signals for larger follicles and fibres. In other words, do the cells at the centre of the aggregation provide anything more that support and are only the cells at the outside of the papilla necessary for interaction and signalling? This could be investigated by growing papilla cells ex vivo to form layers around an appropriate neutral biomaterial of varying papilla sizes and shapes. The capacity of these chimeric structures to induce and maintain growth in established induction assays could then be tested directly. However, this question also touches on issues of papilla cell heterogeneity, which is apparent in some follicles [101, 102]. Microniches within the dermal papilla can direct specific fate specification of adjacent epithelial cells during hair growth [102], indicating that even within a single papilla the cells have different capacities for reprogramming. This however also assumes that cells maintain a static position within the papilla, whereas in some cases at least, there may be cell movement.

What Changes Occur in the Papilla in Follicles Affected by Androgenetic Alopecia?

The hair follicle is an excellent model to study regenerative behaviour of adult tissues, and the inductive properties of the hair follicle and their potential use in the regenerative medicine field are in two main areas: hair follicle regeneration/replacement in alopecia and the production of skin models with appendages.

When trying to understand translational strategies for human hair loss, it is perhaps helpful to discuss the biology of human hair loss which, rather remarkable, is still poorly understood at the cellular level in particular. In human male androgenetic alopecia follicles miniaturise and possess correspondingly diminished dermal papillae containing reduced extracellular matrix and cell numbers. Of the possible cellular mechanisms underpinning these events, there is no evidence for apoptosis amongst the papilla cells, indeed they possess anti-apoptotic mechanisms [103]. Clearly there must be a strong hormonal influence on whatever cellular mechanisms are involved in androgenetic alopecia and there has been a strong body of work investigating the effects of the androgen pathway on papilla cell division [104, 105]. However cell division is relatively rare in the papilla [106] and there still needs to be an explanation of why cells are lost. We therefore previously proposed loss of dermal cell aggregation and migration of cells out of the papilla, either into the dermal sheath or via the sheath to the surrounding dermis as a likely theory [107] (Fig. 1.6). Germane to this, a link has subsequently be shown between migration of skin fibroblasts and estrogen [108]. We also hypothesised that displacement of

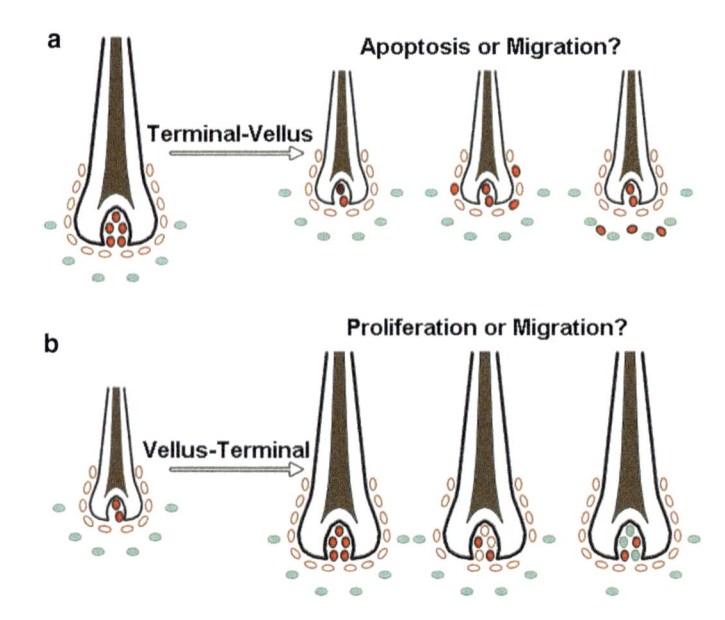

Fig. 1.6 Changes occurring in the dermal papilla with age. **a** Miniaturization of adult hair follicles occurs in androgenetic alopecia, characterized by a terminal to vellus transition. However, whether the dermal papilla miniaturizes due to cell death, or migration of cells into the surrounding sheath or interfollicular dermis remains unknown. **b** With puberty, a vellus to terminal transition of follicles is observed, associated with an increase in the cell number in the dermal papilla. This could be due to cell proliferation in the papilla, migration of cells in from the surrounding sheath or interfollicular dermis, or a combination of the two

papilla cells would most likely take place at a stage of the cycle when the dermal papilla is not enclosed by the follicle epithelium (telogen, catagen, or the very start of anagen) and more probably during transition between these stages, when dynamic movement of the follicle dermis are taking place as a function of the follicle cycle [107, 109]. It used to be thought that follicle miniaturisation or so called terminal to vellus transition was occurring gradually over multiple follicle cycles, however that view subsequently has been changed [110] and it is now proposed that miniaturisation occurs during a single hair follicle cycle. Why cells should reduce cohesiveness and migrate out of the follicle remains an open question, but the loss of aggregative behaviour of human cells in a culture system may serve as a good model for disease progression.

Another idea relating to androgenetic alopecia follicles focuses on oxidative stress, and suggests that 'balding' dermal papilla are more sensitive to environmental stresses compared to dermal papilla which do not undergo miniaturisation [111], but while dermal papilla cells cultured after isolation from balding and non-balding follicles exhibit differential expression profiles [112], their aggregative capacities have not been described.

Loss of cells from the dermal papilla is also confounded by another issue, namely there is a lack of dermal papilla repopulation at the start of a new anagen in androgenetic alopecia which needs to be considered. Work from Morgan's group at Harvard University demonstrated that mouse backskin follicle papillae, partially depleted of papilla cells were able to recover in size in subsequent hair cycles [96], raising the question; what regulates the intrinsic size of a papilla and is this dysregulated in androgenetic alopecia; do follicles no longer recall the size they should be? Biernaskie's group in Calgary later demonstrated that papilla cell repopulation was fuelled by a dermal sheath stem cell, located in the dermal cup beneath the papilla [97, 113]. Intriguingly, in these mouse follicles dermal sheath stem cells only contribute to repopulation of the lower half of the dermal papilla, perhaps helping to explain the genetic makeup of the papilla microniches mentioned earlier [102]. Looking at natural ageing of hair follicles in mice, Biernaskie's group also demonstrated that there is dysfunction in the dermal sheath leading to reduced repopulation of the dermal papilla in a new anagen [113]. This body of evidence therefore, reemphasises the crucial link and importance of crossover between sheath and papilla cells [114] and supports the concept that the sheath acts as a reservoir of reserve cells for the papilla which, if depleted or unable to function, results in papilla cell diminution [107]. It is likely that this mechanism contributes to miniaturisation in human androgenetic follicles as well. The mechanism for cell loss in both compartments is as yet unknown, however a candidate must now be cellular senescence the importance of which is being increasingly apparent in multiple systems [115].

More recently, bulge and secondary hair germ progenitor cells were also quantified in androgenetic, and normal human scalp hair follicles. Garza et al. found a reduction in the number of CD200+ ITGA6+ and CD34+ epithelial progenitor cells in alopecic skin, while the bulge stem cell numbers remained unaffected [116]. This would suggest a key defect in alopecic skin is the conversion of bulge cells into progenitor cells at the start of the hair growth phase. As to whether such an epithelial change is the initial trigger, or one mediated by modified signalling from a depleted dermal papilla compartment remains under debate. However, when evaluating the primary event, perhaps a clue is in the name androgenetic alopecia, since androgen activity requires androgen receptors, which are found in the dermal compartments of follicles [117], while specific genetic variants in the androgen receptor are key risk factors for early onset hair loss [118].

The opposite of miniaturisation in androgenetic alopecia is observed during puberty in adolescents, when follicles switch from a vellus to a terminal fate. In mouse skin, single hair follicles have been seen to switch between hair subtypes, with different numbers of dermal papilla cells [96]. This brings us back to the cellular mechanism enabling papilla enlargement; are papilla cells undergoing cell division, or are fibroblasts, dermal sheath, or dermal stem cells migrating in from the surrounding tissue (Fig. 1.6) to contribute to the dermal papilla? Comparatively, analysis of human follicles suggests 2% of cells in the papilla are dividing in a snapshot of time [119], while murine data suggests that the early anagen environment is conducive to recruitment of dermal sheath stem cells into the papilla, resulting in enlargement [97, 106, 113]. Intriguingly, in humans, the vellus to terminal transition during puberty only occurs in specialized cell types. Adult dermis does maintain an intrinsic HOX code, which sets regionalisation boundaries [120], and it would be interesting to evaluate if the dermis around specialized follicles is of a more plastic or embryonic composition, and in some way enables the transition from a vellus to a terminal fate on these sites.

Clinical Challenges for Reversing Androgenetic Alopecia

Work directed in this area can be subdivided into two strategies; rejuvenation/restoration of the normal size of a bald follicle and new follicle induction. In both cases, a requirement is cells that are papilla-like, or capable of becoming papilla cells under epithelial influence when transplanted. To that end researchers have generally used human hair follicle dermal cells (papilla or sheath). However, given the limitations in the amounts of follicular dermis available, and the difficulties in isolation of papilla and sheath cells another recent strategy has been to make new papilla cell populations, for example from IPSCs [87, 121].

One problem, historically, has been the interpretation of animal experiments so they are applicable to humans. While essential for understanding key biological principles, much animal work is not directly transferable. With regards to dermal papillae, there are differences in composition that are observed between rodents and humans that can influence subsequent experiments/analysis. Rodent papillae isolated in vivo are relatively easy to enzymatically digest, to achieve a suspension, however, human papilla cells contain more fibrous type 1 Collagen and will not digest easily [122]. Paradoxically, after the challenges observed in dispersing human papillae, once in culture the cells do not aggregate autonomously.

There are other clinical challenges associated with the key biological objective which is to achieve human follicle regeneration/rejuvenation. These relate primarily to the required cosmetic effects, such as direction and length of fibre growth, fibre colour and curl, and not least follicle cycling. Some of these will require solutions that are outside of biological expertise. We hope that collaboration across disciplines, between scientists, engineers, and clinicians, who each bring unique insight and problem solving to the field, will one day facilitate clinical translation and use of the hair follicle mesenchyme to benefit patients.

Acknowledgements This work was supported in part by grants from the North American Hair Research Society and the British Skin Foundation (to CAH), and by a grant from the Medical Research Council (to CABJ).

Code of Interest The authors have no conflicts of interest to declare.

References

1. Driskell RR, Lichtenberger BM, Hoste E, Kretzschmar K, Simons BD, Charalambous M, Ferron SR, Herault Y, Pavlovic G, Ferguson-Smith AC, Watt FM (2013) Distinct fibroblast lineages determine dermal architecture in skin development and repair. Nature 504:277–281
2. Sennett R, Rendl M (2012) Mesenchymal-epithelial interactions during hair follicle morphogenesis and cycling. Semin Cell Dev Biol 23:917–927
3. Saxena N, Mok KW, Rendl M (2019) An updated classification of hair follicle morphogenesis. Exp Dermatol 28:332–344
4. Botchkarev VA, Kishimoto J (2003) Molecular control of epithelial-mesenchymal interactions during hair follicle cycling. J Invest Dermatol Symp Proc 8:46–55
5. Higgins CA, Westgate GE, Jahoda CA (2009) From telogen to exogen: mechanisms underlying formation and subsequent loss of the hair club fiber. J Invest Dermatol 129:2100–2108
6. Plikus MV (2012) New activators and inhibitors in the hair cycle clock: targeting stem cells' state of competence. J Invest Dermatol 132:1321–1324
7. Heitman N, Sennett R, Mok KW, Saxena N, Srivastava D, Martino P, Grisanti L, Wang Z, Ma'ayan A, Rompolas P, Rendl M (2020) Dermal sheath contraction powers stem cell niche relocation during hair cycle regression. Science 367:161–166
8. Plikus MV, Baker RE, Chen CC, Fare C, de la Cruz D, Andl T, Maini PK, Millar SE, Widelitz R, Chuong CM (2011) Self-organizing and stochastic behaviors during the regeneration of hair stem cells. Science 332:586–589
9. Plikus MV, Chuong CM (2014) Macroenvironmental regulation of hair cycling and collective regenerative behavior. Cold Spring Harb Perspect Med 4:a015198
10. Jaks V, Kasper M, Toftgard R (2010) The hair follicle-a stem cell zoo. Exp Cell Res 316:1422–1428
11. Joost S, Annusver K, Jacob T, Sun X, Dalessandri T, Sivan U, Sequeira I, Sandberg R, Kasper M (2020) The molecular anatomy of mouse skin during hair growth and rest. Cell Stem Cell 26:441–57 e7
12. Chuong CM, Chodankar R, Widelitz RB, Jiang TX (2000) Evo-devo of feathers and scales: building complex epithelial appendages. Curr Opin Genet Dev 10:449–456
13. Dhouailly D, Sengel P (1972) Morphogenesis of feather and hair, studied by means of heterospecific associations of dermis and epidermis between chicken and mice. C R Acad Sci Hebd Seances Acad Sci D 275:479–482
14. Dhouailly D, Sengel P (1975) Feather- and hair-forming properties of dermal cells of glabrous skin from bird and mammals. C R Acad Sci Hebd Seances Acad Sci D 281:1007–1010
15. Sengel P (1976) Morphogenesis of skin. Cambridge, Eng.; New York, Cambridge University Press
16. Dhouailly D (1973) Dermo-epidermal interactions between birds and mammals: differentiation of cutaneous appendages. J Embryol Exp Morphol 30:587–603
17. Kanzler B, Prin F, Thelu J, Dhouailly D (1997) CHOXC-8 and CHOXD-13 expression in embryonic chick skin and cutaneous appendage specification. Dev Dyn 210:274–287
18. Danforth CH (1925) Hair in its relation to questions of homology and phylogeny. Am J Anat 86:342–356
19. Oliver RF (1966) Whisker growth after removal of the dermal papilla and lengths of follicle in the hooded rat. J Embryol Exp Morphol 15:331–347
20. Wrenn JT, Wessells NK (1984) The early development of mystacial vibrissae in the mouse. J Embryol Exp Morphol 83:137–156
21. Viallet JP, Dhouailly D (1994) Retinoic acid and mouse skin morphogenesis. I. Expression pattern of retinoic acid receptor genes during hair vibrissa follicle, plantar, and nasal gland development. J Invest Dermatol 103:116–121
22. Viallet JP, Ruberte E, du Manoir S, Krust A, Zelent A, Dhouailly D (1991) Retinoic acid-induced glandular metaplasia in mouse skin is linked to the dermal expression of retinoic acid receptor beta mRNA. Dev Biol 144:424–428

23. Holbrook KA, Minami SI (1991) Hair follicle embryogenesis in the human characterization of events in vivo and in vitro. Ann N Y Acad Sci 642:167–196
24. Lillie FR, Wang H (1941) Physiology of development of the feather V experimental morphogenesis. Physiol Zool 14:103–135
25. Oliver RF (1967) The experimental induction of whisker growth in the hooded rat by implantation of dermal papillae. J Embryol Exp Morphol 18:43–51
26. Oliver RF (1970) The induction of hair follicle formation in the adult hooded rat by vibrissa dermal papillae. J Embryol Exp Morphol 23:219–236
27. Jahoda CA, Reynolds AJ, Oliver RF (1993) Induction of hair growth in ear wounds by cultured dermal papilla cells. J Invest Dermatol 101:584–590
28. Jahoda CA (1992) Induction of follicle formation and hair growth by vibrissa dermal papillae implanted into rat ear wounds: vibrissa-type fibres are specified. Development 115:1103–1109
29. Jahoda CA, Horne KA, Oliver RF (1984) Induction of hair growth by implantation of cultured dermal papilla cells. Nature 311:560–562
30. Kishimoto J, Burgeson RE, Morgan BA (2000) Wnt signaling maintains the hair-inducing activity of the dermal papilla. Genes Dev 14:1181–1185
31. Rendl M, Polak L, Fuchs E (2008) BMP signaling in dermal papilla cells is required for their hair follicle-inductive properties. Genes Dev 22:543–557
32. Watson SA, Pisansarakit P, Moore GP (1994) Sheep vibrissa dermal papillae induce hair follicle formation in heterotypic skin equivalents. Br J Dermatol 131:827–835
33. Zheng Y, Nace A, Chen W, Watkins K, Sergott L, Homan Y, Vandeberg JL, Breen M, Stenn K (2010) Mature hair follicles generated from dissociated cells: a universal mechanism of folliculoneogenesis. Dev Dyn 239:2619–2626
34. Ferraris C, Bernard BA, Dhouailly D (1997) Adult epidermal keratinocytes are endowed with pilosebaceous forming abilities. Int J Dev Biol 41:491–498
35. Reynolds AJ, Jahoda CA (2004) Cultured human and rat tooth papilla cells induce hair follicle regeneration and fiber growth. Differentiation 72:566–575
36. Toyoshima KE, Asakawa K, Ishibashi N, Toki H, Ogawa M, Hasegawa T, Irie T, Tachikawa T, Sato A, Takeda A, Tsuji T (2012) Fully functional hair follicle regeneration through the rearrangement of stem cells and their niches. Nat Commun 3:784
37. Bonfanti P, Claudinot S, Amici AW, Farley A, Blackburn CC, Barrandon Y (2010) Microenvironmental reprogramming of thymic epithelial cells to skin multipotent stem cells. Nature 466:978–982
38. Ferraris C, Chaloin-Dufau C, Dhouailly D (1994) Transdifferentiation of embryonic and postnatal rabbit corneal epithelial cells. Differentiation 57:89–96
39. Fliniaux I, Viallet JP, Dhouailly D, Jahoda CA (2004) Transformation of amnion epithelium into skin and hair follicles. Differentiation 72:558–565
40. Horne KA, Jahoda CA (1992) Restoration of hair growth by surgical implantation of follicular dermal sheath. Development 116:563–571
41. Reynolds AJ, Lawrence C, Cserhalmi-Friedman PB, Christiano AM, Jahoda CA (1999) Transgender induction of hair follicles. Nature 402:33–34
42. Jung HS, Francis-West PH, Widelitz RB, Jiang TX, Ting-Berreth S, Tickle C, Wolpert L, Chuong CM (1998) Local inhibitory action of BMPs and their relationships with activators in feather formation: implications for periodic patterning. Dev Biol 196:11–23
43. Pispa J, Mustonen T, Mikkola ML, Kangas AT, Koppinen P, Lukinmaa PL, Jernvall J, Thesleff I (2004) Tooth patterning and enamel formation can be manipulated by misexpression of TNF receptor Edar. Dev Dyn 231:432–440
44. Dhouailly D, Olivera-Martinez I, Fliniaux I, Missier S, Viallet JP, Thelu J (2004) Skin field formation: morphogenetic events. Int J Dev Biol 48:85–91
45. Wessells NK (1965) Morphology and proliferation during early feather development. Dev Biol 12:131–153
46. Chen D, Jarrell A, Guo C, Lang R, Atit R (2012) Dermal beta-catenin activity in response to epidermal Wnt ligands is required for fibroblast proliferation and hair follicle initiation. Development 139:1522–1533

47. Michon F, Charveron M, Dhouailly D (2007) Dermal condensation formation in the chick embryo: requirement for integrin engagement and subsequent stabilization by a possible notch/integrin interaction. Dev Dyn 236:755–768
48. Michon F, Forest L, Collomb E, Demongeot J, Dhouailly D (2008) BMP2 and BMP7 play antagonistic roles in feather induction. Development 135:2797–2805
49. Wessells NK, Roessner KD (1965) Nonproliferation in dermal condensations of mouse vibrissae and pelage hairs. Dev Biol 12:419–433
50. Stuart ES, Moscona AA (1967) Embryonic morphogenesis: role of fibrous lattice in the development of feathers and feather patterns. Science 157:947–948
51. Murray JD, Oster GF, Harris AK (1983) A mechanical model for mesenchymal morphogenesis. J Math Biol 17:125–129
52. Hughes MW, Wu P, Jiang TX, Lin SJ, Dong CY, Li A, Hsieh FJ, Widelitz RB, Chuong CM (2011) In search of the Golden Fleece: unraveling principles of morphogenesis by studying the integrative biology of skin appendages. Integr Biol (Camb) 3:388–407
53. Armstrong NJ, Painter KJ, Sherratt JA (2006) A continuum approach to modelling cell-cell adhesion. J Theor Biol 243:98–113
54. Glazier JA, Graner F (1993) Simulation of the differential adhesion driven rearrangement of biological cells. Phys Rev E Stat Phys Plasmas Fluids Relat Interdiscip Topics 47:2128–2154
55. Ting-Berreth SA, Chuong CM (1996) Local delivery of TGF beta2 can substitute for placode epithelium to induce mesenchymal condensation during skin appendage morphogenesis. Dev Biol 179:347–359
56. Jiang TX, Chuong CM (1992) Mechanism of skin morphogenesis. I. Analyses with antibodies to adhesion molecules tenascin, N-CAM, and integrin. Dev Biol 150:82–98
57. Davies J (2013) Condensation of Cells. In: Mechanisms of morphogenesis. Elsevier Ltd
58. Underhill CB (1993) Hyaluronan is inversely correlated with the expression of CD44 in the dermal condensation of the embryonic hair follicle. J Invest Dermatol 101:820–826
59. Lo CM, Wang HB, Dembo M, Wang YL (2000) Cell movement is guided by the rigidity of the substrate. Biophys J 79:144–152
60. Lin CM, Jiang TX, Baker RE, Maini PK, Widelitz RB, Chuong CM (2009) Spots and stripes: pleomorphic patterning of stem cells via p-ERK-dependent cell chemotaxis shown by feather morphogenesis and mathematical simulation. Dev Biol 334:369–382
61. Song HK, Lee SH, Goetinck PF (2004) FGF-2 signaling is sufficient to induce dermal condensations during feather development. Dev Dyn 231:741–749
62. Widelitz RB, Jiang TX, Noveen A, Chen CW, Chuong CM (1996) FGF induces new feather buds from developing avian skin. J Invest Dermatol 107:797–803
63. Huh SH, Närhi K, Lindfors PH, Haara O, Yang L, Ornitz DM, Mikkola ML (2013) Fgf20 governs formation of primary and secondary dermal condensations in developing hair follicles. Genes Dev 27:450–458
64. Goetinck PF, Sekellick MJ (1972) Observations on collagen synthesis, lattice formation, and morphology of scaleless and normal embryonic skin. Dev Biol 28:636–648
65. Wells KL, Hadad Y, Ben-Avraham D, Hillel J, Cahaner A, Headon DJ (2012) Genome-wide SNP scan of pooled DNA reveals nonsense mutation in FGF20 in the scaleless line of featherless chickens. BMC Genomics 13:257
66. Mammoto T, Mammoto A, Torisawa YS, Tat T, Gibbs A, Derda R, Mannix R, de Bruijn M, Yung CW, Huh D, Ingber DE (2011) Mechanochemical control of mesenchymal condensation and embryonic tooth organ formation. Dev Cell 21:758–769
67. Millar SE (2002) Molecular mechanisms regulating hair follicle development. J Invest Dermatol 118:216–225
68. Linsenmayer TF (1972) Control of integumentary patterns in the chick. Dev Biol 27:244–271
69. Chuong CM, Widelitz RB, Ting-Berreth S, Jiang TX (1996) Early events during avian skin appendage regeneration: dependence on epithelial-mesenchymal interaction and order of molecular reappearance. J Invest Dermatol 107:639–646
70. Novel G (1973) Feather pattern stability and reorganization in cultured skin. J Embryol Exp Morphol 30:605–633

71. Glover JD, Wells KL, Matthaus F, Painter KJ, Ho W, Riddell J, Johansson JA, Ford MJ, Jahoda CAB, Klika V, Mort RL, Headon DJ (2017) Hierarchical patterning modes orchestrate hair follicle morphogenesis. PLoS Biol 15:e2002117
72. Shyer AE, Rodrigues AR, Schroeder GG, Kassianidou E, Kumar S, Harland RM (2017) Emergent cellular self-organization and mechanosensation initiate follicle pattern in the avian skin. Science 357:811–815
73. Driskell RR, Giangreco A, Jensen KB, Mulder KW, Watt FM (2009) Sox2-positive dermal papilla cells specify hair follicle type in mammalian epidermis. Development 136:2815–2823
74. Sennett R, Wang Z, Rezza A, Grisanti L, Roitershtein N, Sicchio C, Mok KW, Heitman NJ, Clavel C, Ma'ayan A, Rendl M (2015) An integrated transcriptome atlas of embryonic hair follicle progenitors their niche, and the developing skin. Dev Cell 34:577–591
75. Jiang TX, Jung HS, Widelitz RB, Chuong CM (1999) Self-organization of periodic patterns by dissociated feather mesenchymal cells and the regulation of size, number and spacing of primordia. Development 126:4997–5009
76. Xavier SP, Gordon-Thomson C, Wynn PC, McCullagh P, Thomson PC, Tomkins L, Mason RS, Moore GP (2013) Evidence that Notch and Delta expressions have a role in dermal condensate aggregation during wool follicle initiation. Exp Dermatol 22:659–662
77. Mok KW, Saxena N, Heitman N, Grisanti L, Srivastava D, Muraro MJ, Jacob T, Sennett R, Wang Z, Su Y, Yang LM, Ma'ayan A, Ornitz DM, Kasper M, Rendl M (2019) Dermal condensate niche fate specification occurs prior to formation and is placode progenitor dependent. Dev Cell 48:32–48 e5
78. DeStefano GM, Fantauzzo KA, Petukhova L, Kurban M, Tadin-Strapps M, Levy B, Warburton D, Cirulli ET, Han Y, Sun X, Shen Y, Shirazi M, Jobanputra V, Cepeda-Valdes R, Cesar Salas-Alanis J, Christiano AM (2013) Position effect on FGF13 associated with X-linked congenital generalized hypertrichosis. Proc Natl Acad Sci U S A 110:7790–7795
79. Fantauzzo KA, Kurban M, Levy B, Christiano AM (2012) Trps1 and its target gene Sox9 regulate epithelial proliferation in the developing hair follicle and are associated with hypertrichosis. PLoS Genet 8:e1003002
80. Collins CA, Kretzschmar K, Watt FM (2011) Reprogramming adult dermis to a neonatal state through epidermal activation of beta-catenin. Development 138:5189–5199
81. Collins CA, Jensen KB, MacRae EJ, Mansfield W, Watt FM (2012) Polyclonal origin and hair induction ability of dermal papillae in neonatal and adult mouse back skin. Dev Biol 366:290–297
82. Driskell RR, Juneja VR, Connelly JT, Kretzschmar K, Tan DW, Watt FM (2012) Clonal growth of dermal papilla cells in hydrogels reveals intrinsic differences between Sox2-positive and -negative cells in vitro and in vivo. J Invest Dermatol 132:1084–1093
83. Duverger O, Morasso MI (2009) Epidermal patterning and induction of different hair types during mouse embryonic development. Birth Defects Res C Embryo Today 87:263–272
84. Fantauzzo KA, Bazzi H, Jahoda CA, Christiano AM (2008) Dynamic expression of the zinc-finger transcription factor Trps1 during hair follicle morphogenesis and cycling. Gene Expr Patterns 8:51–57
85. Lo Celso C, Prowse DM, Watt FM (2004) Transient activation of beta-catenin signalling in adult mouse epidermis is sufficient to induce new hair follicles but continuous activation is required to maintain hair follicle tumours. Development 131:1787–1799
86. Ramos R, Guerrero-Juarez CF, Plikus MV (2013) Hair follicle signaling networks: a dermal papilla-centric approach. J Invest Dermatol 133:2306–2308
87. Veraitch O, Mabuchi Y, Matsuzaki Y, Sasaki T, Okuno H, Tsukashima A, Amagai M, Okano H, Ohyama M (2017) Induction of hair follicle dermal papilla cell properties in human induced pluripotent stem cell-derived multipotent LNGFR(+)THY-1(+) mesenchymal cells. Sci Rep 7:42777
88. Gay D, Kwon O, Zhang Z, Spata M, Plikus MV, Holler PD, Ito M, Yang Z, Treffeisen E, Kim CD, Nace A, Zhang X, Baratono S, Wang F, Ornitz DM, Millar SE, Cotsarelis G (2013) Fgf9 from dermal gammadelta T cells induces hair follicle neogenesis after wounding. Nat Med 19:916–923

89. Rognoni E, Gomez C, Pisco AO, Rawlins EL, Simons BD, Watt FM, Driskell RR (2016) Inhibition of beta-catenin signalling in dermal fibroblasts enhances hair follicle regeneration during wound healing. Development 143:2522–2535
90. Jacobson CM (1966) A comparative study of the mechanisms by which X-irradiation and genetic mutation cause loss of vibrissae in embryo mice. J Embryol Exp Morphol 16:369–379
91. Lehman JM, Laag E, Michaud EJ, Yoder BK (2009) An essential role for dermal primary cilia in hair follicle morphogenesis. J Invest Dermatol 129:438–448
92. Mill P, Mo R, Fu H, Grachtchouk M, Kim PC, Dlugosz AA, Hui CC (2003) Sonic hedgehog-dependent activation of Gli2 is essential for embryonic hair follicle development. Genes Dev 17:282–294
93. Crounse RG, Stengle JM (1959) Influence of the dermal papilla on survival of isolated human scalp hair roots in an heterologous host. J Invest Dermatol 32:477–479
94. Hendrix S, Handjiski B, Peters EM, Paus R (2005) A guide to assessing damage response pathways of the hair follicle: lessons from cyclophosphamide-induced alopecia in mice. J Invest Dermatol 125:42–51
95. Panteleyev AA, Botchkareva NV, Sundberg JP, Christiano AM, Paus R (1999) The role of the hairless (hr) gene in the regulation of hair follicle catagen transformation. Am J Pathol 155:159–171
96. Chi W, Wu E, Morgan BA (2013) Dermal papilla cell number specifies hair size, shape and cycling and its reduction causes follicular decline. Development 140:1676–1683
97. Rahmani W, Abbasi S, Hagner A, Raharjo E, Kumar R, Hotta A, Magness S, Metzger D, Biernaskie J (2014) Hair follicle dermal stem cells regenerate the dermal sheath, repopulate the dermal papilla, and modulate hair type. Dev Cell 31:543–558
98. Jahoda CA, Oliver RF, Reynolds AJ, Forrester JC, Horne KA (1996) Human hair follicle regeneration following amputation and grafting into the nude mouse. J Invest Dermatol 107:804–807
99. Rompolas P, Deschene ER, Zito G, Gonzalez DG, Saotome I, Haberman AM, Greco V (2012) Live imaging of stem cell and progeny behaviour in physiological hair-follicle regeneration. Nature 487:496–499
100. Greco V, Chen T, Rendl M, Schober M, Pasolli HA, Stokes N, Dela Cruz-Racelis J, Fuchs E (2009) A two-step mechanism for stem cell activation during hair regeneration. Cell Stem Cell 4:155–169
101. Kaushal GS, Rognoni E, Lichtenberger BM, Driskell RR, Kretzschmar K, Hoste E, Watt FM (2015) Fate of prominin-1 expressing dermal papilla cells during homeostasis, wound healing and Wnt activation. J Invest Dermatol
102. Yang H, Adam RC, Ge Y, Hua ZL, Fuchs E (2017) Epithelial-mesenchymal micro-niches govern stem cell lineage choices Cell 169:483–96 e13
103. Morgan MB, Rose P (2003) An investigation of apoptosis in androgenetic alopecia. Ann Clin Lab Sci 33:107–112
104. Randall VA, Hibberts NA, Hamada K (1996) A comparison of the culture and growth of dermal papilla cells from hair follicles from non-balding and balding (androgenetic alopecia) scalp. Br J Dermatol 134:437–444
105. Thornton MJ, Messenger AG, Elliott K, Randall VA (1991) Effect of androgens on the growth of cultured human dermal papilla cells derived from beard and scalp hair follicles. J Invest Dermatol 97:345–348
106. Chi WY, Enshell-Seijffers D, Morgan BA (2010) De novo production of dermal papilla cells during the anagen phase of the hair cycle. J Invest Dermatol 130:2664–2666
107. Jahoda CA (1998) Cellular and developmental aspects of androgenetic alopecia. Exp Dermatol 7:235–248
108. Stevenson S, Taylor AH, Meskiri A, Sharpe DT, Thornton MJ (2008) Differing responses of human follicular and nonfollicular scalp cells in an in vitro wound healing assay: effects of estrogen on vascular endothelial growth factor secretion. Wound Repair Regen 16:243–253
109. Pantelireis N, Higgins CA (2018) A bald statement - Current approaches to manipulate miniaturisation focus only on promoting hair growth. Exp Dermatol 27:959–965

110. Whiting DA (2001) Possible mechanisms of miniaturization during androgenetic alopecia or pattern hair loss. J Am Acad Dermatol 45:S81–S86
111. Upton JH, Hannen RF, Bahta AW, Farjo N, Farjo B, Philpott MP (2015) Oxidative stress-associated senescence in dermal papilla cells of men with androgenetic alopecia. J Invest Dermatol 135:1244–1252
112. Chew EG, Tan JH, Bahta AW, Ho BS, Liu X, Lim TC, Sia YY, Bigliardi PL, Heilmann S, Wan AC, Nothen MM, Philpott MP, Hillmer AM (2016) Differential expression between human dermal papilla cells from balding and non-balding scalps reveals new candidate genes for androgenetic alopecia. J Invest Dermatol 136:1559–1567
113. Shin W, Rosin NL, Sparks H, Sinha S, Rahmani W, Sharma N, Workentine M, Abbasi S, Labit E, Stratton JA, Biernaskie J (2020) Dysfunction of hair follicle mesenchymal progenitors contributes to age-associated hair loss. Dev Cell 53:185–98 e7
114. Jahoda CA (2003) Cell movement in the hair follicle dermis—more than a two-way street? J Invest Dermatol 121:ix–xi
115. Di Micco R, Krizhanovsky V, Baker D, d'Adda di Fagagna F (2021) Cellular senescence in ageing: from mechanisms to therapeutic opportunities. Nat Rev Mol Cell Biol 22:75–95
116. Garza LA, Yang CC, Zhao T, Blatt HB, Lee M, He H, Stanton DC, Carrasco L, Spiegel JH, Tobias JW, Cotsarelis G (2011) Bald scalp in men with androgenetic alopecia retains hair follicle stem cells but lacks CD200-rich and CD34-positive hair follicle progenitor cells. J Clin Invest 121:613–622
117. Choudhry R, Hodgins MB, Van der Kwast TH, Brinkmann AO, Boersma WJ (1992) Localization of androgen receptors in human skin by immunohistochemistry: implications for the hormonal regulation of hair growth, sebaceous glands and sweat glands. J Endocrinol 133:467–475
118. Hillmer AM, Hanneken S, Ritzmann S, Becker T, Freudenberg J, Brockschmidt FF, Flaquer A, Freudenberg-Hua Y, Jamra RA, Metzen C, Heyn U, Schweiger N, Betz RC, Blaumeiser B, Hampe J, Schreiber S, Schulze TG, Hennies HC, Schumacher J, Propping P, Ruzicka T, Cichon S, Wienker TF, Kruse R, Nothen MM (2005) Genetic variation in the human androgen receptor gene is the major determinant of common early-onset androgenetic alopecia. Am J Hum Genet 77:140–148
119. Tobin DJ, Gunin A, Magerl M, Paus R (2003) Plasticity and cytokinetic dynamics of the hair follicle mesenchyme during the hair growth cycle: implications for growth control and hair follicle transformations. J Investig Dermatol Symp Proc 8:80–86
120. Johansson JA, Headon DJ (2014) Regionalisation of the skin. Semin Cell Dev Biol 25–26:3–10
121. Gnedeva K, Vorotelyak E, Cimadamore F, Cattarossi G, Giusto E, Terskikh VV, Terskikh AV (2015) Derivation of hair-inducing cell from human pluripotent stem cells. PLoS One 10:e0116892
122. Topouzi H, Logan NJ, Williams G, Higgins CA (2017) Methods for the isolation and 3D culture of dermal papilla cells from human hair follicles. Exp Dermatol 26: 491–496

Chapter 2
Hair Regeneration and Rejuvenation: Pipeline of Medical and Technical Strategies

Summik Limbu and Paul Kemp

Abstract *Introduction*: The original discovery of the hair neogenesis ability of adult mouse dermal follicular cells sparked a great interest in developing an alternative therapeutic product for hair loss especially for androgenetic alopecia. While significant research has been conducted to understand fundamental processes such as hair follicle morphogenesis, cycling, and regeneration, there is currently no cell therapy in the market as a treatment for androgenetic alopecia. In this chapter, we discuss the medical and technical strategies that have changed over the years to treat androgenetic alopecia by utilising follicular cells, in particular, the dermal papilla and dermal sheath. *Methods*: We used Pubmed database to select published literature with keywords such as androgenetic alopecia, cell therapy, hair regeneration and hair rejuvenation. *Results*: Hair follicle dermal papilla and dermal sheath cells possess inductive ability to create new hair follicles. Clinical trials in the early 2000s was conducted by Intercytex and Aderans Research Institute to evaluate the efficacy of using follicular cells for hair regeneration as a treatment for androgenetic alopecia; however, although some hair neogenesis was observed, the trails failed to show expected cosmetic efficacy. Currently, the focus has moved away from regeneration and instead companies developing a cell therapy for androgenetic alopecia are focusing on rejuvenation of existing, miniaturising hair follicles. *Conclusions*: There is a clear need for alternative treatments for androgenetic alopecia and cell therapies using dermal papilla and dermal sheath cells show potential to reverse hair follicle miniaturisation.

Keywords Androgenetic alopecia · Cell therapy · Dermal papilla · Dermal sheath · Hair regeneration · Hair rejuvenation · Hair follicle miniaturisation

---

S. Limbu (✉)
Department of Bioengineering, Imperial College London, London, UK
e-mail: s.limbu17@imperial.ac.uk

P. Kemp
HairClone, Manchester, UK

Introduction

Androgenetic alopecia (AGA) is a common hair loss disorder impacting both genders. At least 50% of men are affected by AGA by the time they reach 50 years old and up to 40% of women are affected by age 70 [1]. In males, AGA is characterised by hair loss from the temples, vertex, and frontal scalp [2], while in females, it is characterised by overall hair thinning and hair loss especially from the central, frontal, and parietal scalp regions [3]. Although AGA is not considered a serious medical condition, hair loss can have a significant psychological impact. It is well-known that androgens, especially dihydrotestosterone (DHT), can affect the growth of androgen-sensitive follicles, leading to the production of small, fine hairs and eventually hair loss. Androgens exert their effects through the dermal papilla (DP) as androgen-sensitive follicles have higher levels of androgen receptors [4]. Therefore, the DP has become a target of emerging cell therapies for AGA.

Role of the Dermal Papilla in Androgenetic Alopecia

In AGA, affected hair follicles undergo a process known as miniaturisation which is the transformation of thick, pigmented terminal hairs into thin, unpigmented vellus hairs (Fig. 2.1), resulting in less coverage and appearance of a bald scalp [5]. The precise mechanism behind miniaturisation is not well understood; however, androgens have been shown to have a crucial role in AGA [6] and for the induction of miniaturisation [5]. In androgen-sensitive follicles, circulating androgens are converted to DHT by 5 α-reductase in the DP. DHT binds to androgen receptors (AR) and activates androgen-responsive genes such as TGF-β1, TGF-β2, DKK-1, and IL-6. TGF-β1 inhibits proliferation of keratinocytes in vitro [7] while TGF-β2 activates caspase-9 and caspase-3 leading to apoptosis of keratinocytes and suppression of hair growth in follicle organ cultures [8]. DKK-1 induces apoptosis of keratinocytes in vitro and in cultured hair follicles [9]. Similarly, IL-6 inhibits hair growth in vitro and injection of recombinant IL-6 into the hypodermis induces catagen onset in mice [10]. This suggests that activation of androgen-responsive genes promotes transition of the DP from a growth state to a regression state which contributes to the onset of miniaturisation.

In miniaturised follicles, there are also changes to the hair cycle. The duration of the anagen phase is shorter while the telogen phase is extended. This results in a higher percentage of follicles in the telogen phase [11, 12]. Previously, miniaturisation was thought to occur gradually over many hair cycles with a normal anagen phase lasting for years. However, miniaturisation is observed within 6–12 months clinically. This has led to the suggestion that miniaturisation may occur abruptly in one hair cycle [12]. Along with changes in the hair cycle, a decrease in both the volume of the DP and the number of cells within the DP has been observed in miniaturised follicles [11]. This is important as the volume of the DP is proportionally linked to the size

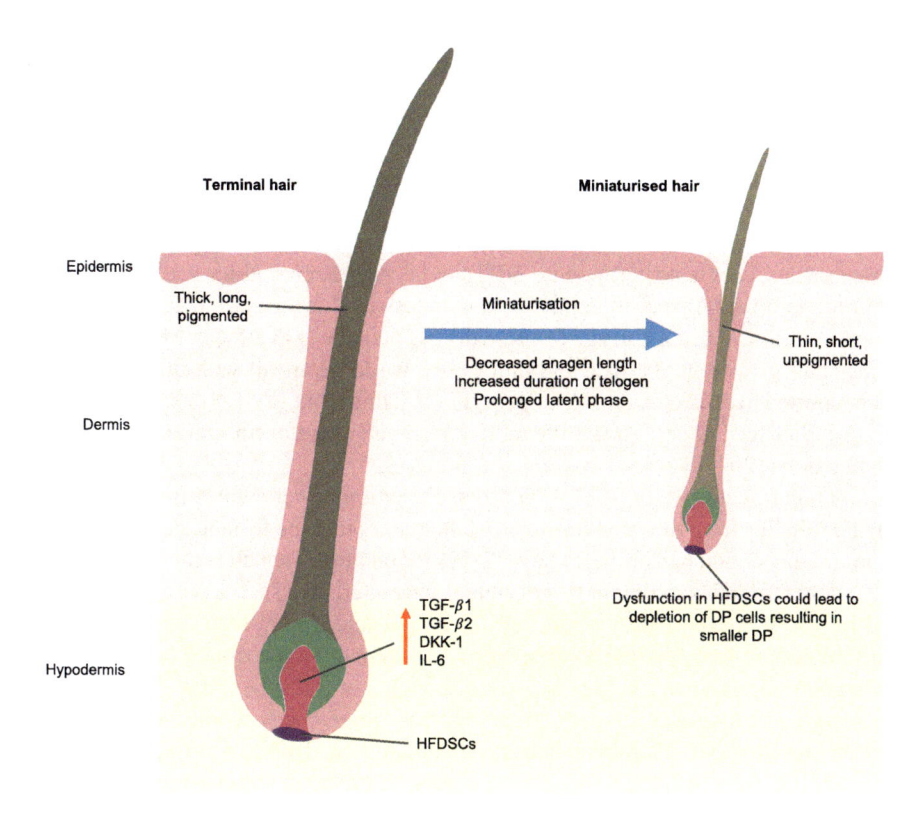

Fig. 2.1 Miniaturisation of a hair follicle
In the balding scalp, the androgen-sensitive terminal follicles transition into miniaturised vellus-like hairs. This transition is accompanied by changes in the hair shaft as well as the duration of different stages of the hair cycle. The anagen phase decreases, the telogen phase increases with a prolonged latent phase. In response to androgen-mediated signaling, there is an increase in secretion of TGF-β1, TGF-β2, DKK-1 and IL-6 from the DP, which all contribute to shorten hair growth. Finally, it has been hypothesised that the impairment in differentiation of hair follicle dermal stem cells (HFDSCs) could lead to decreased number of DP cells resulting in a smaller DP and production of a smaller hair shaft (Pantelireis and Higgins, 2018).

of the hair matrix and the width of the hair shaft produced [13]. The reduction in DP cell numbers could be due to a larger than normal number of cells moving out of the DP during the anagen to catagen phase or due to a lower number of cells moving into the DP from the dermal sheath (DS) during the telogen to anagen phase [14]. If the DP cell number cannot be replenished at early anagen, then this will result in a smaller DP and miniaturisation of a follicle in one hair cycle. This suggests that it might be possible to reverse miniaturisation quickly by stimulating an increase in DP cell numbers or by inhibiting cell movements out of the DP at the end of anagen.

Current Treatments for Androgenetic Alopecia

There are only two FDA-approved pharmacological treatments available for AGA: minoxidil and finasteride. Minoxidil can be used by both men and women. It is a potassium channel opener that triggers follicles in telogen to enter anagen phase, and also increases the duration of the anagen phase which increases the length of the hair produced [15]. Finasteride is an inhibitor of 5_α-reductase type II enzyme, which metabolises testosterone into DHT, and is only approved for use by men [3]. Both drugs must be used indefinitely as hair loss can occur upon discontinuation. Moreover, in some cases, usage of these drugs can have side effects. Minoxidil can induce hair growth on the face and limbs [16], while treatment with finasteride has been reported to cause sexual side effects [17–19]; this remains controversial as many other studies have found no significant link between usage of finasteride and sexual dysfunctions [20–22].

A popular treatment for AGA is hair transplantation surgery which involves transferring follicles from the occipital scalp (donor zone) into the frontal scalp (recipient zone). The occipital hair follicles have "donor dominance" which is the concept that these donor follicles maintain their original characteristics such as texture, colour, and length of anagen phase when transferred into the frontal scalp. As the occipital follicles are less sensitive to androgens, they remain in the frontal scalp without undergoing miniaturisation [23, 24]. Even though hair transplantation surgeries are performed with local anaesthesia and are relatively quick, many people have concerns undergoing surgeries. Moreover, hair transplantation involves one-to-one relocation, and therefore, it is difficult to use for patients who are beginning to notice hair loss as the eventual extent of hair miniaturisation and receding is unknown. Also, some women exhibit diffuse unpatterned alopecia (DUPA) in which hair miniaturisation occurs throughout the entire scalp including the donor area making hair transplantation unsuitable as a treatment [25]. Thus, alternative therapies are required which can give similar results to hair transplantations without the need to undergo surgery.

Effective therapy for AGA would address two main concerns: (1) replacing lost hair follicles through follicle regeneration, and (2) converting existing miniaturised hairs into terminal ones through rejuvenation (Figs. 2.2a, b). A cell therapy treatment would be beneficial for people with advanced AGA with insufficient amounts of donor hairs for transplantation. It would be permanent, like transplantations, but would not require the use of long-term medications. Also, it would be useful for both men and women making cell therapy an attractive alternative treatment for AGA.

Development of a Commercial Cell Therapeutic Product for Androgenetic Alopecia

As discussed extensively in Chap. 1, studies have shown that the DP is critical for the induction of new hair follicles [26, 27]. The inductive ability retained by cultured DP cells raised the possibility of using DP cells to treat AGA [28, 29]. Currently, several

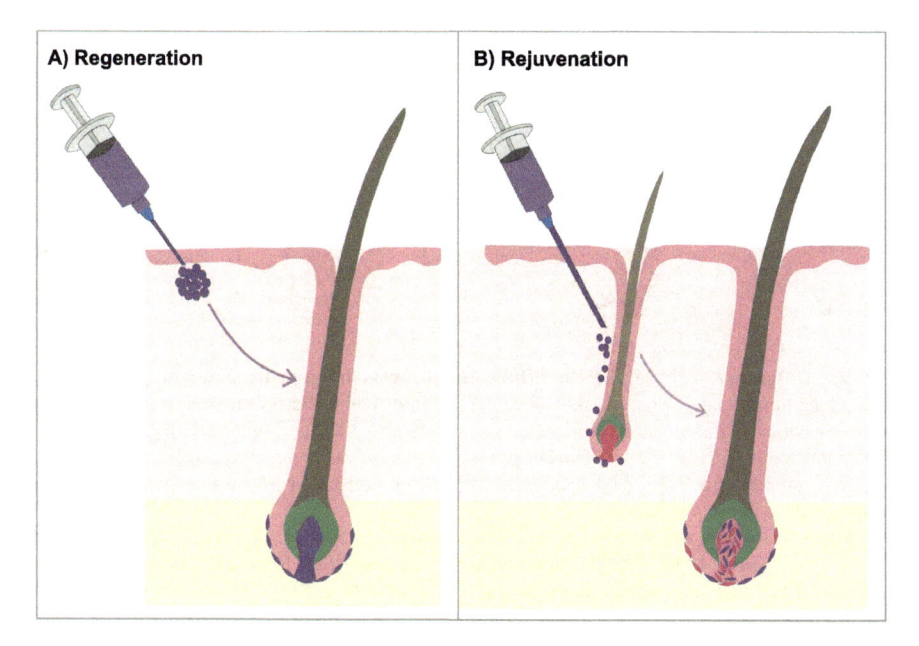

Fig. 2.2 Regeneration versus Rejuvenation
A) Regeneration is the process of producing new hair follicles through injection of DP cells. These cells induce the overlying epidermis into a follicular state, thereby producing a new hair follicle.
B) During rejuvenation of a hair follicle, the injected cells are thought to migrate to miniaturised follicles and replenish the DP. This increases DP size resulting in transition of vellus follicles into terminal ones.

companies are working on developing commercial products for AGA using DP cells as well as other follicular cells such as the DS cup cells. One of the earliest companies involved in developing cell therapeutic products was Intercytex. From its founding in 1997, Intercytex was interested in taking advantage of the inductive capability of cultured DP cells to treat hair loss. Their process would involve taking biopsies of few hair follicles from the occipital scalp, expanding DP cells to obtain sufficient number of cells, and then injecting the cells back into the balding scalp where the intention was that they would interact with local epithelial cells to induce the formation of new hairs as seen in rodent model systems. This process is known as "hair multiplication" as small numbers of hair follicles could be used to create indefinite number of new hair follicles in theory (Fig. 2.3a) and is different to hair transplantation surgeries which requires good quality and intact hair follicles (Fig. 2.3b).

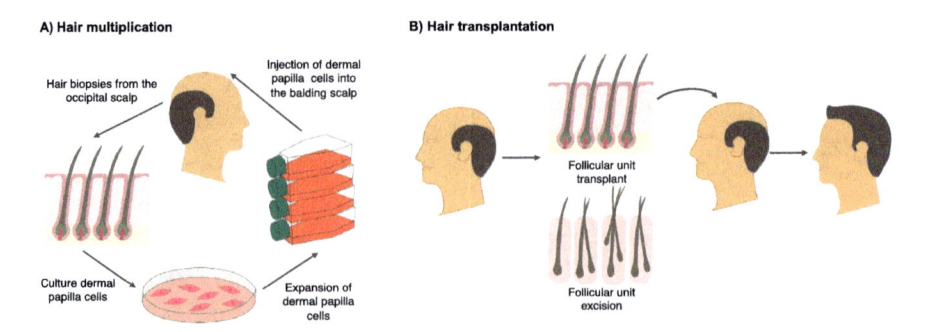

Fig. 2.3 Differences between hair multiplication process and hair transplantation

A) For hair multiplication process, skin biopsies containing hair follicles are taken from the occipital scalp of a patient. The DP is micro-dissected out of the hair follicle and cultured in a good manufacturing practice (GMP) laboratory. The DP cells are expanded to obtain large number of cells. After sufficient number of cells have been cultured, DP cells are injected back into the patient's balding scalp to induce hair regeneration.

B) During the surgery, hair follicles are harvested from the occipital scalp either as a strip of donor skin (strip harvesting follicular unit transplantation) or as individual follicular units (follicular unit excision). The donor follicular units are inserted into the balding scalp leading to hair growth after few months of surgery.

Autologous Versus Allogeneic Cell Therapy Considerations at Intercytex

There are advantages and disadvantages of using autologous or allogeneic cells to treat hair loss. Allogeneic cell therapy would involve the culture and expansion of cells from one donor to treat all patients. This process would be easier for patients as they do not have to wait for the expansion of their cells. Reynolds et al. [30] showed that it was possible to induce allogeneic hair growth using freshly isolated DS. For Intercytex, there were few concerns regarding the use of allogeneic DP cells. Injection of allogeneic cells could potentially elicit the host's immune system to attack donor cells which could trigger inflammation in the scalp surrounding the injection site and hair follicles. There was a worry that this could lead to the development of a common autoimmune disorder called alopecia areata. The source of allogeneic DP cells would be taken post-mortem which may not be acceptable by the consumers. Moreover, several hundred to thousands of follicles would have been required to generate a large, useable cell bank as regulators require a large bank of adventitious pathogen testing and the large costs of this would have to be spread over a large number of patients. This would in turn have been problematic as there are major limitations to overcome in maintaining the inductivity of cultured human DP cells for an extended period as cultured DP cells are not inductive after long-term passage [31]. Later, Inamatsu et al. [32] showed that the addition of keratinocyte-growth media to cultured vibrissae DP cells can maintain their inductivity for up to 70 passages. Even though this work was adapted for cultured human DP cells [33] the population doublings required to generate a sizeable cell bank may have affected the phenotype and inductive potential of DP cells. Also, in general, producing a large cell bank is technically challenging. With all these issues in mind, the decision to use autologous cells was made by Intercytex (personal communication with Intercytex).

Clinical Trials Involving Follicular Cells for Hair Regeneration

Intercytex licensed the technology from Hiroshima University, where the work by Mutsumi Inamatsu was carried out. In 2002, they built their manufacturing facility and developed a good manufacturing practice (GMP) process to culture autologous human DP cells. Several millions of cultured DP cells would be injected intradermally into the patient's scalp. Phase I clinical trials started in 2003 intending to regenerate new hair follicles in completely balding regions of patients with AGA. Results showed an increase in hair numbers with no adverse side effects (personal communication with Farjo Hair Institute). Phase II trials conducted in the UK with DP cell injections showed some efficacy and produced new hair follicles [34]; however, these were mainly fine vellus-type hairs which did not offer a cosmetic benefit. In 2010, Intercytex stopped research into the hair multiplication process and sold their assets to Aderans Research Institute (ARI).

ARI was founded in 2002 as a subsidiary of Aderans, a leading global wig manufacturer. Their original intellectual property (IP) was based on research from Zheng et al. [35] in which mice neonatal epidermal and dermal cells are injected together to form new hair follicles. ARI conducted Phase II clinical trials using autologous, cultured dermal cells combined with epidermal cells which were injected into the balding scalp of men and women. Their Phase II trials involved comparing the efficacy of injecting dermal and epidermal cells with injecting dermal cells only or with topical minoxidil and number of injections required [36]. No results of the clinical trials have been published and Aderans stopped funding this research in 2013 (Fig. 2.4).

Spheroid Versus Monolayer Cultures

Even though two-dimensional (2D) culture is widely used to culture various cell types, it changes both the morphology of cells and its' extracellular environment [37]. DP cells in their native environment rarely divide, while in culture, they proliferate and undergo several rounds of division [38, 39]. Also, cultured DP cells express smooth muscle $_\alpha$-actin, while this marker is not present in intact DP [40] indicating that 2D culture induces genotypic and phenotypic changes to the cells. Transcriptomic profiling of intact and 2D cultured DP cells showed that the DP cells undergo large changes in gene expression when cultured. Culturing DP cells as spheroids

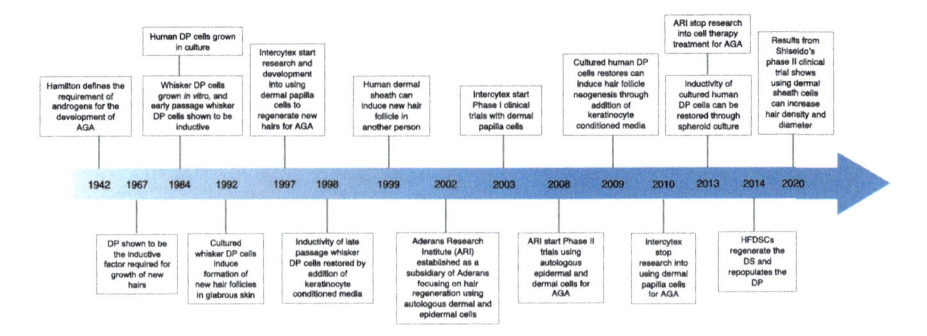

Fig. 2.4 Timeline of major research conducted which helped to understand the basis of hair regeneration and the companies involved in the development of a cell therapeutic product for androgenetic alopecia

mimic their environment in intact tissue and partially restores their transcriptome profile making them inductive again [41]. This finding highlighted the importance of the DP cell environment for inductivity and the possibility of using DP spheroids instead of dissociated cells. However, from a commercial perspective using DP spheroids would create an extra step before patient treatment. This step would involve creating spheroids from cultured cells which increases culture time and expenses. From a clinical perspective, additional time would be required to collect spheres before injecting them into the scalp of patients and injecting spheroids under the epidermis to induce hair growth may be difficult. Also, clinical trial results from Intercytex and Aderans indicated that the injection of DP cells into the balding scalp does not lead to formation of new hairs. Instead, they seemed to migrate into existing miniaturised hair follicles and reactivate them suggesting rejuvenation may be easier and more cost-effective rather than trying to create new follicles in the balding scalp.

Moving on- Focus on Rejuvenation

Rejuvenation is the transition of vellus into terminal hairs with a likely increase in the DP cell numbers resulting in a bigger DP. For rejuvenation to occur, the injected cells are expected to migrate into the DS and then to the DP which would lead to a reversal of miniaturisation. During a normal hair cycle, the DP is replenished by the migration of cells from the DS [39, 42, 43]. Recently, Yoshida et al. [44] have shown that intradermally injected human DS cup cells can migrate to hair follicles and were present in the DS and DS cup. Even though injected DS cells were not shown to be present in the DP, a clinical study by Shiseido, in collaboration with RepliCel Life Sciences, showed promising results for using DS cup cells to reverse miniaturisation. The clinical trial involved 65 patients with AGA (50 males and 15 females) who were injected with autologous DS cup cells, and their progress was monitored for 12 months. Results of the clinical trial showed that the injection of DS cup cells increased total hair density and cumulative hair diameter with the largest hair growth seen at 9 months post-injection, in both men and women [45]. They suggest that the increase in total hair density and cumulative hair diameter is due to the movement of injected cells into the DS and DP of miniaturised follicles and anagen initiation in these hairs. The clinical trial showed promising results and shows that rejuvenation can be a viable strategy to treat AGA.

Conclusion

Early research into the inductive property of DP cells using mouse models showed great promise as a potential cell therapy for hair loss and several labs and commercial biotechnology companies demonstrated follicular neogenesis. Although a lot more research has been conducted since then to utilise the inductive abilities of DP and DS to interact with epithelial cells and form early new hair follicular structures, a cell therapeutic treatment for AGA that regenerates follicles is still not here. Results from Intercytex and Aderans showed that follicle neogenesis by simply injecting single cell suspensions of DP and DS cells is not feasible as a commercial product for follicular neogenesis as cosmetically acceptable follicles would have to be produced in sufficient numbers and issues such as the direction of hair growth, hair cycle, and the formation of secondary structures (arrector pili muscle and sebaceous glands) would need to be addressed. There are several groups still pursuing this and looking at the in vitro formation of organoid structures that could then be implanted but the commercial costs of such extensive in vitro culture would be considerable. The cost of the therapy is often less important in other life changing diseases where a cost–benefit to the quality of life can be used to justify payment of the treatment from national health bodies or health insurers. In this case, it will be the patient themselves that pays for this therapy, and this must be taken into account when designing new therapies. The early results into the use of injected suspensions of human DP or DS cells have also pointed to a new avenue of rebuilding existing miniaturised follicles. Although, with current GMP cell culture systems, this will, at least initially, be somewhat expensive to produce but it will be orders of magnitude less than the cost of producing and delivering multi-cell type organoid structures. Looking ahead, the need for a cosmetically efficacious treatment for AGA is still there and there is a massive pent up commercial demand as well. This guarantees that solutions will continue to be pursued and advances in other areas of cellular therapy that will address manufacturing cost issues will, in future, also be adapted for this application.

Acknowledgements We would like to thank Claire Higgins, Nilofer Farjo and Bessam Farjo for their insights.

Conflict of Interest Summik Limbu is funded through an EPSRC iCASE Ph.D. studentship sponsored by HairClone. Paul Kemp is a founder, shareholder, and CEO of HairClone and was previously the CEO of Intercytex.

References

1. Trüeb RM (2002) Molecular mechanisms of androgenetic alopecia. Exp Gerontol 37(8–9):981–990
2. Cranwell W, Sinclair R (2016) Male androgenetic alopecia. Hair, Hair Growth Hair Disord 159–170. https://doi.org/10.1007/978-3-540-46911-7_9
3. Fabbrocini G, Cantelli M, Masarà A, Annunziata MC, Marasca C, Cacciapuoti S (2018) Female pattern hair loss: a clinical, pathophysiologic, and therapeutic review. Int J Women's Dermatol 4(4):203–211. https://doi.org/10.1016/j.ijwd.2018.05.001
4. Hibberts NA, Howell AE, Randall VA (1998) Balding hair follicle dermal papilla cells contain higher levels of androgen receptors than those from non-balding scalp. J Endocrinol 156(1):59–65. https://doi.org/10.1677/joe.0.1560059
5. Sinclair RD (2004) Male androgenetic alopecia. JMHG 1(4):319–327. https://doi.org/10.1007/978-3-540-46911-7_9
6. Hamilton JB (1942) Male hormone stimulation is a prerequisite and incitant in common baldness. J Invest Dermatol 5(6):451–480. https://doi.org/10.1038/jid.1942.62
7. Inui S, Fukuzato Y, Nakajima T, Yoshikawa K, Itami S (2002) Androgen-inducible TGF-beta1 from balding dermal papilla cells inhibits epithelial cell growth: a clue to understand paradoxical effects of androgen on human hair growth. FASEB J 16: 1967–9
8. Hibino T, Nishiyama T (2004) Role of TGF-β2 in the human hair cycle. J Dermatol Sci 35: 9–18 https://doi.org/10.1016/j.jdermsci.2003.12.003
9. Kwack MH, Sung YK, Chung EJ, Im SU, Ahn JS, Kim MK, Kim JC (2008) Dihydrotestosterone-inducible dickkopf 1 from balding dermal papilla cells causes apoptosis in follicular keratinocytes. J Invest Dermatol 128(2):262–269. https://doi.org/10.1038/sj.jid.5700999
10. Kwack MH, Ahn JS, Kim MK, Kim JC, Sung YK (2012) Dihydrotestosterone-inducible IL-6 inhibits elongation of human hair shafts by suppressing matrix cell proliferation and promotes regression of hair follicles in mice. J Invest Dermatol 132(1):43–49. https://doi.org/10.1038/JID.2011.274
11. Jahoda CAB (1998) Cellular and developmental aspects of androgenetic alopecia. Exp Dermatol 7(5):235–248. https://doi.org/10.1111/j.1600-0625.2007.00666.x
12. Whiting DA (2001) Possible mechanisms of miniaturization during androgenetic alopecia or pattern hair loss. J Am Acad Dermatol 45(3 SUPPL.):81–86. https://doi.org/10.1067/mjd.2001.117428
13. Chi W, Wu E, Morgan BA (2013) Dermal papilla cell number specifies hair size, shape and cycling and its reduction causes follicular decline. Development (Cambridge) 140(8):1676–1683. https://doi.org/10.1242/dev.090662
14. Pantelireis N, Higgins CA (2018) A bald statement—current approaches to manipulate miniaturisation focus only on promoting hair growth. Exp Dermatol 27(9):959–965. https://doi.org/10.1111/exd.13690
15. Messenger AG, Rundegren J (2004) Minoxidil: mechanisms of action on hair growth. Br J Dermatol 150(2):186–194. https://doi.org/10.1111/j.1365-2133.2004.05785.x
16. Peluso AM, Misciali C, Vincenzi C, Tosti A (1997) Diffuse hypertrichosis during treatment with 5% topical minoxidil. Br J Dermatol 136:118–120. https://doi.org/10.1046/J.1365-2133.1997.D01-1156.X
17. Irwig MS, Kolukula S (2011) Persistent sexual side effects of finasteride for male pattern hair loss. J Sex Med 8(6):1747–1753. https://doi.org/10.1111/J.1743-6109.2011.02255.X
18. Mella JM, Perret MC, Manzotti M, Catalano HN, Guyatt G (2010) Efficacy and safety of finasteride therapy for androgenetic alopecia: a systematic review. Arch Dermatol 146(10):1141–1150. https://doi.org/10.1001/archdermatol.2010.256
19. Traish AM (2020) Post-finasteride syndrome: a surmountable challenge for clinicians. Fertil Steril 113(1):21–50. https://doi.org/10.1016/J.FERTNSTERT.2019.11.030

20. Haber RS, Gupta AK, Epstein E, Carviel JL, Foley KA (2019) Finasteride for androgenetic alopecia is not associated with sexual dysfunction: a survey-based, single-centre, controlled study. J Eur Acad Dermatol Venereol 33(7):1393–1397. https://doi.org/10.1111/JDV.15548
21. Pallotti F, Senofonte G, Pelloni M, Cargnelutti F, Carlini T, Radicioni AF, Rossi A, Lenzi A, Paoli D, Lombardo F (2020) Androgenetic alopecia: effects of oral finasteride on hormone profile, reproduction and sexual function. Endocrine 68:688–694. https://doi.org/10.1007/s12020-020-02219-2
22. Tosti A, Piraccini B, Soli M (2001) Evaluation of sexual function in subjects taking finasteride for the treatment of androgenetic alopecia. J Eur Acad Dermatol Venereol 15(5):418–421. https://doi.org/10.1046/J.1468-3083.2001.00315.X
23. Dinh HV, Sinclair RD, Martinick J (2008) Donor site dominance in action: transplanted hairs retain their original pigmentation long term. Dermatol Surg 34(8):1108–1111. https://doi.org/10.1111/j.1524-4725.2008.34228.x
24. Orentreich N (1959) Autografts in alopecias and other selected dermatological conditions. Ann N Y Acad Sci 83(3):463–479. https://doi.org/10.1111/J.1749-6632.1960.TB40920.X
25. Jimenez F, Alam M, Vogel JE, Avram M (2021) Hair transplantation: basic overview. J Am Acad Dermatol 85(4): 803–814. https://doi.org/10.1016/j.jaad.2021.03.124
26. Oliver RF (1966) Histological studies of whisker regeneration in the hooded rat. J Embryol Exp Morphol 16(2):231–244
27. Oliver RF (1967) The experimental induction of whisker growth in the hooded rat by implantation of dermal papillae. J Embryol Exp Morphol 18(1):43–51
28. Jahoda CAB, Oliver RF (1984) Vibrissa dermal papilla cell aggregative behaviour in vivo and in vitro. J Embryol Exp Morphol 79:211–224.
29. Jahoda CAB, Reynolds AJ, Oliver RF (1993) Induction of hair growth in ear wounds by cultured dermal papilla cells. J Invest Dermatol 101(4):584–590. https://doi.org/10.1111/1523-1747.ep12366039
30. Reynolds AJ, Lawrence C, Cserhalmi-Friedman PB, Christiano AM, Jahoda CAB (1999) Trans-gender induction of hair follicles. Nature 402(6757):33–34. https://doi.org/10.1038/46938
31. Jahoda CAB, Horne KA, Oliver RF (1984) Induction of hair growth by implantation of cultured dermal papilla cells. Nature 311(5986):560–562. https://doi.org/10.1038/311560a0
32. Inamatsu M, Matsuzaki T, Iwanari H, Yoshizato K (1998) Establishment of rat dermal papilla cell lines that sustain the potency to induce hair follicles from afollicular skin. J Invest Dermatol 111(5):767–775. https://doi.org/10.1046/j.1523-1747.1998.00382.x
33. Qiao J, Zawadzka A, Philips E, Turetsky A, Batchelor S, Peacock J, Durrant S, Garlick D, Kemp P, Teumer J (2009) Hair follicle neogenesis induced by cultured human scalp dermal papilla cells. Regen Med. https://doi.org/10.2217/rme.09.50
34. Adams JU (2011) Raising hairs. Nat Biotechnol 29(6):474–476. https://doi.org/10.1038/nbt0611-474
35. Zheng Y, Du X, Wang W, Boucher M, Parimoo S, Stenn KS (2005) Organogenesis from dissociated cells: generation of mature cycling hair follicles from skin-derived cells. J Investig Dermatol 124(5):867–876. https://doi.org/10.1111/J.0022-202X.2005.23716.X
36. de Sousa ICVD, Tosti A (2013) New investigational drugs for androgenetic alopecia. Expert Opin Invest Drugs 22(5):573–589. https://doi.org/10.1517/13543784.2013.784743
37. Kapałczyńska M, Kolenda T, Przybyła W, Zajączkowska M, Teresiak A, Filas V, Ibbs M, Bliźniak R, Łuczewski Ł, Lamperska K (2018) 2D and 3D cell cultures—a comparison of different types of cancer cell cultures. Arch Med Sci. https://doi.org/10.5114/aoms.2016.63743
38. Messenger A (1984) The culture of dermal papilla cells from human hair follicles. Br J Dermatol 110:685–689. https://doi.org/10.1111/j.1365-2133.1984.tb04705.x
39. Tobin DJ, Gunin A, Magerl M, Paus R (2003) Plasticity and cytokinetic dynamics of the hair follicle mesenchyme during the hair growth cycle: Implications for growth control and hair follicle transformations. J Invest Dermatol Symp Proc 80–86. https://doi.org/10.1046/j.1523-1747.2003.12177.x

40. Jahoda CAB, Reynolds AJ, Chaponnier C, Forester JC, Gabbiani G (1991) Smooth muscle α-actin is a marker for hair follicle dermis in vivo and in vitro. J Cell Sci 99: 627–636
41. Higgins CA, Chen JC, Cerise JE, Jahoda CAB, Christiano AM (2013) Microenvironmental reprogramming by threedimensional culture enables dermal papilla cells to induce de novo human hair-follicle growth. Proc Natl Acad Sci USA 110(49):19679–19688. https://doi.org/10.1073/pnas.1309970110
42. McElwee KJ, Kissling S, Wenzel E, Huth A, Hoffmann R (2003) Cultured peribulbar dermal sheath cells can induce hair follicle development and contribute to the dermal sheath and dermal papilla. J Invest Dermatol 121:1267–1275. https://doi.org/10.1111/j.1523-1747.2003.12568.x
43. Rahmani W, Abbasi S, Hagner A, Raharjo E, Kumar R, Hotta A, Magness S, Metzger D, Biernaskie J (2014) Hair follicle dermal stem cells regenerate the dermal sheath, repopulate the dermal papilla, and modulate hair type. Dev Cell 31(5):543–558. https://doi.org/10.1016/j.devcel.2014.10.022
44. Yoshida Y, Soma T, Matsuzaki T, Kishimoto J (2019) Wnt activator CHIR99021-stimulated human dermal papilla spheroids contribute to hair follicle formation and production of reconstituted follicle-enriched human skin. Biochem Biophys Res Commun 516:599–605. https://doi.org/10.1016/j.bbrc.2019.06.038
45. Tsuboi R, Niiyama S, Irisawa R, Harada K, Nakazawa Y, Kishimoto J (2020) Autologous cell-based therapy for male and female pattern hair loss using dermal sheath cup cells: a randomized placebo-controlled double-blinded dose finding clinical study. J Am Acad Dermatol 83: 109–116. https://doi.org/10.1016/j.jaad.2020.02.033

Part II
Cells and Structures Involved in Hair Follicle Regeneration

Chapter 3
Cells and Structures Involved in Hair Follicle Regeneration: An Introduction

Jerry Tsai and Luis A. Garza

Abstract *Introduction* The hair follicle spans across the skin epidermis, dermis, and hypodermis and undergoes repeated cycles of growth (anagen), regression (catagen), quiescence (telogen), and shedding (exogen). Research over the past decades have increasingly shown that hair regeneration not only depends on hair follicle stem cells (HFSCs), but also involves coordinated interactions with surrounding cells and structures. *Methods* PubMed, Web of Science, and Google Scholar were used to find peer-reviewed articles examining HFSCs, fibroblasts, adipocytes, immune cells, lymphatics, blood vessels, nerves, and exosomes in the context of hair regeneration. *Results* Distinct populations of HFSCs exist in the bulge and secondary germ of the hair follicle. During hair cycling, HFSCs interact with specialized dermal fibroblasts called dermal papilla cells and dermal sheath cells, which have hair-inducing properties. Dermal adipocyte differentiation and remodeling, including fluctuations in dermal adipose thickness, are associated with changes in HFSC activity. Perifollicular macrophages, mast cells, cytotoxic T cells, and regulatory T cells modulate hair growth and have been implicated in various hair disorders. The organization of lymphatic and blood vessels relative to the hair follicle may be correlated with phases of the hair cycle. Substance P and cutaneous norepinephrine signaling may also regulate hair growth. Intercellular communication in the hair follicle microenvironment may be mediated by extracellular vesicles called exosomes, which deliver a variety of cargo in a cell-specific manner. *Conclusions* Understanding the interdependence of cells and structures involved in the hair growth cycle, as well as mediators of intercellular signaling, are essential for the study of hair regeneration.

Keywords Hair follicle stem cell · Bulge · Secondary germ · Fibroblast · Dermal papilla · Dermal sheath · Preadipocyte · Adipocyte · Immune cell · Macrophage · Mast cell · Lymphocyte · T cell · Lymphatic vessel · Blood vessel · Nerve · Extracellular vesicle · Exosome

J. Tsai · L. A. Garza (✉)
Department of Dermatology, Johns Hopkins University School of Medicine, Baltimore, MD, USA
e-mail: LAG@jhmi.edu

© The Author(s), under exclusive license to Springer Nature Switzerland AG 2022
F. Jimenez and C. Higgins (eds.), *Hair Follicle Regeneration*, Stem Cell Biology and Regenerative Medicine 72, https://doi.org/10.1007/978-3-030-98331-4_3

Summary

In this chapter, we introduce major cells and structures involved in hair cycling and regeneration, including hair follicle stem cells, fibroblasts, adipocytes, immune cells, lymphatics, blood vessels, and nerves (Fig. 3.1). We also introduce the concept of exosomes and describe their role in intercellular communication. While substantial insight has arisen from the study of mouse models, differences between murine and human hair biology, such as hair cycle duration, synchronicity of hair growth,

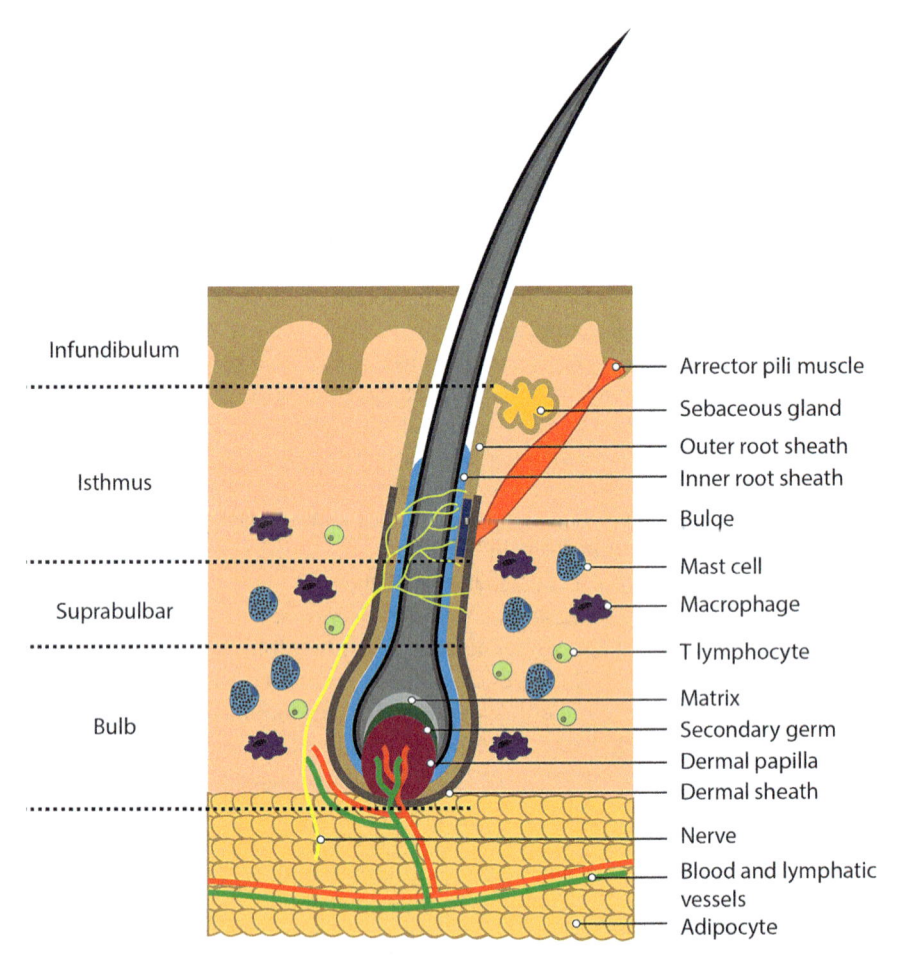

Fig. 3.1 Schematic diagram of cells and structures involved in hair follicle regeneration (anagen hair follicle). Hair growth and cycling involve coordination between various cell types within and outside the hair follicle. These include distinct populations of hair follicle stem cells in the bulge and secondary germ, fibroblasts in the dermal papilla and dermal sheath, dermal adipocytes, immune cells (e.g., macrophages, mast cells, and T lymphocytes), lymphatics, blood vessels, and nerves. Some intercellular interactions may be mediated through signaling by exosomes (discussed in the text).

and response to hormones like estrogen and androgen [1], may prevent direct translation of findings from one species to another. Corresponding findings in human hair follicles, clinical correlates, and clinical applications are emphasized whenever possible.

Key points:

- Hair regeneration depends on the activity of multiple distinct populations of hair follicle stem cells (HFSCs).
- Hair growth and cycling may also be modulated by fibroblasts in the dermal papilla and dermal sheath, dermal adipocytes, immune cells, lymphatic and blood vessels, and nerves.
- Understanding the interdependence of these cells and structures, as well as mediators of intercellular signaling, are essential for the study of hair regeneration.

Introduction

The skin may be divided into the epidermis, dermis, and hypodermis. The epidermis is avascular and composed mostly of keratinocytes. The dermis contains collagen and elastin fibers, fibroblasts, immune cells, lymphatic and blood vessels, and nerves. The hypodermis includes adipose tissue, as well as larger nerves and vessels. The hair follicle spans across these skin layers and are found in association with the sebaceous gland and arrector pili muscle as part of the pilosebaceous unit.

The hair has many functions in animals including mechanoreception, thermoregulation, and piloerection [2, 3]. The hair follicle also participates in epidermal and dermal wound healing [4–6]. In humans, the hair serves a cosmetic role in social interactions, and both abnormal hair growth (hypertrichosis) and hair loss (alopecia) may cause significant psychological distress. While numerous pharmacologic (e.g., 5α-reductase inhibitors, minoxidil) and procedural (e.g., hair transplantation, laser therapy) treatments for hair loss have been developed, novel strategies for hair regeneration may continue to arise with deeper understanding of hair follicle biology.

The hair follicle is a mini organ featuring well-coordinated cellular interactions within its internal compartments and with external structures. During hair development, epithelial-mesenchymal signaling involving molecules like Wnt, Shh, EDA, BMP, FGF, and TNF result in sequential development of the hair placode, hair germ, hair peg, bulbous peg, and mature follicle along with the adjacent dermal papilla [3, 7]. The mature hair is composed of multiple concentric layers, including the outermost connective tissue sheath, followed by the outer root sheath, companion layer, inner root sheath, hair cuticle, hair cortex, and hair medulla. The upper segment of the hair follicle is permanent and includes the infundibulum and isthmus. The remaining lower segment degrades and regenerates with each hair cycle, and it includes the hair bulb located at the base of the hair follicle. Matrix keratinocytes in the hair bulb undergo proliferation and differentiation during the growth (anagen) phase of the hair cycle to drive hair elongation, and the duration of anagen (and possibly keratinization rate) determines hair length. The hair may subsequently undergo apoptosis-driven regression (catagen), relative quiescence (telogen), and shedding (exogen), before reentering the cycle autonomously [1, 3, 8, 9]. Dysfunction of hair cycling is seen in various hair loss disorders such as telogen effluvium, anagen effluvium, and androgenetic alopecia [2].

Hair Follicle Stem Cells

Hair regeneration depends on the activity of hair follicle stem cells (HFSCs) that exist in distinct compartments of the hair follicle. Stem cells responsible for hair regeneration were traditionally thought to reside exclusively in the hair bulb at the base of the hair follicle. Using pulse-chase methods, however, a population of slow-cycling stem cells were identified at the outer root sheath of the lower isthmus in mice, in a histologically identifiable outpouching called the bulge where the arrector pili muscle attaches [10]. Subsequent studies showed that bulge stem cells contribute to formation of all epithelial lineages of the hair follicle [11–13] and the interfollicular epidermis following wounding [14]. While adult human hair follicles do not contain a morphologically distinct bulge [15, 16], label-retaining bulge stem cells still localize to a similar region near the insertion of the arrector pili [17, 18]. Whether bulge stem cells are affected may underlie the reversibility of hair loss disorders. Cicatricial (scarring) alopecia is thought to be permanent due to destruction of bulge stem cells [19]. On the other hand, alopecia areata is a nonscarring hair loss disorder that involves lymphocyte-driven inflammation at the hair bulb and relative sparing of the bulge, and it may be improved with treatments like corticosteroids, minoxidil, and JAK inhibitors [17, 20–22].

The bulge also gives rise to a population of progenitor cells in the secondary germ, located at the base of the hair follicle and above the dermal papilla in telogen hair follicles [23, 24]. In the telogen to anagen transition, successive activation of secondary germ cells followed by bulge cells contribute to formation of the new hair follicle. The secondary germ cells, in particular, give rise to the hair matrix that generates the anagen hair shaft [23, 25, 26].

Gene expression and immunohistochemical studies have identified several stem cell markers in human hair, including keratin 15 that localizes to the bulge and CD200 that localizes to the bulge and secondary germ [16, 18, 27, 28]. The Ber-EP4 antibody, which binds epithelial adhesion molecule (EpCAM), has also been used as a specific marker for the secondary germ in human follicles [27, 29, 30]. While CD34 is a bulge cell marker in mice [31], it is not expressed in human bulge cells [17]. Instead, CD34 positive cells are found in the lower outer root sheath in human hair and may represent descendants of keratin 15 positive bulge cells [17, 28, 32]. Bald scalp of individuals with androgenetic alopecia (AGA), also a nonscarring hair loss disorder, has been shown to retain keratin 15 positive bulge cells but lack progenitor cells derived from the bulge, and the observation that bulge stem cells are relatively undisrupted in AGA may support potential reversibility of the disorder [27]. Lgr5, a target of Wnt signaling, has also been identified as a HFSC marker in mice [33]. Decreased expression of Lgr5 was observed in bald scalp affected by AGA [27], but the importance of Lgr5, along with the closely related murine HFSC marker Lgr6 [34], both remain to be studied in human hair follicle.

Activation and suppression of hair growth depend on the interaction of multiple signaling pathways between HFSCs and surrounding structures. Studies using mouse models have revealed opposite roles of the Wnt/β-catenin and BMP pathways on HFSC activity, which promote and inhibit hair growth, respectively [23, 35–39]. Additional molecular regulators of hair cycling that have been identified in mice include FGF, Shh, and Noggin [23, 40–43]. The long duration of anagen (several years), small proportion of hair follicles in other phases (<10%), and asynchronous growth of human hair significantly increase the difficulty of observing the entire human hair cycle and underlying signaling events in culture [1–3]. However, a recent study successfully isolated telogen and early-anagen hair follicles from hair grafts of patients undergoing hair transplantation and confirmed activation the Wnt pathway during the telogen to anagen transition [44]. Differences in the expression pattern of individual Wnt proteins and Wnt inhibitors in comparison to mice were noted [44], highlighting the need to reexamine other signaling pathways in human hair.

Dermal Papilla and Dermal Sheath

Hair follicle regeneration also relies on interactions of HFSCs with closely associated fibroblasts in the dermis. The dermal papilla (DP) is a population of mesenchymal cells located at the base of the hair follicle and serves as one of the major sources of signaling for hair growth [2, 3, 45]. During catagen, degeneration of the lower segment of the hair follicle causes upward migration of the DP towards the bulge, where DP cells may interact with bulge HFSCs through telogen, eventually leading to initiation of anagen. As the lower segment of the hair follicle regenerates during anagen, the DP is pushed downward and away from bulge HFSCs but continue to interact with HFSCs in the secondary germ [1, 3, 45].

Experimentally implanted DP cells can induce hair follicle formation and growth in mice [46, 47] and depend on Wnt, BMP, and FGF signaling [23, 48, 49]. Human DP cells often lose their hair-inducing property in culture, but use of three-dimensional spheroid cultures of DP cells have been shown to prevent this, allowing de novo formation of hair follicles upon implantation of DP aggregates in human skin [50]. High expression of androgen receptors have been also found in human frontal scalp DP cells, which may explain androgen-induced hair follicle miniaturization in androgenetic alopecia (AGA) and mitigation of hair loss in AGA with 5α-reductase inhibitors (e.g., finasteride, dutasteride) [45, 51, 52].

A separate population of dermal sheath (DS) cells encapsulate the DP, contributing to both the maintenance of DP cells and hair growth. Experimental implantation of cultured DS cells in mice induces the formation of new DP cells and hair follicles

[53, 54]. Conversely, ablation of the DP and adjacent DS in telogen follicle has been found to impair initiation of hair growth in mice [25]. The hair growth and DP-inducing properties of DS cells may arise from a subset of self-renewing cells within the DS that have been called hair follicle dermal stem cells [55]. The DS also expresses α-smooth muscle actin (αSMA), exhibits contractile function in mice, and is responsible for pushing the DP toward the hair follicle bulge during catagen to allow DP-bulge HFSC crosstalk [56], but whether this function is present in human DS remains unknown. An intriguing human study in 1999 observed formation of new hair follicles around one month after transplanting human DS tissue from terminal follicle of a male donor to the inner forearm of a genetically unrelated female recipient [57]. The use of DS cells for treating androgenetic alopecia was also studied in a recent randomized controlled trial, which observed increased hair density and diameter following autologous injection of DS cells, although improvements were no longer present at 12 months after treatment [58]. Together, studies on DP and DS cells suggest that they are promising targets for modulating hair regeneration.

Adipose Cells

White adipose tissue has traditionally been recognized for its roles in energy balance, thermal insulation, and mechanical cushioning. Brown adipose tissue allows thermogenesis through uncoupling of oxidative phosphorylation but are rare in adult humans [59], and the following discussions will focus on white adipose tissue. Although adipose tissue is often associated with the hypodermis, a histologically distinct population of adipocytes has been identified in the dermis [60]. These dermal adipocytes are closely associated with and may undergo reciprocal signaling with the hair follicle to coordinate adipocyte remodeling and hair cycling [61].

Dermal adipose tissue in mammalian skin undergoes structural changes during the hair cycle and modulates HFSC activation through shared signaling pathways. Dermal adipose in mice thickens during transition to anagen, with both hyperplasia and hypertrophy of adipocytes, followed by decrease in thickness upon transition to catagen [62–66]. Related findings have been observed in human skin, with regions affected by alopecia having decreased adipose thickness [62]. Studies using mouse models found that the coordination of dermal adipocyte proliferation and differentiation with hair growth may be driven by epidermal Wnt/β-catenin signaling [62] and Shh signaling from the hair follicle [67], with the latter acting through peroxisome proliferator-activated receptor gamma (PPAR-gamma) in dermal preadipocytes to stimulate their proliferation and differentiation to adipocytes [67]. Preadipocytes secrete platelet-derived growth factors (PDGF) to promote both adipogenesis [65] and transition to anagen in hair follicles [63]. In contrast, adipocytes promote quiescence of HFSCs through bone morphogenetic protein 2 (Bmp2) that increases during telogen and decreases during anagen [38]. The importance of adipose in human hair growth has been suggested in the context of scarring alopecia, which may arise from dysfunction of PPAR-gamma signaling [68]. Further human studies of adipocyte remodeling and their association with HFSC activity will help clarify whether findings in mice are applicable to human, and whether they may be targeted for hair regeneration.

Immune Cells and Responses

The skin functions as a barrier and may respond to foreign substance through immune cells such as granulocytes (i.e., neutrophils, mast cells, basophils, eosinophils), macrophages, natural killer cells, dendritic cells, and lymphocytes. Besides their well-established roles in the innate and adaptive immune responses, several of these immune cells may also affect hair growth, loss, and regeneration [69, 70]. These include macrophages, mast cells, and T lymphocytes, which are commonly found near the hair follicle in mice and humans [69, 71, 72].

In mice, perifollicular macrophages have been found to maintain hair in the telogen phase [73, 74]. This may occur through secretion of oncostatin M by TREM2$^+$ macrophages, which acts through JAK-STAT5 signaling to promote HFSC quiescence [74]. Consistent with these findings, induction of macrophage apoptosis promotes early transition to anagen through upregulation of Wnt/β-catenin ligands, including Wnt7b and Wnt10a [73]. Ex vivo studies of human hair follicles have similarly demonstrated secretion of Wnt7b and Wnt10a by macrophages during anagen [75]. Macrophages also express the fibroblast growth factor proteins Fgf-5 and Fgf-5S to promote and inhibit the transition to catagen, respectively [76, 77]. The catagen-promoting characteristic of Fgf5 is supported by findings of abnormally long body hair in mice (angora phenotype) and eyelashes in humans that occur with dysfunction of Fgf5 [78, 79].

Mast cells may also regulate the hair cycle and contribute to the development of pathologic hair loss. Stimulation of mast cell degranulation with compound 48/80 and adrenocorticotropic hormone (ACTH) induces anagen in telogen mouse hair follicles in vivo, whereas administration of mast cell stabilizers and selective inhibition of mast cell mediators with antihistamines and serotonin antagonists slow the development of anagen [80]. On the other hand, stimulation of mast cell degranulation with substance P and ACTH also promote anagen to catagen transition in mice in vivo, while inhibition of mast cell degranulation delays catagen development [81], suggesting contrasting roles of mast cells at different phases of the murine hair cycle. Mast cells are present in human hair follicles [82], and increased mast cell degranulation induced by the stress-related neuropeptide substance P has been associated with the development of catagen [83]. Individuals with androgenetic alopecia (AGA) show prominent mast cell degranulation in hair follicles [84]. Subsequent gene expression studies found upregulation of mast cell markers in AGA including prostaglandin D$_2$ synthase (PTGDS) [85], in agreement with the association of PTGDS and prostaglandin D$_2$ with AGA [86]. Mast cells' inflammatory activity and interaction with CD8 + T cells have also been implicated in the development of alopecia areata [87].

Multiple classes of T cells may affect hair growth in mice and humans. Studies using mouse models and in patients with alopecia areata have identified infiltration of hair follicles by cytotoxic CD8 + NKG2D + T cells as important for the development of alopecia areata [88, 89]. CD8 + NKG2D + T cells secrete interferon gamma to activate the JAK-STAT pathway and maintain quiescence of HFSCs, and the use of JAK inhibitors to target this pathway has seen success in treating alopecia areata [22, 89, 90]. FOXP3 + regulatory T cells, on the other hand, promote anagen entry in mice through expression of Jagged 1, which activates the Notch signaling pathway [91]. Ablation of regulatory T cells reduces hair growth in mice [91], and single nucleotide polymorphisms and dysfunction of regulatory T cells have been associated with development of alopecia areata [88, 92].

Immune cells and responses contribute to wound-induced hair neogenesis (WIHN) in mice, in which full thickness wounding of skin causes de novo hair formation [93]. These include dermal γδ T cells that increase Wnt signaling through Fgf9, as well as macrophages that activate AKT/β-catenin signaling through cytokines like TNF-α and TGF-β1 [94–96]. Double strand RNA (dsRNA) released by damaged cells also trigger hair regeneration through the innate immune system receptor Toll-Like Receptor 3 (TLR3) in mice [97, 98]. Despite consistent observations in mice, reports of WIHN in humans remain scarce [99–102].

Lymphatics

Cutaneous lymphatic vessels are located in the dermis in close association with blood vessels. Their typical functions include drainage of interstitial fluid and proteins to systemic circulation. Lymphatic vessels also provide a route for the transport of immune cells, such as lymphocytes and dendritic cells, between the skin and regional lymph nodes during immune responses [103]. Recent studies have shown that lymphatic vessels may also influence HFSC activity. Using three-dimensional deep imaging, one group visualized close association of lymphatic capillaries, marked by lymphatic vessel endothelial hyaluronan receptor-1 (LYVE1) and vascular endothelial growth factor tyrosine kinase receptor-3 (VEGFR3), with the hair follicle bulge in both mouse and human hair follicles [104]. They also identified angiopoietin-like (Angptl) factors released by bulge stem cells as potential regulators of stem cell and lymphatic dynamics [104]. Angptl7 was associated with increased lymphatic drainage, association of lymphatic vessels with the bulge, and telogen phase, while Angptl4 was associated with decreased lymphatic drainage, dissociation of lymphatic vessels from the bulge, and anagen phase [104]. Another study found proximity of lymphatic capillaries to HFSCs at the anterior side of mouse hair follicles maintained by HFSC-derived Wnt ligands, with each lymphatic vessel associated with a triad of hair follicles [105]. Disruption of lymphatic vessels in mice have yielded varying outcomes in these studies ranging from earlier entry to anagen to inhibition of hair growth [104, 105]. Further studies are needed to better characterize interactions between lymphatic vessels and HFSCs, as well as the extent to which these processes occur independently from surrounding cells and structures, such as the closely associated blood vessels.

Cutaneous Blood Vessels

Cutaneous blood vessels are present in the dermis, and their major functions include maintaining perfusion, nutrition, thermoregulation, and transport of immune cells. Studies in mice have identified associations between growth of blood vessels and activity of HFSCs [106, 107]. Increase in diameter, length, and density of cutaneous vasculature with evidence of endothelial cell proliferation has been observed during anagen in mice, with the opposite seen during catagen and telogen [106, 107]. Induction of angiogenesis through vascular endothelial growth factor (VEGF) overexpression in outer root sheath keratinocytes enhances hair growth [107], whereas administration of angiogenesis inhibitors including TNP-470 [106] and anti-VEGF antibody [107] results in delayed hair growth. While interactions between cutaneous vasculature and HFSCs remain uncertain in humans, the association of minoxidil with hypertrichosis and its established efficacy for treating androgenetic alopecia and alopecia areata may offer some insights [21, 52, 108]. VEGF is highly expressed in human DP cells [109], with increased expression during anagen and decreased expression during catagen and telogen [110], which supports the temporal coordination of angiogenesis and hair cycling in human hair. Minoxidil might promote angiogenesis by stimulating the production of VEGF in DP cells [110, 111] and also causes vasodilation in the scalp [112, 113]. Whether increased vascularization and blood flow may fully explain minoxidil-induced hair growth remains to be determined, but it supports the involvement of cutaneous vasculature in hair regeneration.

Nerves

Peripheral nerves may carry out sensory, motor, and autonomic functions in the skin. Cutaneous nerves are abundant near the hair follicle, including the bulge region, and they may affect hair cycling through the release of neuropeptides [114, 115]. One neuropeptide that has received scrutiny is substance P, which is associated with states of emotional stress and neuroinflammation [83, 116]. One study found that mice exposed to stress in the form of noise develop catagen prematurely, with upregulation of substance P-positive nerve fibers [116]. Inhibition of hair growth was also seen with direct injection of substance P in non-stressed mice and blocked with concurrent administration of an NK1 receptor antagonist [116]. Notably, stress was also associated with increased number of perifollicular macrophages and dermal mast cell degranulation [116], suggesting how cutaneous nerves may indirectly modulate hair growth through immune cells. Administration of substance P to organ-cultured human hair follicles resulted in similar findings, including inhibition of hair elongation, early transition to catagen, and mast cell degranulation [83]. Together, these findings provide attractive explanations for how stress may contribute to hair loss in conditions like telogen effluvium and alopecia areata [83]. Additional types of innervation have been linked to hair growth. One group found expression of the olfactory receptor OR2AT4 in the outer root sheath of human hair follicles, with stimulation of OR2AT4 leading to prolonged anagen ex vivo that was reversible through both pharmacologic antagonism and gene silencing of OR2AT4 [117]. Recent studies have also shown that sympathetic innervation, which is responsible for piloerection, may modulate HFSC activity and hair growth [118, 119]. Specifically, activation of cutaneous norepinephrine signaling with visible light stimulation of the eyes [118] and cold temperature [119] caused early entry to anagen in mice in vivo. Whether these findings are applicable to hair regeneration in human remains to be investigated.

Exosomes

Recent discoveries on extracellular vesicles have further enhanced our understanding of intercellular communication in the hair follicle microenvironment. Extracellular vesicles are membrane-bound particles secreted by most, if not all cells, and they have been evolutionarily conserved in organisms ranging from bacteria and archaea to eukaryotes [120–122]. Extracellular vesicles have been classified into three subtypes – exosomes, microvesicles, and apoptotic bodies – which vary in morphology, biogenesis, and function [122]. Exosomes, in particular, have received attention for their potential role in regulating skin and hair regeneration [120].

Exosomes range in diameter from 40 to 160 nm and originate from the endocytic pathway. Normally, endocytosis of extracellular molecules and fluid generates endosomes that may interact with organelles such as the endoplasmic reticulum, trans-Golgi network, mitochondria, and lysosome. Endosomes, however, may undergo further inward budding to become multivesicular bodies (MVB) containing intraluminal vesicles (ILV). The MVBs may dock and fuse with the plasma membrane, causing secretion of ILVs out of the cell as membrane-bound exosomes [123]. In this manner, exosomes may transport a variety of cargo, such as proteins, lipids, and metabolites, to specific recipient cells [121, 123–125]. Biogenesis and secretion of exosomes likely involves Rab GTPases and ESCRT (Endosomal Sorting Complexes Required for Transport) proteins, among others, while uptake of exosomes by recipient cells may occur through endocytosis mediated by cell surface receptors, lipid rafts, clathrin, and calveolin, as well as phagocytosis and macropinocytosis [120, 122, 123, 126]. Exosomes have been implicated in a wide range of biological processes such as cell growth and differentiation, innate and adaptive immune responses, angiogenesis, homeostasis, metabolism, and neoplasia [122, 123, 127].

The ability to facilitate intercellular communication makes exosomes strong candidates for studying and manipulating cellular processes related to hair growth. Active Wnt proteins have been identified in exosomes isolated from human cells that are capable of inducing Wnt signaling in vitro [128]. Multiple studies have also shown increased occurrence and duration of anagen upon injection of mesenchymal stem cell (MSC) or DP cell-derived exosomes in mice, with evidence of increased Wnt/β-catenin or Shh activity [129–132]. Technical challenges remain in the isolation of exosomes [122, 133, 134], and further human clinical trials will be necessary to clarify whether exosomes may be used safely and efficiently for hair regeneration. Nevertheless, purification of exosomes with cargo of interest for future injection and bioengineering of exosomes for cell-specific delivery of therapeutics both signify promising applications in the field of regenerative medicine.

Conclusion

Remarkable advances in skin regeneration have occurred in recent years, including multiple studies that have used stem cell-based procedures and genetic engineering to restore epidermis in mice [135–137] and in patients with epidermolysis bullosa [138, 139]. Another recent study used human pluripotent stem cells to reconstruct skin organoids with epidermis, dermis, sebaceous glands, adipocytes, neurons, and hair follicles including the dermal papilla and dermal sheath [140]. The study provides a promising model for future study of human hair regeneration in vitro, although the presence of additional cellular components such as immune cells would further enhance its representation of the hair follicle microenvironment [141]. As discussed in this chapter, the hair follicle involves complex interactions between cells and structures that are temporally and spatially coordinated. Understanding the interdependence of these cellular processes is essential to study and achieve hair regeneration—a feat that would not only benefit patients suffering from hair loss, but also strengthen efforts to regenerate other human appendages and organs.

References

1. Oh JW, Kloepper J, Langan EA, Kim Y, Yeo J, Kim MJ, Hsi TC, Rose C, Yoon GS, Lee SJ, Seykora J, Kim JC, Sung YK, Kim M, Paus R, Plikus MV (2016) A guide to studying human hair follicle cycling in vivo. J Invest Dermatol 136(1):34–44. https://doi.org/10.1038/JID.2015.354
2. Paus R, Cotsarelis G (1999) The biology of hair follicles. N Engl J Med 341(7):491–497. https://doi.org/10.1056/NEJM199908123410706
3. Schneider MR, Schmidt-Ullrich R, Paus R (2009) The hair follicle as a dynamic miniorgan. Curr Biol 19(3):R132-142. https://doi.org/10.1016/j.cub.2008.12.005
4. Bishop GH (1945) Regeneration after experimental removal of skin in man. Am J Anat 76(2):153–181. https://doi.org/10.1002/aja.1000760202
5. Jahoda CA, Reynolds AJ (2001) Hair follicle dermal sheath cells: unsung participants in wound healing. The Lancet 358(9291):1445–1448. https://doi.org/10.1016/s0140-6736(01)06532-1

6. Ansell DM, Kloepper JE, Thomason HA, Paus R, Hardman MJ (2011) Exploring the "hair growth-wound healing connection": anagen phase promotes wound re-epithelialization. J Invest Dermatol 131(2):518–528. https://doi.org/10.1038/jid.2010.291
7. Garza L (2019) Developmental biology of the skin. In Kang S, Amagai M, Bruckner AL et al. (eds) Fitzpatrick's Dermatology, 9e. McGraw-Hill Education, New York, NY
8. Milner Y, Sudnik J, Filippi M, Kizoulis M, Kashgarian M, Stenn K (2002) Exogen, shedding phase of the hair growth cycle: characterization of a mouse model. J Invest Dermatol 119(3):639–644. https://doi.org/10.1046/j.1523-1747.2002.01842.x
9. Paus R, Foitzik K (2004) In search of the "hair cycle clock": a guided tour. Differentiation 72(9–10):489–511. https://doi.org/10.1111/j.1432-0436.2004.07209004.x
10. Cotsarelis G, Sun TT, Lavker RM (1990) Label-retaining cells reside in the bulge area of pilosebaceous unit: implications for follicular stem cells, hair cycle, and skin carcinogenesis. Cell 61(7):1329–1337. https://doi.org/10.1016/0092-8674(90)90696-c
11. Oshima H, Rochat A, Kedzia C, Kobayashi K, Barrandon Y (2001) Morphogenesis and renewal of hair follicles from adult multipotent stem cells. Cell 104(2):233–245. https://doi.org/10.1016/s0092-8674(01)00208-2
12. Morris RJ, Liu Y, Marles L, Yang Z, Trempus C, Li S, Lin JS, Sawicki JA, Cotsarelis G (2004) Capturing and profiling adult hair follicle stem cells. Nat Biotechnol 22(4):411–417. https://doi.org/10.1038/nbt950
13. Taylor G, Lehrer MS, Jensen PJ, Sun T-T, Lavker RM (2000) Involvement of follicular stem cells in forming not only the follicle but also the epidermis. Cell 102(4):451–461. https://doi.org/10.1016/s0092-8674(00)00050-7
14. Ito M, Liu Y, Yang Z, Nguyen J, Liang F, Morris RJ, Cotsarelis G (2005) Stem cells in the hair follicle bulge contribute to wound repair but not to homeostasis of the epidermis. Nat Med 11(12):1351–1354. https://doi.org/10.1038/nm1328
15. Akiyama M, Dale BA, Sun T-T, Holbrook KA (1995) Characterization of hair follicle bulge in human fetal skin: the human fetal bulge is a pool of undifferentiated keratinocytes. J Invest Dermatol 105(6):844–850. https://doi.org/10.1111/1523-1747.ep12326649
16. Lyle S, Christofidou-Solomidou M, Liu Y, Elder DE, Albelda S, Cotsarelis G (1998) The C8/144B monoclonal antibody recognizes cytokeratin 15 and defines the location of human hair follicle stem cells. J Cell Sci 111(Pt 21):3179–3188
17. Cotsarelis G (2006) Epithelial stem cells: a folliculocentric view. J Invest Dermatol 126(7):1459–1468. https://doi.org/10.1038/sj.jid.5700376
18. Ohyama M (2005) Characterization and isolation of stem cell-enriched human hair follicle bulge cells. J Clin Invest 116(1):249–260. https://doi.org/10.1172/jci26043
19. Harries MJ, Paus R (2010) The pathogenesis of primary cicatricial alopecias. Am J Pathol 177(5):2152–2162. https://doi.org/10.2353/ajpath.2010.100454
20. Whiting DA (2003) Histopathologic features of alopecia areata. Arch Dermatol 139(12). doi:https://doi.org/10.1001/archderm.139.12.1555
21. Alkhalifah A, Alsantali A, Wang E, Mcelwee KJ, Shapiro J (2010) Alopecia areata update. J Am Acad Dermatol 62(2):191–202. https://doi.org/10.1016/j.jaad.2009.10.031
22. Wang EHC, Sallee BN, Tejeda CI, Christiano AM (2018) JAK inhibitors for treatment of alopecia areata. J Invest Dermatol 138(9):1911–1916. https://doi.org/10.1016/j.jid.2018.05.027
23. Greco V, Chen T, Rendl M, Schober M, Pasolli HA, Stokes N, Dela Cruz-Racelis J, Fuchs E (2009) A two-step mechanism for stem cell activation during hair regeneration. Cell Stem Cell 4(2):155–169. https://doi.org/10.1016/j.stem.2008.12.009
24. Ito M, Cotsarelis G, Kizawa K, Hamada K (2004) Hair follicle stem cells in the lower bulge form the secondary germ, a biochemically distinct but functionally equivalent progenitor cell population, at the termination of catagen. Differentiation 72(9–10):548–557. https://doi.org/10.1111/j.1432-0436.2004.07209008.x
25. Rompolas P, Deschene ER, Zito G, Gonzalez DG, Saotome I, Haberman AM, Greco V (2012) Live imaging of stem cell and progeny behaviour in physiological hair-follicle regeneration. Nature 487(7408):496–499. https://doi.org/10.1038/nature11218

26. Hsu YC, Li L, Fuchs E (2014) Emerging interactions between skin stem cells and their niches. Nat Med 20(8):847–856. https://doi.org/10.1038/nm.3643
27. Garza LA, Yang CC, Zhao T, Blatt HB, Lee M, He H, Stanton DC, Carrasco L, Spiegel JH, Tobias JW, Cotsarelis G (2011) Bald scalp in men with androgenetic alopecia retains hair follicle stem cells but lacks CD200-rich and CD34-positive hair follicle progenitor cells. J Clin Invest 121(2):613–622. https://doi.org/10.1172/JCI44478
28. Inoue K, Aoi N, Sato T, Yamauchi Y, Suga H, Eto H, Kato H, Araki J, Yoshimura K (2009) Differential expression of stem-cell-associated markers in human hair follicle epithelial cells. Lab Invest 89(8):844–856. https://doi.org/10.1038/labinvest.2009.48
29. Ozawa M, Aiba S, Kurosawa M, Tagami H (2004) Ber-EP4 antigen is a marker for a cell population related to the secondary hair germ. Exp Dermatol 13(7):401–405. https://doi.org/10.1111/j.0906-6705.2004.00153.x
30. Purba TS, Haslam IS, Poblet E, Jiménez F, Gandarillas A, Izeta A, Paus R (2014) Human epithelial hair follicle stem cells and their progeny: Current state of knowledge, the widening gap in translational research and future challenges. BioEssays 36(5):513–525. https://doi.org/10.1002/bies.201300166
31. Trempus CS, Morris RJ, Bortner CD, Cotsarelis G, Faircloth RS, Reece JM, Tennant RW (2003) Enrichment for living murine keratinocytes from the hair follicle bulge with the cell surface marker CD34. J Invest Dermatol 120(4):501–511. https://doi.org/10.1046/j.1523-1747.2003.12088.x
32. Poblet E, Jiménez F, Godínez JM, Pascual-Martín A, Izeta A (2006) The immunohistochemical expression of CD34 in human hair follicles: a comparative study with the bulge marker CK15. Clin Exp Dermatol 31(6):807–812. https://doi.org/10.1111/j.1365-2230.2006.02255.x
33. Jaks V, Barker N, Kasper M, Van Es JH, Snippert HJ, Clevers H, Toftgård R (2008) Lgr5 marks cycling, yet long-lived, hair follicle stem cells. Nat Genet 40(11):1291–1299. https://doi.org/10.1038/ng.239
34. Snippert HJ, Haegebarth A, Kasper M, Jaks V, Van Es JH, Barker N, Van De Wetering M, Van Den Born M, Begthel H, Vries RG, Stange DE, Toftgard R, Clevers H (2010) Lgr6 marks stem cells in the hair follicle that generate all cell lineages of the skin. Science 327(5971):1385–1389. https://doi.org/10.1126/science.1184733
35. Blanpain C, Lowry WE, Geoghegan A, Polak L, Fuchs E (2004) Self-renewal, multipotency, and the existence of two cell populations within an epithelial stem cell niche. Cell 118(5):635–648. https://doi.org/10.1016/j.cell.2004.08.012
36. Choi S, Yeon, Zhang Y, Xu M, Yang Y, Ito M, Peng T, Cui Z, Nagy A, Hadjantonakis A-K, Lang A, Richard, Cotsarelis G, Andl T, Morrisey E, Edward, Millar E, Sarah (2013) Distinct functions for Wnt/β-catenin in hair follicle stem cell proliferation and survival and interfollicular epidermal homeostasis. Cell Stem Cell 13(6):720–733. doi:https://doi.org/10.1016/j.stem.2013.10.003
37. Myung PS, Takeo M, Ito M, Atit RP (2013) Epithelial Wnt ligand secretion is required for adult hair follicle growth and regeneration. J Invest Dermatol 133(1):31–41. https://doi.org/10.1038/jid.2012.230
38. Plikus MV, Mayer JA, De La Cruz D, Baker RE, Maini PK, Maxson R, Chuong C-M (2008) Cyclic dermal BMP signalling regulates stem cell activation during hair regeneration. Nature 451(7176):340–344. https://doi.org/10.1038/nature06457
39. Reddy S, Andl T, Bagasra A, Lu MM, Epstein DJ, Morrisey EE, Millar SE (2001) Characterization of Wnt gene expression in developing and postnatal hair follicles and identification of Wnt5a as a target of Sonic hedgehog in hair follicle morphogenesis. Mech Dev 107(1–2):69–82. https://doi.org/10.1016/s0925-4773(01)00452-x
40. Botchkarev VA, Botchkareva NV, Roth W, Nakamura M, Chen L-H, Herzog W, Lindner G, Mcmahon JA, Peters C, Lauster R, Mcmahon AP, Paus R (1999) Noggin is a mesenchymally derived stimulator of hair-follicle induction. Nat Cell Biol 1(3):158–164. https://doi.org/10.1038/11078
41. Brownell I, Guevara E, Bai CB, Loomis CA, Joyner AL (2011) Nerve-derived sonic hedgehog defines a niche for hair follicle stem cells capable of becoming epidermal stem cells. Cell Stem Cell 8(5):552–565. https://doi.org/10.1016/j.stem.2011.02.021

42. Kimura-Ueki M, Oda Y, Oki J, Komi-Kuramochi A, Honda E, Asada M, Suzuki M, Imamura T (2012) Hair cycle resting phase is regulated by cyclic epithelial FGF18 signaling. J Invest Dermatol 132(5):1338–1345. https://doi.org/10.1038/jid.2011.490
43. Sun X, Are A, Annusver K, Sivan U, Jacob T, Dalessandri T, Joost S, Fullgrabe A, Gerling M, Kasper M (2020) Coordinated hedgehog signaling induces new hair follicles in adult skin. Elife 9. https://doi.org/10.7554/eLife.46756
44. Hawkshaw NJ, Hardman JA, Alam M, Jimenez F, Paus R (2020) Deciphering the molecular morphology of the human hair cycle: Wnt signalling during the telogen–anagen transformation. Br J Dermatol 182(5):1184–1193. https://doi.org/10.1111/bjd.18356
45. Driskell RR, Clavel C, Rendl M, Watt FM (2011) Hair follicle dermal papilla cells at a glance. J Cell Sci 124(Pt 8):1179–1182. https://doi.org/10.1242/jcs.082446
46. Jahoda CA, Horne KA, Oliver RF (1984) Induction of hair growth by implantation of cultured dermal papilla cells. Nature 311(5986):560–562. https://doi.org/10.1038/311560a0
47. Reynolds AJ, Jahoda CA (1992) Cultured dermal papilla cells induce follicle formation and hair growth by transdifferentiation of an adult epidermis. Development 115(2):587–593
48. Kishimoto J, Burgeson RE, Morgan BA (2000) Wnt signaling maintains the hair-inducing activity of the dermal papilla. Genes Dev 14(10):1181–1185
49. Rendl M, Polak L, Fuchs E (2008) BMP signaling in dermal papilla cells is required for their hair follicle-inductive properties. Genes Dev 22(4):543–557. https://doi.org/10.1101/gad.1614408
50. Higgins CA, Chen JC, Cerise JE, Jahoda CAB, Christiano AM (2013) Microenvironmental reprogramming by three-dimensional culture enables dermal papilla cells to induce de novo human hair-follicle growth. Proc Natl Acad Sci 110(49):19679–19688. https://doi.org/10.1073/pnas.1309970110
51. Hibberts N, Howell A, Randall V (1998) Balding hair follicle dermal papilla cells contain higher levels of androgen receptors than those from non-balding scalp. J Endocrinol 156(1):59–65. https://doi.org/10.1677/joe.0.1560059
52. Adil A, Godwin M (2017) The effectiveness of treatments for androgenetic alopecia: a systematic review and meta-analysis. J Am Acad Dermatol 77(1):136-141.e135. https://doi.org/10.1016/j.jaad.2017.02.054
53. Horne KA, Jahoda CA (1992) Restoration of hair growth by surgical implantation of follicular dermal sheath. Development 116(3):563–571
54. McElwee KJ, Kissling S, Wenzel E, Huth A, Hoffmann R (2003) Cultured peribulbar dermal sheath cells can induce hair follicle development and contribute to the dermal sheath and dermal papilla. J Invest Dermatol 121(6):1267–1275. https://doi.org/10.1111/j.1523-1747.2003.12568.x
55. Rahmani W, Abbasi S, Hagner A, Raharjo E, Kumar R, Hotta A, Magness S, Metzger D, Biernaskie J (2014) Hair follicle dermal stem cells regenerate the dermal sheath, repopulate the dermal papilla, and modulate hair type. Dev Cell 31(5):543–558. https://doi.org/10.1016/j.devcel.2014.10.022
56. Heitman N, Sennett R, Mok KW, Saxena N, Srivastava D, Martino P, Grisanti L, Wang Z, Ma'ayan A, Rompolas P, Rendl M (2020) Dermal sheath contraction powers stem cell niche relocation during hair cycle regression. Science 367(6474):161–166. https://doi.org/10.1126/science.aax9131
57. Reynolds AJ, Lawrence C, Cserhalmi-Friedman PB, Christiano AM, Jahoda CAB (1999) Trans-gender induction of hair follicles. Nature 402(6757):33–34. https://doi.org/10.1038/46938
58. Tsuboi R, Niiyama S, Irisawa R, Harada K, Nakazawa Y, Kishimoto J (2020) Autologous cell-based therapy for male and female pattern hair loss using dermal sheath cup cells: a randomized placebo-controlled double-blinded dose-finding clinical study. J Am Acad Dermatol 83(1):109–116. https://doi.org/10.1016/j.jaad.2020.02.033
59. Virtanen KA, Lidell ME, Orava J, Heglind M, Westergren R, Niemi T, Taittonen M, Laine J, Savisto N-J, Enerbäck S, Nuutila P (2009) Functional brown adipose tissue in healthy adults. N Engl J Med 360(15):1518–1525. https://doi.org/10.1056/nejmoa0808949

60. Driskell RR, Jahoda CAB, Chuong C-M, Watt FM, Horsley V (2014) Defining dermal adipose tissue. Exp Dermatol 23(9):629–631. https://doi.org/10.1111/exd.12450
61. Guerrero-Juarez CF, Plikus MV (2018) Emerging nonmetabolic functions of skin fat. Nat Rev Endocrinol 14(3):163–173. https://doi.org/10.1038/nrendo.2017.162
62. Donati G, Proserpio V, Lichtenberger BM, Natsuga K, Sinclair R, Fujiwara H, Watt FM (2014) Epidermal Wnt/ -catenin signaling regulates adipocyte differentiation via secretion of adipogenic factors. Proc Natl Acad Sci 111(15):E1501–E1509. https://doi.org/10.1073/pnas.1312880111
63. Festa E, Fretz J, Berry R, Schmidt B, Rodeheffer M, Horowitz M, Horsley V (2011) Adipocyte lineage cells contribute to the skin stem cell niche to drive hair cycling. Cell 146(5):761–771. https://doi.org/10.1016/j.cell.2011.07.019
64. Hansen LS, Coggle JE, Wells J, Charles MW (1984) The influence of the hair cycle on the thickness of mouse skin. Anat Rec 210(4):569–573. https://doi.org/10.1002/ar.1092100404
65. Rivera-Gonzalez GC, Shook BA, Andrae J, Holtrup B, Bollag K, Betsholtz C, Rodeheffer MS, Horsley V (2016) Skin Adipocyte Stem Cell Self-Renewal Is Regulated by a PDGFA/AKT-Signaling Axis. Cell Stem Cell 19(6):738–751. https://doi.org/10.1016/j.stem.2016.09.002
66. Rodeheffer MS, Birsoy K, Friedman JM (2008) Identification of white adipocyte progenitor cells in vivo. Cell 135(2):240–249. https://doi.org/10.1016/j.cell.2008.09.036
67. Zhang B, Tsai P-C, Gonzalez-Celeiro M, Chung O, Boumard B, Perdigoto CN, Ezhkova E, Hsu Y-C (2016) Hair follicles' transit-amplifying cells govern concurrent dermal adipocyte production through Sonic Hedgehog. Genes Dev 30(20):2325–2338. https://doi.org/10.1101/gad.285429.116
68. Karnik P, Tekeste Z, Mccormick TS, Gilliam AC, Price VH, Cooper KD, Mirmirani P (2009) Hair follicle stem cell-specific PPARγ deletion causes scarring alopecia. J Invest Dermatol 129(5):1243–1257. https://doi.org/10.1038/jid.2008.369
69. Rahmani W, Sinha S, Biernaskie J (2020) Immune modulation of hair follicle regeneration. NPJ Regen Med 5(1). doi:https://doi.org/10.1038/s41536-020-0095-2
70. Wang ECE, Higgins CA (2020) Immune cell regulation of the hair cycle. Exp Dermatol 29(3):322–333. https://doi.org/10.1111/exd.14070
71. Paus R, Van Der Veen C, Eichmüller S, Kopp T, Hagen E, Müller-Röver S, Hofmann U (1998) Generation and cyclic remodeling of the hair follicle immune system in mice. J Invest Dermatol 111(1):7–18. https://doi.org/10.1046/j.1523-1747.1998.00243.x
72. Christoph T, Müller-Röver S, Audring H, Tobin DJ, Hermes B, Cotsarelis G, Rückert R, Paus R (2000) The human hair follicle immune system: cellular composition and immune privilege. Br J Dermatol 142(5):862–873. https://doi.org/10.1046/j.1365-2133.2000.03464.x
73. Castellana D, Paus R, Perez-Moreno M (2014) Macrophages contribute to the cyclic activation of adult hair follicle stem cells. PLoS Biol 12(12). https://doi.org/10.1371/journal.pbio.1002002
74. Wang ECE, Dai Z, Ferrante AW, Drake CG, Christiano AM (2019) A subset of TREM2(+) dermal macrophages secretes oncostatin m to maintain hair follicle stem cell quiescence and inhibit hair growth. Cell Stem Cell 24(4):654–669 e656. doi:https://doi.org/10.1016/j.stem.2019.01.011
75. Hardman JA, Muneeb F, Pople J, Bhogal R, Shahmalak A, Paus R (2019) Human perifollicular macrophages undergo apoptosis, express wnt ligands, and switch their polarization during catagen. J Invest Dermatol 139(12):2543-2546.e2549. https://doi.org/10.1016/j.jid.2019.04.026
76. Suzuki S, Kato T, Takimoto H, Masui S, Oshima H, Ozawa K, Suzuki S, Imamura T (1998) Localization of rat FGF-5 protein in skin macrophage-like cells and FGF-5S protein in hair follicle: possible involvement of twoFgf-5 gene products in hair growth cycle regulation. J Invest Dermatol 111(6):963–972. https://doi.org/10.1046/j.1523-1747.1998.00427.x
77. Suzuki S, Ota Y, Ozawa K, Imamura T (2000) Dual-mode regulation of hair growth cycle by two Fgf-5 gene products. J Invest Dermatol 114(3):456–463. https://doi.org/10.1046/j.1523-1747.2000.00912.x

78. Hébert JM, Rosenquist T, Götz J, Martin GR (1994) FGF5 as a regulator of the hair growth cycle: evidence from targeted and spontaneous mutations. Cell 78(6):1017–1025. https://doi.org/10.1016/0092-8674(94)90276-3
79. Higgins CA, Petukhova L, Harel S, Ho YY, Drill E, Shapiro L, Wajid M, Christiano AM (2014) FGF5 is a crucial regulator of hair length in humans. Proc Natl Acad Sci 111(29):10648–10653. https://doi.org/10.1073/pnas.1402862111
80. Paus R, Maurer M, Slominski A, Czarnetzki BM (1994) Mast cell involvement in murine hair growth. Dev Biol 163(1):230–240. https://doi.org/10.1006/dbio.1994.1139
81. Maurer M, Fischer E, Handjiski B, von Stebut E, Algermissen B, Bavandi A, Paus R (1997) Activated skin mast cells are involved in murine hair follicle regression (catagen). Lab Invest 77(4):319–332
82. Weber A, Knop J, Maurer M (2003) Pattern analysis of human cutaneous mast cell populations by total body surface mapping. Br J Dermatol 148(2):224–228. https://doi.org/10.1046/j.1365-2133.2003.05090.x
83. Peters EMJ, Liotiri S, Bodó E, Hagen E, Bíró T, Arck PC, Paus R (2007) Probing the effects of stress mediators on the human hair follicle. Am J Pathol 171(6):1872–1886. https://doi.org/10.2353/ajpath.2007.061206
84. Jaworsky C, Kligman AM, Murphy GF (1992) Characterization of inflammatory infiltrates in male pattern alopecia: implications for pathogenesis. Br J Dermatol 127(3):239–246. https://doi.org/10.1111/j.1365-2133.1992.tb00121.x
85. Michel L, Reygagne P, Benech P, Jean-Louis F, Scalvino S, Ly Ka So S, Hamidou Z, Bianovici S, Pouch J, Ducos B, Bonnet M, Bensussan A, Patatian A, Lati E, Wdzieczak-Bakala J, Choulot JC, Loing E, Hocquaux M (2017) Study of gene expression alteration in male androgenetic alopecia: evidence of predominant molecular signalling pathways. Br J Dermatol 177(5):1322–1336. https://doi.org/10.1111/bjd.15577
86. Garza LA, Liu Y, Yang Z, Alagesan B, Lawson JA, Norberg SM, Loy DE, Zhao T, Blatt HB, Stanton DC, Carrasco L, Ahluwalia G, Fischer SM, Fitzgerald GA, Cotsarelis G (2012) Prostaglandin D2 inhibits hair growth and is elevated in bald scalp of men with androgenetic alopecia. Sci Transl Med 4(126):126ra134–126ra134. doi:https://doi.org/10.1126/scitranslmed.3003122
87. Bertolini M, Zilio F, Rossi A, Kleditzsch P, Emelianov VE, Gilhar A, Keren A, Meyer KC, Wang E, Funk W, Mcelwee K, Paus R (2014) Abnormal interactions between perifollicular mast cells and CD8+ T-cells may contribute to the pathogenesis of alopecia areata. PLoS ONE 9(5). https://doi.org/10.1371/journal.pone.0094260
88. Petukhova L, Duvic M, Hordinsky M, Norris D, Price V, Shimomura Y, Kim H, Singh P, Lee A, Chen WV, Meyer KC, Paus R, Jahoda CAB, Amos CI, Gregersen PK, Christiano AM (2010) Genome-wide association study in alopecia areata implicates both innate and adaptive immunity. Nature 466(7302):113–117. https://doi.org/10.1038/nature09114
89. Xing L, Dai Z, Jabbari A, Cerise JE, Higgins CA, Gong W, De Jong A, Harel S, Destefano GM, Rothman L, Singh P, Petukhova L, Mackay-Wiggan J, Christiano AM, Clynes R (2014) Alopecia areata is driven by cytotoxic T lymphocytes and is reversed by JAK inhibition. Nat Med 20(9):1043–1049. https://doi.org/10.1038/nm.3645
90. Harel S, Higgins CA, Cerise JE, Dai Z, Chen JC, Clynes R, Christiano AM (2015) Pharmacologic inhibition of JAK-STAT signaling promotes hair growth. Sci Adv 1(9). https://doi.org/10.1126/sciadv.1500973
91. Ali N, Zirak B, Rodriguez RS, Pauli ML, Truong H-A, Lai K, Ahn R, Corbin K, Lowe MM, Scharschmidt TC, Taravati K, Tan MR, Ricardo-Gonzalez RR, Nosbaum A, Bertolini M, Liao W, Nestle FO, Paus R, Cotsarelis G, Abbas AK, Rosenblum MD (2017) Regulatory T cells in skin facilitate epithelial stem cell differentiation. Cell 169(6):1119-1129.e1111. https://doi.org/10.1016/j.cell.2017.05.002
92. Tembhre MK, Sharma VK (2013) T-helper and regulatory T-cell cytokines in the peripheral blood of patients with active alopecia areata. Br J Dermatol 169(3):543–548. https://doi.org/10.1111/bjd.12396

93. Wier EM, Garza LA (2020) Through the lens of hair follicle neogenesis, a new focus on mechanisms of skin regeneration after wounding. Semin Cell Dev Biol 100:122–129. https://doi.org/10.1016/j.semcdb.2019.10.002
94. Kasuya A, Ito T, Tokura Y (2018) M2 macrophages promote wound-induced hair neogenesis. J Dermatol Sci 91(3):250–255. https://doi.org/10.1016/j.jdermsci.2018.05.004
95. Rahmani W, Liu Y, Rosin NL, Kline A, Raharjo E, Yoon J, Stratton JA, Sinha S, Biernaskie J (2018) Macrophages promote wound-induced hair follicle regeneration in a CX3CR1- and TGF-beta1-dependent manner. J Invest Dermatol 138(10):2111–2122. https://doi.org/10.1016/j.jid.2018.04.010
96. Wang X, Chen H, Tian R, Zhang Y, Drutskaya MS, Wang C, Ge J, Fan Z, Kong D, Wang X, Cai T, Zhou Y, Wang J, Wang J, Wang S, Qin Z, Jia H, Wu Y, Liu J, Nedospasov SA, Tredget EE, Lin M, Liu J, Jiang Y, Wu J (2017) Macrophages induce AKT/β-catenin-dependent Lgr5+ stem cell activation and hair follicle regeneration through TNF. Nat Commun 8(1):14091. https://doi.org/10.1038/ncomms14091
97. Nelson AM, Reddy SK, Ratliff TS, Hossain MZ, Katseff AS, Zhu AS, Chang E, Resnik SR, Page C, Kim D, Whittam AJ, Miller LS, Garza LA (2015) dsRNA released by tissue damage activates TLR3 to drive skin regeneration. Cell Stem Cell 17(2):139–151. https://doi.org/10.1016/j.stem.2015.07.008
98. Zhu AS, Li A, Ratliff TS, Melsom M, Garza LA (2017) After skin wounding, noncoding dsRNA coordinates prostaglandins and wnts to promote regeneration. J Invest Dermatol 137(7):1562–1568. https://doi.org/10.1016/j.jid.2017.03.023
99. Kligman AM, Strauss JS (1956) The formation of vellus hair follicles from human adult epidermis. J Invest Dermatol 27(1):19–23. https://doi.org/10.1038/jid.1956.71
100. Sun Z, Diao J, Guo S, Yin G (2009) A very rare complication: new hair growth around healing wounds. J Int Med Res 37(2):583–586. https://doi.org/10.1177/147323000903700236
101. Beachkofsky TM, Henning JS, Hivnor CM (2011) Induction of de novo hair regeneration in scars after fractionated carbon dioxide laser therapy in three patients. Dermatol Surg 37(9):1365–1368. https://doi.org/10.1111/j.1524-4725.2011.01934.x
102. Wong T-W, Hughes M, Wang S-H (2018) Never too old to regenerate? Wound induced hair follicle neogenesis after secondary intention healing in a geriatric patient. J Tissue Viability 27(2):114–116. https://doi.org/10.1016/j.jtv.2018.01.001
103. Skobe M, Detmar M (2000) Structure, function, and molecular control of the skin lymphatic system. J Invest Dermat Symp Proc 5(1):14–19. https://doi.org/10.1046/j.1087-0024.2000.00001.x
104. Gur-Cohen S, Yang H, Baksh SC, Miao Y, Levorse J, Kataru RP, Liu X, De La Cruz-Racelis J, Mehrara BJ, Fuchs E (2019) Stem cell–driven lymphatic remodeling coordinates tissue regeneration. Science 366(6470):1218–1225. https://doi.org/10.1126/science.aay4509
105. Peña-Jimenez D, Fontenete S, Megias D, Fustero-Torre C, Graña-Castro O, Castellana D, Loewe R, Perez-Moreno M (2019) Lymphatic vessels interact dynamically with the hair follicle stem cell niche during skin regeneration in vivo. EMBO J 38(19). doi:https://doi.org/10.15252/embj.2019101688
106. Mecklenburg L, Tobin DJ, Müller-Röver S, Handjiski B, Wendt G, Peters EMJ, Pohl S, Moll I, Paus R (2000) Active hair growth (Anagen) is associated with angiogenesis. J Invest Dermatol 114(5):909–916. https://doi.org/10.1046/j.1523-1747.2000.00954.x
107. Yano K, Brown LF, Detmar M (2001) Control of hair growth and follicle size by VEGF-mediated angiogenesis. J Clin Invest 107(4):409–417. https://doi.org/10.1172/jci11317
108. Burton JL, Marshall A (1979) Hypertrichosis due to minoxidil. Br J Dermatol 101(5):593–595. https://doi.org/10.1111/j.1365-2133.1979.tb15106.x
109. Kozlowska U, Blume-Peytavi U, Kodelja V, Sommer C, Goerdt S, Majewski S, Jablonska S, Orfanos CE (1998) Expression of vascular endothelial growth factor (VEGF) in various compartments of the human hair follicle. Arch Dermatol Res 290(12):661–668. https://doi.org/10.1007/s004030050370
110. Lachgar C, Gall B (1998) Minoxidil upregulates the expression of vascular endothelial growth factor in human hair dermal papilla cells. Br J Dermatol 138(3):407–411. https://doi.org/10.1046/j.1365-2133.1998.02115.x

111. Messenger AG, Rundegren J (2004) Minoxidil: mechanisms of action on hair growth. Br J Dermatol 150(2):186–194. https://doi.org/10.1111/j.1365-2133.2004.05785.x
112. Wester RC, Maibach HI, Guy RH, Novak E (1984) Minoxidil stimulates cutaneous blood flow in human balding scalps: pharmacodynamics measured by laser Doppler velocimetry and photopulse plethysmography. J Invest Dermatol 82(5):515–517. https://doi.org/10.1111/1523-1747.ep12261084
113. Bunker CB, Dowd PM (1987) Alterations in scalp blood flow after the epicutaneous application of 3% minoxidil and 0.1% hexyl nicotinate in alopecia. Br J Dermatol 117(5):668–669. doi:https://doi.org/10.1111/j.1365-2133.1987.tb07505.x
114. Botchkarev VA, Eichmuller S, Johansson O, Paus R (1997) Hair cycle-dependent plasticity of skin and hair follicle innervation in normal murine skin. J Comp Neurol 386(3):379–395. https://doi.org/10.1002/(sici)1096-9861(19970929)386:3%3c379::aid-cne4%3e3.0.co;2-z
115. Paus R, Peters EM, Eichmuller S, Botchkarev VA (1997) Neural mechanisms of hair growth control. J Invest Dermatol Symp Proc 2(1):61–68. https://doi.org/10.1038/jidsymp.1997.13
116. Arck PC, Handjiski B, Peters EMJ, Peter AS, Hagen E, Fischer A, Klapp BF, Paus R (2003) Stress inhibits hair growth in mice by induction of premature catagen development and deleterious perifollicular inflammatory events via neuropeptide substance P-dependent pathways. Am J Pathol 162(3):803–814. https://doi.org/10.1016/s0002-9440(10)63877-1
117. Chéret J, Bertolini M, Ponce L, Lehmann J, Tsai T, Alam M, Hatt H, Paus R (2018) Olfactory receptor OR2AT4 regulates human hair growth. Nat Commun 9(1). doi:https://doi.org/10.1038/s41467-018-05973-0
118. Fan SM-Y, Chang Y-T, Chen C-L, Wang W-H, Pan M-K, Chen W-P, Huang W-Y, Xu Z, Huang H-E, Chen T, Plikus MV, Chen S-K, Lin S-J (2018) External light activates hair follicle stem cells through eyes via an ipRGC–SCN–sympathetic neural pathway. Proc Natl Acad Sci 115(29):E6880–E6889. https://doi.org/10.1073/pnas.1719548115
119. Shwartz Y, Gonzalez-Celeiro M, Chen C-L, Pasolli HA, Sheu S-H, Fan SM-Y, Shamsi F, Assaad S, Lin ET-Y, Zhang B, Tsai P-C, He M, Tseng Y-H, Lin S-J, Hsu Y-C (2020) Cell types promoting goosebumps form a niche to regulate hair follicle stem cells. Cell 182(3):578–593.e519. https://doi.org/10.1016/j.cell.2020.06.031
120. Carrasco E, Soto-Heredero G, Mittelbrunn M (2019) The role of extracellular vesicles in cutaneous remodeling and hair follicle dynamics. Int J Mol Sci 20(11):2758. https://doi.org/10.3390/ijms20112758
121. Van Niel G, D'Angelo G, Raposo G (2018) Shedding light on the cell biology of extracellular vesicles. Nat Rev Mol Cell Biol 19(4):213–228. https://doi.org/10.1038/nrm.2017.125
122. Wiklander OPB, Brennan MÁ, Lötvall J, Breakefield XO, El Andaloussi S (2019) Advances in therapeutic applications of extracellular vesicles. Sci Transl Med 11(492):eaav8521. doi:https://doi.org/10.1126/scitranslmed.aav8521
123. Kalluri R, LeBleu VS (2020) The biology, function, and biomedical applications of exosomes. Science 367(6478). doi:https://doi.org/10.1126/science.aau6977
124. Jeppesen DK, Fenix AM, Franklin JL, Higginbotham JN, Zhang Q, Zimmerman LJ, Liebler DC, Ping J, Liu Q, Evans R, Fissell WH, Patton JG, Rome LH, Burnette DT, Coffey RJ (2019) Reassessment of exosome composition. Cell 177(2):428-445.e418. https://doi.org/10.1016/j.cell.2019.02.029
125. Keerthikumar S, Chisanga D, Ariyaratne D, Al Saffar H, Anand S, Zhao K, Samuel M, Pathan M, Jois M, Chilamkurti N, Gangoda L, Mathivanan S (2016) ExoCarta: a web-based compendium of exosomal cargo. J Mol Biol 428(4):688–692. https://doi.org/10.1016/j.jmb.2015.09.019
126. Riazifar M, Pone EJ, Lötvall J, Zhao W (2017) Stem cell extracellular vesicles: extended messages of regeneration. Annu Rev Pharmacol Toxicol 57(1):125–154. https://doi.org/10.1146/annurev-pharmtox-061616-030146
127. Théry C, Ostrowski M, Segura E (2009) Membrane vesicles as conveyors of immune responses. Nat Rev Immunol 9(8):581–593. https://doi.org/10.1038/nri2567
128. Gross JC, Chaudhary V, Bartscherer K, Boutros M (2012) Active Wnt proteins are secreted on exosomes. Nat Cell Biol 14(10):1036–1045. https://doi.org/10.1038/ncb2574

129. Hu S, Li Z, Lutz H, Huang K, Su T, Cores J, Dinh P-UC, Cheng K (2020) Dermal exosomes containing miR-218–5p promote hair regeneration by regulating β-catenin signaling. Science Advances 6(30):eaba1685. doi:https://doi.org/10.1126/sciadv.aba1685
130. Kwack MH, Seo CH, Gangadaran P, Ahn BC, Kim MK, Kim JC, Sung YK (2019) Exosomes derived from human dermal papilla cells promote hair growth in cultured human hair follicles and augment the hair-inductive capacity of cultured dermal papilla spheres. Exp Dermatol 28(7):854–857. https://doi.org/10.1111/exd.13927
131. Rajendran RL, Gangadaran P, Bak SS, Oh JM, Kalimuthu S, Lee HW, Baek SH, Zhu L, Sung YK, Jeong SY, Lee S-W, Lee J, Ahn B-C (2017) Extracellular vesicles derived from MSCs activates dermal papilla cell in vitro and promotes hair follicle conversion from telogen to anagen in mice. Sci Rep 7(1). doi:https://doi.org/10.1038/s41598-017-15505-3
132. Zhou L, Wang H, Jing J, Yu L, Wu X, Lu Z (2018) Regulation of hair follicle development by exosomes derived from dermal papilla cells. Biochem Biophys Res Commun 500(2):325–332. https://doi.org/10.1016/j.bbrc.2018.04.067
133. Mcbride JD, Rodriguez-Menocal L, Badiavas EV (2017) Extracellular vesicles as biomarkers and therapeutics in dermatology: a focus on exosomes. J Invest Dermatol 137(8):1622–1629. https://doi.org/10.1016/j.jid.2017.04.021
134. Théry C, Witwer KW, Aikawa E, Alcaraz MJ, Anderson JD, Andriantsitohaina R, Antoniou A, Arab T, Archer F, Atkin-Smith GK, Ayre DC, Bach J-M, Bachurski D, Baharvand H, Balaj L, Baldacchino S, Bauer NN, Baxter AA, Bebawy M, Beckham C, Bedina Zavec A, Benmoussa A, Berardi AC, Bergese P, Bielska E, Blenkiron C et al (2018) Minimal information for studies of extracellular vesicles 2018 (MISEV2018): a position statement of the International Society for Extracellular Vesicles and update of the MISEV2014 guidelines. J Extracell Vesic 7(1):1535750. doi:https://doi.org/10.1080/20013078.2018.1535750
135. Sebastiano V, Zhen HH, Haddad B, Bashkirova E, Melo SP, Wang P, Leung TL, Siprashvili Z, Tichy A, Li J, Ameen M, Hawkins J, Lee S, Li L, Schwertschkow A, Bauer G, Lisowski L, Kay MA, Kim SK, Lane AT, Wernig M, Oro AE (2014) Human COL7A1-corrected induced pluripotent stem cells for the treatment of recessive dystrophic epidermolysis bullosa. Sci Transl Med 6(264):264ra163. doi:https://doi.org/10.1126/scitranslmed.3009540
136. Umegaki-Arao N, Pasmooij AM, Itoh M, Cerise JE, Guo Z, Levy B, Gostynski A, Rothman LB, Jonkman MF, Christiano AM (2014) Induced pluripotent stem cells from human revertant keratinocytes for the treatment of epidermolysis bullosa. Sci Transl Med 6(264):264ra164. doi:https://doi.org/10.1126/scitranslmed.3009342
137. Wenzel D, Bayerl J, Nystrom A, Bruckner-Tuderman L, Meixner A, Penninger JM (2014) Genetically corrected iPSCs as cell therapy for recessive dystrophic epidermolysis bullosa. Sci Transl Med 6(264):264ra165. doi:https://doi.org/10.1126/scitranslmed.3010083
138. Hirsch T, Rothoeft T, Teig N, Bauer JW, Pellegrini G, De Rosa L, Scaglione D, Reichelt J, Klausegger A, Kneisz D, Romano O, Secone Seconetti A, Contin R, Enzo E, Jurman I, Carulli S, Jacobsen F, Luecke T, Lehnhardt M, Fischer M, Kueckelhaus M, Quaglino D, Morgante M, Bicciato S, Bondanza S, De Luca M (2017) Regeneration of the entire human epidermis using transgenic stem cells. Nature 551(7680):327–332. https://doi.org/10.1038/nature24487
139. Siprashvili Z, Nguyen NT, Gorell ES, Loutit K, Khuu P, Furukawa LK, Lorenz HP, Leung TH, Keene DR, Rieger KE, Khavari P, Lane AT, Tang JY, Marinkovich MP (2016) Safety and wound outcomes following genetically corrected autologous epidermal grafts in patients with recessive dystrophic epidermolysis bullosa. JAMA 316(17):1808–1817. https://doi.org/10.1001/jama.2016.15588
140. Lee J, Rabbani CC, Gao H, Steinhart MR, Woodruff BM, Pflum ZE, Kim A, Heller S, Liu Y, Shipchandler TZ, Koehler KR (2020) Hair-bearing human skin generated entirely from pluripotent stem cells. Nature 582(7812):399–404. https://doi.org/10.1038/s41586-020-2352-3
141. Wang LL, Cotsarelis G (2020) Regenerative medicine could pave the way to treating baldness. Nature 582(7812):343–344. https://doi.org/10.1038/d41586-020-01568-2

Chapter 4
The Dermal Papilla and Hair Follicle Regeneration: Engineering Strategies to Improve Dermal Papilla Inductivity

Nikolaos Pantelireis, Gracia Goh, and Carlos Clavel

Abstract *Introduction*: The Dermal Papilla (DP) has been the focus of many studies regarding the hair follicle due to its critical role in morphogenesis, hair cycling, and de novo induction. These studies have led to the current ambition of harnessing DP cells to provide a clinical cell-based therapy which overcomes the major limitation in hair follicle transplantation, a lack of donor material. *Methods*: Revision of peer-reviewed published literature using different databases including Pubmed. *Results*: Early work using explanted DP demonstrated a capacity to induce de novo hair follicles at implantation sites. In order to overcome limited donor material, cells need to be expanded in vitro. However, DP cells lose their inductive potential when cultured. In order to address this issue, researchers have attempted to maintain, restore, and improve inductivity in a number of ways. The most successful of these studies to date have combined elements of the 3 pillars of Tissue Engineering: Cells, Signals and Scaffolds. *Conclusions*: In the past, most approaches have utilised single pillar approaches in order to improve DP inductivity with limited success. More recent bioengineered solutions which are carefully designed to factor all three pillars of tissue engineering are starting to show real promise. Challenges remain, largely surrounding the quality of induced hair follicles and the efficiency to which they are induced as well as time and cost to deliver these therapies. However, if these issues can be overcome, tissue engineered hair follicle restoration would present a significant step-change in hair follicle restoration.

Keywords Dermal papilla · Inductivity · Hair follicle · AGA · Tissue engineering · Balding

---

N. Pantelireis
Department of Bioengineering, Imperial College London, London, UK

N. Pantelireis · G. Goh · C. Clavel (✉)
Skin Research Institute of Singapore, Agency for Science, Technology and Research, Singapore, Singapore
e-mail: carlos_clavel@asrl.a-star.edu.sg

© The Author(s), under exclusive license to Springer Nature Switzerland AG 2022
F. Jimenez and C. Higgins (eds.), *Hair Follicle Regeneration*, Stem Cell Biology and Regenerative Medicine 72, https://doi.org/10.1007/978-3-030-98331-4_4

Summary

In this chapter will introduce the Dermal Papilla (DP) and why it has been the focus of so much study regarding the hair follicle. This will then be followed by a transition into the current ambition of harnessing the DP's inductive potential to provide a clinical cell-based alternative to the current yardstick of hair follicle restoration, hair follicle transplantation. First, we will summarise the work done in trying to increase DP inductivity, starting with the early work demonstrating the capacity of the DP to induce de novo hair follicle morphogenesis and then proceeding to explain how DP cells lose their inductive potential when cultured. Then, we will cover how researchers have attempted to maintain, restore and improve inductivity to date. Here we outline the numerous studies using the 3 pillars of Tissue Engineering: Cells, Signals and Scaffolds. Most approaches have utilised single pillar approaches in order to improve DP inductivity, however, in recent years bioengineered solutions which factor all three pillars are starting to show promise. In engineering solutions, factoring all three pillars, we can create multicellular micro-organoids, capable of directing the proliferation and differentiation of multipotent epidermal cells semi-autonomously into functional hair follicles.

Key Points

1. Translation of DP inductive studies to the clinic has been limited due to loss of inductive capacity in DP cells in vitro.
2. Numerous studies have attempted to restore inductivity, with the greatest success coming through multidisciplinary approaches.
3. In order to create, clinically acceptable induced hair follicles which function, all three pillars of tissue engineering (Cells, Signals, and Scaffolds) need to be considered in conjunction.

The Dermal Papilla—Introducing the Conductor of the Hair Follicle

The DP is an unassuming structure, small compared to the entire hair follicle, which is mostly formed by extracellular matrix proteins and relatively few mesenchymal cells. DP cells show little cell death and low proliferation levels but nevertheless, the DP is critical in regulating the hair follicle. Throughout the hair follicle's life, the DP and its progenitor are pivotal in hair follicle morphogenesis and hair follicle cycling. DP cells also control the spatial differentiation of hair follicle epidermal cells, hair shaft diameter and hair follicle miniaturisation.

Don't Simply Condense All Cells Together—Cellular Identity Matters

The DP forms from the dermal condensate, a condensation of early mesenchymal progenitors which migrate underneath epidermal placodes during morphogenesis in response to a paracrine signal from these epidermal structures. Placodes are discrete thickened loci in the epidermis which form in response to broad signalling from the dermis [1–3]. While we know that a broad signal from the dermis is required for placode specification, this signal remains elusive. Recent studies have shown that prior to the placode forming a morphologically distinct thickened locus, pre-placodal epidermal cells at the future placodal site upregulate FGF20 [4]. In response to this pre-placodal signal, dermal condensate precursor cells are induced in the surrounding embryonic dermis [4–6]. These induced cells are randomly dispersed within the fibroblasts beneath the pre-placode, showing a distinct molecular profile and a post-mitotic behaviour, linking a loss of proliferation to a gain in hair follicle inductivity [4, 7]. Why certain cells are indued and others are not remains unclear, however, a fibroblast subpopulation closely related to pre-dermal condensate cells was observed, indicating a possible precursor cell type which responds to FGF20 signalling [4]. Under continued FGF20 stimulation from the placode, pre-dermal condensate cells begin to migrate underneath the placode to form the dermal condensate [2]. The dermal condensate is crucial for hair follicle morphogenesis as it signals the overlying placodal cells to form the downward growing hair germ which proliferates and differentiates forming the various epidermal layers of the hair follicle, eventually engulfing the condensate to form the adult hair follicle. During early morphogenesis, the cells which form the dermal condensate move through a developmental trajectory with three distinct states: a pre-condensate state, an early condensate state, and late condensate state [4]. At each of these stages, condensate cells show distinct transcription factor and signalling molecule expression patterns which are proposed to play a role in dermal condensate fate acquisition, aggregation, initiation, and maintenance of hair follicle morphogenesis [4].

The mature hair follicle's DP retains its regulatory role throughout the adult hair follicle's lifetime. Unlike most tissues, the hair follicle undergoes a continuous cycle through a series of stages known as anagen (growth phase), catagen (destructive phase), and telogen (rest phase) [8, 9]. The DP is particularly important in maintaining the stem cell niche in the hair follicle while also stimulating the hair follicle's re-entry into anagen during cycling. During anagen, crosstalk between the DP and the overlying epithelium regulates the hair bulb micro-niche at the hair follicle base. Heterogeneous cells within the DP lie close to overlying epithelial transit-amplifying cells (TACs) which show a similar heterogeneity. Distinct lineages are observed within the TAC population, where they are organised spatially and adjacent to equally organised and heterogenous DP cells [10]. Each unique TAC progenitor gives rise to the various differentiated cells which make up each of the distinct inner hair follicle and hair shaft layers. Essentially, one TAC type gives rise to an entire sub-structure within the hair follicle (Fig. 4.1). The surprising presence of such heterogeneity in

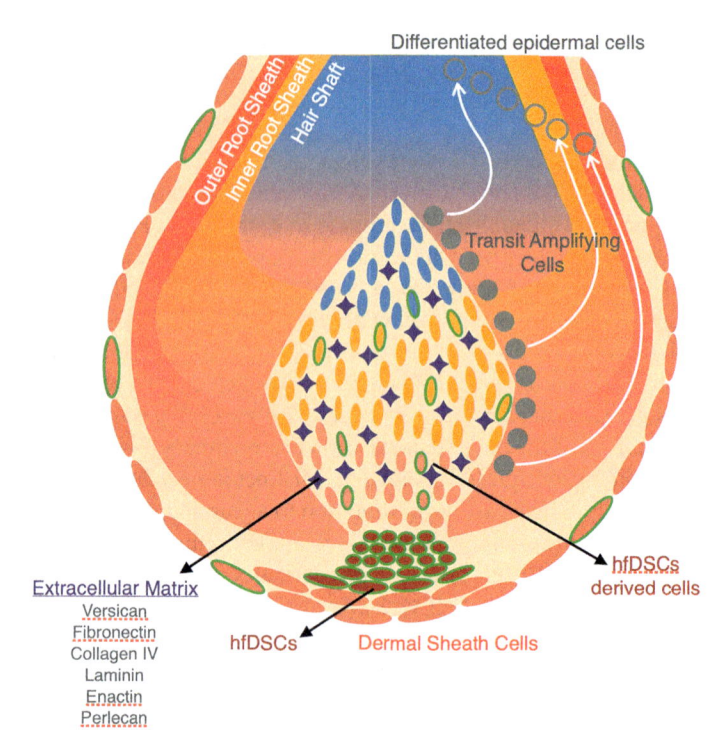

Fig. 4.1 The Dermal Papilla and its niche. The DP has a natural scaffold rich in basement membrane proteins such as collagen IV, laminin, and entactin. This is in addition to a number of proteoglycans such as versican, fibronectin and perelcan. Heterogeneity within the cells of the DP regulates the overlying transit amplifying cells to form the various layers of the hair follicle. While hfDSCs from the dermal sheath replenish the DP during cycling

the hair follicle bulb's epithelial portion and the adjacent DP's corresponding pattern suggests a delicate cellular crosstalk between the two structures. Thus, the DP plays a role in maintaining the cellular identity of epithelial TAC progenitors during anagen in a spatially organised manner, resulting in the overlying concentric pattern of the hair follicle inner root sheath and hair shaft.

Hence, as structures, the dermal condensate and the DP, are vital in morphogenesis and niche maintenance. Moreover, the interspersed response to placodal FGF20 signalling, the temporal heterogeneity within the dermal condensate during morphogenesis and the spatial heterogeneity of the mature DP shows how specific sub populations of cells play a crucial role in regulating the hair follicle niche. This something we need to be conscious of when looking to engineer strategies to improve hair follicle inductivity. The choice of cells, the aggregation of these cells, and spatial organisation within these structures are all vital.

Round and Round We Go Unless There is no DP

Further to the dermal condensate's role in morphogenesis, and the DP's in maintaining epidermal progenitor identity, the DP's role in the cycling of the hair follicle has also been well established. This role was previously characterised in a study where telogen stage DP and dermal sheath (DS) were laser ablated [11]. Ablated hair follicles were unable to continue cycling, stagnating in telogen while non-ablated follicles progressed into anagen. In the same study, partial ablation of the hair germ, but not the DP and DS did not perturb the hair cycle's progression [11]. Here, despite the removal of progenitors that eventually form the anagen hair follicle, the DP's presence allowed the hair germ to regenerate; hair follicles still entered anagen in synchronisation with non-ablated follicles. What we see in telogen is reminiscent of morphogenesis, where the DP signals overlying epidermal cells to re-form the hair follicle. Similar to morphogenesis, removing the dermal component stops the process [12]. Inductive signals that the dermal condensate and DP provide to the hair follicle's epidermal component are essential to both morphogenesis and hair cycling. Later on, we will detail what is known about these signals and what is being done to recreate them.

The Larger the Root the Larger the Fruit

Knowing that transcriptionally heterogeneous cells within the DP help direct TACs to form the overlying hair follicle layers, it may seem obvious that a larger DP volume and an associated increase in DP cells is associated with a thicker hair shaft diameter [10, 13]. Nevertheless, it is another significant example of the DP's regulatory role. We see this correlation throughout the body, and it is most notable when comparing course male facial hair with large DP (mean volume—1,878 × 10^3 μm^3) to that of near-invisible vellus (peach fuzz) female facial hairs with DP (mean volume—47 × 10^3 μm^3) which are almost 40-fold smaller [14]. However, it isn't just hair shaft diameter that is correlated to the DP size. In mice we see that DP size also dictates the type of hair follicle formed [15]. Furthermore, the size of the DP and hair follicle do not only differ from location to location, a reduction in DP is also observed in androgenic alopecia (AGA) [14].

Miniaturisation—The Diaspora of Dermal Papilla Cells

In AGA hair follicles undergo miniaturisation [16]. Here thick terminal hair follicles drastically shrink into thin, almost indistinguishable vellus hairs [17, 18]. This process happens during transitions in the hair cycle where the follicle undergoes comprehensive remodelling. The DP itself changes between anagen and telogen shrinking in size [19]. However, due to limited apoptosis and proliferation in the DP, it is hypothesised that the reduction in DP size in the anagen to telogen transition is due to cell migration into the DS [19–21]. In the telogen to anagen transition, replenishment of the DP cell number has been shown to come from hair follicle dermal stem cells (hfDSCs) which reside in the DS directly underneath the DP (Fig. 4.1). However, with age, DP homeostasis is compromised due to a loss of stemness in hfDSCs [21–23]. In miniaturised hair follicles, like those seen in androgenic alopecia or associated with ageing, it is likely that this loss of function ultimately results in a net reduction in the DP's size and the subsequent inability to stimulate re-entry into anagen for terminal hair follicles. Instead, only short, white vellus hairs are supported, which exhibit long telogen phases giving the combined impression of a scalp devoid of hair follicles.

Further implicating the DP in hair loss is the androgen dihydrotestosterone (DHT), whose action results in AGA onset. DHT is a metabolite of testosterone which 5alfa-reductase catalyses. The increased presence of 5alfa-reductase in the DP of balding hair follicles compared to non-balding hair follicles indicates that balding hair follicles precociously catalyse DHT production through the mesenchymal compartment of the hair follicle [24]. It remains to be seen whether DHT may act through the mesenchyme by perturbing the process through which hfDSC replenish the DP. There is no doubt then of the importance of the DP; however, our discussion's primary focus will be whether it can provide the solution to hair loss pathologies.

Dermal Papilla Induction—A Not so New Hope

We have summarised some of the most ground-breaking work done on characterising the DP, focusing on its role in hair follicle morphogenesis and regulation of the hair follicle cycle. Furthermore, we summarised how a net efflux of DP cells from the DP into the DS could play a central role in the miniaturisation of hair follicles. However, possibly the most impressive capability of the DP with potential clinical applications we are yet to touch on. This is its capability to induce the formation of brand-new hair follicles. An ability first demonstrated over 50 years ago [25]. Building on early work done using the DP from feathers and using rodent DP isolation techniques, several experiments were carried out which established the DP's role in hair follicle induction [26–28].

Counter to what we may expect, at first, removing the DP from rat whisker follicles did not stop the generation of new hair shafts [29]. However, deeper histological analysis showed that the excised DP were regenerated before the hair shaft grew [30]. At the time, it was postulated that the DP might be replenished from the DS or the dermis. Since, and as we have already touched on, the DP's replenishment by hfDSC is now established. These first experiments showed that the DP as a structure is always present in hair follicles that can produce a hair shaft and those that can cycle. The DP's requirement was further demonstrated explicitly when the DP was implanted at the base of inactivated non-cycling follicles [31]. Post implantation, the hair follicle re-entered an active state with active hair shaft growth and cycling. Essentially, this confirms what has now become common knowledge in the field today, that the DP is necessary for the follicle to function and cycle.

At first, isolated DP implantation was somewhat unsuccessful in inducing hair follicles [26]. The inability to do so was likely due to a lack of incubation time as small epidermal invaginations did occur. In the same experiment "whole papillae", DP with matrix cells still attached, from whiskers were also implanted. These whole papillae were able to induce large whisker follicles at the recipient site. Due to this differing response, it was thought that if the DP was implanted on its own, it could induce a hair follicle but that the epidermis at the recipient site would dictate the follicle type formed.

In 1970, excised DP were reconstituted with epidermal cells from the scrotal sac and those from the oral epithelium [25]. These recombined structures were then implanted into recipient sites located in the ears where they induced abnormally large hair follicles compared to the local hair follicles.

It is worth noting that both the isolated DP and the whole papillae induced follicle formation but neither produced hair shafts. However, the studies showed that the DP could instruct the early formation of a hair follicle. Furthermore, they could also direct non-follicle bearing epidermal cells to form hair follicles. However, it was almost 20 years later until it was confirmed that rat whisker DP implanted in wounded recipient skin gives rise to rat whisker hair follicles which produce long thick whisker type hair shafts [32]. This work finally confirmed that it is indeed the DP that dictates the hair follicle phenotype and instructs the epidermis to form it. It should also be noted that implanted DP need to be implanted in contact with the epithelial tissue in order to be able to induce de novo hair follicle morphogenesis.

While this work was conducted on rodents, the human dermal compartment can also induce de-novo hair follicle morphogenesis in adult human skin [33]. In a transgenic male to female transplant, DP and DS were isolated and implanted separately into adult skin. Here the larger DS was able to reform the DP and induce de-novo hair follicles. When karyotyped, the hair follicles' dermal compartment possessed XY chromosomes while the epidermal compartment possessed XX chromosomes, confirming the male DS had successfully induced the female epidermis to form brand-new transgenic hair follicles. However, surprisingly, implanted DP were unable to induce hair follicles. This may be due to difficulty maintaining contact between the small DP and the epidermis [33]. It is worth noting that the implanted DS would likely contain hfDSCs capable of regenerating the DP, possibly giving it more capacity to induce. Still, human DP have since been shown to be inductive when they are transplanted into inactive follicles and when recombined with neonatal human epidermal cells and transplanted into mice [34, 35].

The ability to regenerate a whole follicle from a single small dermal structure has tremendous potential, yet, clinically, using excised DP does not provide an advantage to hair transplantation. Currently, hair transplantation is the only therapy which can restore cycling terminal hairs to bald skin. However, its primary limitation is that we can only move hair follicles from one location to another. So, there is a limited number of hair follicles one can harvest from non-balding regions. Each implanted follicle gives rise to a single follicle in the recipient site, coming at the expense of that hair follicle from the donor region. Implanting excised DP individually doesn't address this shortcoming and requires additional steps to excise the DP from the hair follicle. Furthermore, as demonstrated already, not all implanted DP will induce a new hair follicle.

The promising potential comes from the fact that a single hair follicle compartment can induce an entire hair follicle, hence, making cell-based approaches to hair loss more practical. With cell-based approaches, we can overcome limited donor material as we can expand cells in-vitro to levels that will enable more precocious coverage. Furthermore, the usual problems encountered in tissue engineering, namely complexity, are reduced. Typically, in tissue engineering, we need to consider whole complex organs. These usually have many different cells types, associated paracrine signalling, functional and structural requirements, and even vascularisation. With an inductive structure, we can design simpler alternatives that can self-regulate their morphogenesis and form the organ's intricate compartments autonomously. While there is potential, this simplified structure is still incredibly complex, and we will delve into the steps researchers have taken to bring us closer to a translatable inductive therapy.

Dermal Papilla Induction—Cultured Cells Strike Back

One of the first steps in both cell therapies and tissue engineering is to take a small autologous biopsy in order to isolate cells which subsequently will be used for treatment. In most cases these cells are usually expanded in-vitro. For hair-loss, where the limitation is a lack of donor tissue from the occipital scalp, in-vitro cell expansion is a prerequisite. To this end, both human and rodent DP have been successfully excised and explanted in order to grow DP cells in culture [36, 37]. Since then, cultured DP have been characterised extensively.

Characteristics of Cultured DP Cells

Human and rodent DP cells show a number of differences in culture, yet some characteristic traits remain. These differences likely play a role in each cells capacity to induce hair follicle morphogenesis as well as the pathologies unique to human hair.

One of the most unique differences that likely plays a role in inductivity is the ability of rat whisker DP to self-aggregate in-vitro and in-vivo [38]. Self-aggregation is also observed in rat pelage DP cells, albeit to a lesser extent. However, this is seemingly absent from cultured human DP cells [39].

Another difference between human DP and rodent DP cells is in the expression of androgen receptors. In human DP, androgen receptor expression varies according to the location from which the hair is taken [40–42] while in mice, there is an absence of androgen receptors on the whole [39]. This goes someway to explaining why humans exhibit substantial hormone dependant hair growth and loss while mice do not.

A common feature of cultured DP across species is the expression of alpha smooth muscle actin (αSMA), a cytoskeletal protein which is not expressed at high levels in vivo within the DP [43]. Additionally, cultured DP produce distinctive proteoglycan and glycoprotein rich extra cellular matrix similar to that characterised in-vivo [44]. Therefore, DP cells maintain some characteristic traits in-vitro and their assessment does show heterogeneity within different follicle types and across species.

The Loss of Inductive Capacity

Despite the retention of some distinct characteristic traits, DP cells undergo a drastic change when moving from their in vivo environment within the DP to the hard plastic of the culture plate. This is most apparent when we look at how cultured DP cells behave with respect to hair follicle induction. In early work, culturing rat DP cells yielded positive results [45, 46]. Introducing cultured whisker DP cells to inactivated follicles stimulated hair shaft formation in over 50% of hair follicles [46], which is similar to rates observed using intact DP [25]. The cells were implanted in the form of a loose pellet at the base of whisker follicles whose bulb region had been amputated. Fibroblasts were used as a control and showed almost no signs of hair shaft regrowth (3% of follicles showed spontaneous regrowth) [47]. This indicated that DP even when cultured retained their inductive power. However, over progressive culture the DP cells lose their inductive potential. When late passage cells were used (passage 6–15) and implanted none of the 42 follicles targeted showed follicle growth [47]. Nevertheless, the ability to passage cells even to a relatively low passage number still represents an upside as donor material can be amplified. This is especially true when induction efficiency was comparable to that of intact implanted DP.

While most rodent cultured DP are able to induce hair follicle formation, they all lose this capability over progressive passages. With whisker type cells, late passage cultured cells did not aggregate in-vivo post implantation, while their late passage in vitro aggregative behaviour was not commented on [47]. It was thus believed that a loss in inductive capacity may be linked to a loss in self-aggregation. This is of interest when we consider human DP cells, which as we alluded to, do not aggregate in vitro and have not been shown to self-aggregate in vivo post injection [34]. Furthermore, micro-array analysis of gene expression in intact DP and early passage cells showed a drastic change in gene expression from the moment the cells begin to attach to the tissue culture plate [34]. Almost 4000 transcripts were shown to be differentially expressed from intact ex vivo DP, when place in vitro, with this phenotype persisting throughout the passages. Ontology analysis indicated that many cell cycle, DNA replication, and mitotic terms were enriched in cultured cells while Wnt-signalling was precociously down regulated. Despite this, while a large-scale change occurs immediately, DP cells still show distinct differences to fibroblasts in culture. However, it is believed that over progressive passages, human DP cells tend towards a fibroblast-like phenotype losing much of their own characteristics [48]. This could be linked to culture-induced senescence, as some have suggested [49], or it could be a loss of paracrine signalling or contact from surrounding cells and extracellular matrix.

In the following section, we will focus on how researchers have tried to restore a hair follicle inductivity to DP cells. This will be broadly split out into the three pillars of tissue engineering: Cells, Signals and Scaffolds. It is worth noting however, that these pillars do have significant overlap and that a multi-pillar approach may be required and where this has been attempted, we will highlight it.

Dermal Papilla Induction—The Return of Inductivity

Signalling—Does Supplementation Work?

The most straightforward and thus most popular strategy to restore a specific phenotype is to add a specific signal that may be missing from the culture. In this context a signal is related to an extracellular additive, whether organic or inorganic.

With regards to the DP, a number of essential genes related to hair morphogenesis and hair cycling have been characterised. These usually consist of genes within the WNT, BMP and FGF signalling pathways. When additional factors which upregulate these pathways are added to cultured DP cells, DP signature genes are restored and, in some cases, a beneficial effect on induction occurs when testing using in vivo hair induction assay. We summarise some of the most relevant attempts to restore an inductive environment in Table 4.1. Despite the numerous studies, to date no additives on their own have been able to recapture a lost inductive potential in human DP cells. The more successful studies have shown that supplementation of media with extracellular factors can provide an additional benefit by enriching DP characteristics [50–54]. Here, an increase in hair inductive capacity accompanies DP enriching supplementation. While the holy grail may be to determine the exact inductive signal that leads to hair follicle development, it seems apparent that the dermal compartment provides continuous signals to the epidermal compartment of the hair follicle throughout the process of morphogenesis and the adult follicles lifetime. To this end a single factor is likely not sufficient. Therefore, combining extracellular signals with other aspects of tissue engineering to generate inductive mini organs provides the most likely improvement to current hair follicle transplants therapies. To achieve a simpler topical or systemic therapy, researchers need to look at how to replenish the DP of miniaturised hair follicles. To do so we need to answer why hfDSCs lose their ability to replenish the DP and whether this can be restored in-vivo through extracellular signalling. In the following sections we will expand on methods which focus on the implantable mini-organ model.

4 The Dermal Papilla and Hair Follicle Regeneration ... 69

Table 4.1 Signals – Table of factors added to DP cells in culture with the aim to restore inductive phenotype

Signal	Pathway	Effect on DP	Induction demonstrated?	Cell type tested	Ref.
Conditioned medium					
Adipocyte stem cell conditioned medium	NA	I. Alkaline phosphatase levels increase in cultured DP cells after treatment with ASC-CM II. Use of differentiated adipocyte CM did not increase levels of Alkaline Phosphatase	❌ No Induction shown with DP cells treated with ASC-CM	Rat whisker cultured DP	[55]
Keratinocyte conditioned medium	NA	I. Medium collected from rat foot pad keratinocytes II. Culture of rat whisker DP maintained up to passage 90 III. Inductive capacity of whisker DP maintained up to passage 70, 60 more passages than non-supplemented whisker DP	✅ Induced hair follcile up to passage 70	Rat whisker cultured DP	[56]
Extracellular vesicles					
DP-EVs miR-140-5p	BMP Pathway	I. DP extracellular vesicles (DP-EVs) contain high levels of miR-140-5p II. Decreases BMP2 and downstream targets III. Direct inhibition of miR-140-5p restores BMP signalling IV. Addition of DP-EVs to human HF explants results in prolonged anagen V. Injection of DP-EVs into mice post depilation accelerates onset of anagen versus control groups	❌ Prolonged anagen and acceleration of telogen to anagen	Human HF explants and in-vivo murine models	[57, 58]

(continued)

Table 4.1 (continued)

Signal	Pathway	Effect on DP	Induction demonstrated?	Cell type tested	Ref.
DP-Exos Dermal Papilla derived exosomes	Wnt and BMP pathway	I. Human dermal papilla cell exosomes isolated from 3 and 2D cultures and added as a supplement to culture II. 3D-exos promoted proliferation in DP and ORS III. 3D-exos increased expression of IGF1, KGF, and HGF in DP cells IV. 3D-exos increased explant anagen duration V. Local injections in mice accelerated anagen to telogen transition and prolonged anagen VI. Increased number of hairs induced when 3D DP treated with 3D-exos versus non treated 3D DP cells VII. Similar results observed with 2D DP exos	✓ Increased induction in human DP spheroids	Human cultured DP cells Human HF explants model In-vivo murine model Human DP spheres recombined with neonatal murine epidermis	[52]
MAC-EVs Macrophage extracellular vesicles	Wnt pathway	I. Increased proliferation, migration and upregulation of DP-specific markers II. MAC-EVs contain high levels of Wnt3a and Wnt7b resulting in downstream expression of Axin2 and Lef1 III. Treated DP cells showed increased expression of VEGF and KGF	✗ Increased hair growth	DP and extracellular vesicles from human scalp tissues	[59]
MSC-EVs Mesenchymal stem cell extracellular vesicles	Wnt Pathway	I. Increases DP cell proliferation and migration II. Increases levels of Bcl-2. Phosphorylation of Akt and ERK III. Increased expression and secretion of VEGF and IGF-1 IV. Promotion of telogen to anagen transition V. Increased expression of Wnt3a, Wnt5a and Versican in C57BL/6 mice after intradermal injection of MSC-EV	✗ Promotes activation of anagen	Human DP cells In-vivo murine HFs	[60]

(continued)

Table 4.1 (continued)

Signal	Pathway	Effect on DP	Induction demonstrated?	Cell type tested	Ref.
stDF-EVs Stimulated dermal fibroblast extracellular vesicles	Wnt pathway	I. Fibroblasts were treated with bFGF and PDGF-AA prior to st-EV collection II. St-EVs activate secretion of the non-Wnt ligand protein Norrin activating proliferation in overlying epidermal cells III. Increases hair shaft growth IV. Increases DP cell proliferation in vitro	✗ Increased hair growth	Human ex-vivo hair follicles	[61]
Growth factors and others					
Adenosine	Wnt pathway	I. Increased expression of B-catenin and activation of Wnt/ERK pathways II. Increases proliferation in cultured DP cells III. Anagen phase activation and elongation IV. Promotes expression of known hair growth factors such as FGF7, FGF2, IGF1, and VEGF	✗ Increased hair growth	Anagen phase follicles from upper lip pad of mice Human DP cells	[62]
BMP6	BMP pathway	I. BMP6 added to early cultured adult murine DP cells II. In early passages DP able to induce significantly more hairs than control DP III. BMP6 delays onset of culture directed loss of induction but does not stop it entirely	✓ Increased hair follicle induction versus control	Murine early culture DP cells	[63]

(continued)

Table 4.1 (continued)

Signal	Pathway	Effect on DP	Induction demonstrated?	Cell type tested	Ref.
CHIR99021	Wnt pathway	I. Increased expression of DP signature genes such as: LEF1, Versican, BMP4, Sox2 and Noggin in 3D DP spheroids versus non-treated spheroids II. CHIR99021-stimulated DP spheroids can induce de-novo hairs III. No mention made of improved inductivity versus non treated DP spheroids IV. In separate study use of CHIR99021 increased 3D DP inductivity in in-vitro human skin constructs to 50% from 20% efficiency	✓ Increased inductivity of 3D DP cultured cells	Human cultured DP	[51, 53]
DPAC medium	Wnt/BMP/FGF pathway	I. Dermal Papilla activation medium consisting of 6-bromoindirubin-39-oxime (BIO), BMP2, and basic FGF added to 10%FBS DMEM II. Human cells treated with DPAC medium show increased expression of DP like genes III. Alkaline phosphatase activity increased in cultured DP cells IV. Increased level of inductivity in DPAC cultured 3D DP spheres versus non treated 3D DP spheres when recombined with mouse sole epidermis	✓ Increased inductivity of 3D DP cultured cells	Human cultured 3D DP cells recombined with murine paw epidermis	[50]
FGF2	FGF pathway	I. Increases proliferation and maintains population for longer passages II. Late passage cells are unable to induce III. When cultured in spheres the inductivity of FGF2 treated DP cells is restored	✗ Does not alone restore inductivity	Mouse whisker DP	[64]

(continued)

4 The Dermal Papilla and Hair Follicle Regeneration … 73

Table 4.1 (continued)

Signal	Pathway	Effect on DP	Induction demonstrated?	Cell type tested	Ref.
Glucose	Glycolysis	I. 3D DP cells show increased aerobic glycolysis II. When glycolysis is inhibited induction associated genes show decreased expression III. In the presence of high glucose, induction associated genes increase IV. Increased acetylation is also observed in these inductive genes	❌ Increased hair growth No inductive tests	Human cultured DP cells	[65]
Hypoxia	NOX4 mediated ROS generation	I. DP cells cultured in 2% Oxygen conditions show increased proliferation compared to DP cells cultured normally II. Increased inductive capacity shown in late passage DP cells after spheroids formation III. Mild Reactive oxidative species are generated which act through NOX4 to stimulate increased proliferation and a delay in sensecense	✅ Increased inductivity in late passage 3D DP	Human cultured DP cells	[66]
lncRNA-599547	Wnt pathway	I. lncRNA-599547 is highly expressed in cashmere goat during anagen compared to telogen II. Overexpression of lncRNA-599547 increases expression of ALP, LEF1, and Wnt10b gene	❌ qRT-PCR only	Cashmere goat DP	[67]

(continued)

Table 4.1 (continued)

Signal	Pathway	Effect on DP	Induction demonstrated?	Cell type tested	Ref.
Minoxidil	VEGF pathway	I. Minoxidil-treated dermal papilla cells show increased expression of VEGF II. Believed to underly an increased vascularisation response in-vivo	✗ Characterisation only	Human cultured DP	[68]
Platelet rich plasma	NA	I. Addition of 10% PRP to cultured DP cells increased number of hairs induced compared to control II. Morphogenesis also occurred at a quicker rate to control	✓ Increased induction versus control	Cultured Whisker DP cells	[69]
RSPO1 + iPKC + WNT5a + MMP14	Wnt pathway	I. Factors added to adult murine co-cultures grown on submerged culture inserts over a period of 6 days II. Cultured inserts sutured face down into full thickness wounds in-vivo III. Treated cells were able to induce hair morphogenesis where control conditions could not	✓ Induction activated in adult murine dermal cells	Cultured adult murine epidermis and dermis	[54]

(continued)

Table 4.1 (continued)

Signal	Pathway	Effect on DP	Induction demonstrated?	Cell type tested	Ref.
SHH	SHH pathway	I. VersicanGFP + murine DP cells isolated from adult mice II. Co-cultured with shh + chicken embryonic fibroblasts III. Shh signalling cannot maintain GFP expression (marker of anagen phase in DP) IV. Shh co-cultured DP recombined with neonatal murine epidermal cells are not able to induce hair follicles	Adult murine DP not able to induce	Cultured adult murine versicanGFP+DP cells	[70]
Tofacitinib	JAK-STAT pathway	I. JAK-Stat inhibitor Tofacitinib treated human DP spheroids show an increased number of induced hair follicles in murine patch assay II. Human DP spheroid treated with Tofacitinib show increased expression of DP signatures, which are not restored through 3D culture alone	Increased induction versus control	Human DP cultured spheroids recombined with neonatal murine keratinocytes	[71]
Wnt3a	Wnt pathway	I. VersicanGFP+ murine DP cells isolated from adult mice Co-cultured with Wnt3a+ chicken embryonic fibroblasts II. Wnt3a signalling maintains GFP expression (marker of anagen phase in DP) III. Wnt3a co-cultured cells recombined with neonatal murine epidermal cells able to induce hair follicles	Induction activated in adult murine DP	Cultured adult murine versicanGFP+DP cells	[70]

(continued)

Table 4.1 (continued)

Signal	Pathway	Effect on DP	Induction demonstrated?	Cell type tested	Ref.
Wnt10b	Wnt pathway	1. Addition of Wnt10b to human cultured DP in human skin construct. Treated aggregated DP increased the efficiency of hair follicle induction from 20% to close to 50%	✓ Increased inductivity of 3D DP cultured cells	Human Cultured DP	[51]

✓ Increased inductive capacity demonstrated

✗ No increase in inductive capacity demonstrated

Scaffolds and Augmenting the Macroenvironment of the DP

The DP has an extracellular matrix which is rich in basement membrane proteins, such as laminin, collagen IV, and entactin. A large number of glycoproteins are also present such as versican, fibronectin, and perlecan [72, 73] (Fig. 4.1). The proteins act as a scaffold but are also important in cell signalling, in particular cell adhesion, migration, and cytoskeletal organisation and differentiation [72]. The ECM is also dynamically remodelled during hair cycling, with anagen DP showing the greatest level of ECM proteins. The deposition of these ECM proteins also occurs in the culture of rat whiskers DP cells in early passages where these cells self-aggregate, however, their expression is diminished with continued passage [36, 74]. In their place a number of connective tissue collagens such as collagen II and III are expressed [75].

Due to the observation of in-vivo self-aggregation in the highly inductive rat whisker DP cells and the lack thereof in human DP cells, the idea to force aggregation was explored. When looking back at the early cultured mouse models in many cases the cells were recombined with the epidermis in loose pellet forms. In a loose pellet, cells with an affinity to aggregate can do so readily. Since human DP don't readily self-aggregate, the classical hanging drop technique was utilised to stimulate aggregation in human DP [76, 77].

In hanging drop culture gene expression of versican (a marker of DP anagen) and a number of other ECM proteins as well as Wnt signalling factors are restored through 3D culture to levels akin to in vivo DP [34, 77]. Additionally, alkaline phosphatase activity is strongly increased and αSMA expression, which is increased in culture, becomes downregulated. While a number of ECM proteins are restored in 3D DP spheroids, a full intact DP ECM is not generated [77]. Therefore, culturing DP cells with ECM type proteins may yield a positive effect, although this has not been comprehensively studied.

Further to hanging drop culture, 3D DP spheroids can also be generated using hydrophobic polyvinyl alcohol (PVA) coated tubes. DP cells cultured within these tubes form 3D spheroids after 6–24 h of culture [78]. Low bind 96 well plates and hydrocell plates can also be utilised to form DP spheres [50, 79].

Despite the wide array of techniques available, all these aforementioned 3D DP systems have been shown to restore an inductive capacity to human DP cells [34, 50, 78, 79]. However, the size of the DP spheres seems to correlate to the number of hair follicles induced. Using 30,000 cells per spheroid, gives an average efficiency of 60%, while smaller 3,000 cell spheres show on average efficiency of 18.5% [34, 78]. It should be noted that the in-vivo hair induction model selected also plays a role. Most of the studies use simple patch assays recombining human DP spheres with murine epidermal cells. These patches form pelage type hair follicles subcutaneously [50, 78, 79]. The only initial 3D DP study using human epidermal cells shows a much-reduced efficiency [34].

Looking at the molecular signature, microarray analysis of hanging drop DP spheroids shows that around 22% of the intact DP transcriptome is recovered [34]. This partial recovery is somewhat surprising due to the restoration of an inductive phenotype, but the lack of paracrine signalling of surrounding DP niche compartments is likely required to maintain characteristic traits within the DP. By comparing intact DP with plate cultured DP and 3D cultured DP cells, differentially expressed genes were determined. The expression of a number of known DP signatures such as BMP2, HEY1, FGF7, and SFRP2 was restored in spheroid culture, while expression of 2D culture related genes such as a-SMA, GREM1, and FZD6 was lost. However, a number of genes such as SOSTDC1, WIF1, LEF1 and CD133 were not restored through 3D culture of DP cells. This is of note as all these genes have been linked to hair growth previously. The inability to restore these genes to intact DP expression levels indicates that these genes may not play a role in induction. That being said hair follicles which are induced do not represent a fully healthy hair follicle. Some do not include sebaceous glands while in others hair shafts are missing or truncated [80]. Therefore, another possibility is that even though these genes are not completely necessary for hair follicle induction they may increase the quality or efficiency of induction.

The restoration of inductivity through 3D culture recaptures the potential for cell-based therapies. However, there is one more major compromise, the loss of proliferation. DP cells become post mitotic when forced to aggregate, which more closely reproduces the behaviour within the DP in vivo. This restoration of an in vivo environment is what underlies the return of inductivity yet without a proliferative capacity, expanding cells to clinically required levels is not possible. Consequently, cells must be expanded using plate culture first. The issue here is the immediate loss and only partial restoration of DP gene expression subsequent to 3D culture. Despite this, human DP cell spheres have been shown to be inductive up to P10 [79]. Hence in theory a 1000X fold amplification from P0 cells is possible. In an effort to restore traits lost in DP due to culture, spheroid culture can be combined with the simultaneous treatment of DP cells with DP derived exosomes or supplemented media. Here inductivity can be increased or prolonged with the various methods summarised in Table 4.1. This combinatorial approach takes into account both macroenvironmental requirements and supplementation.

Scalability is a further issue of 3D culture. Hanging drops and PVA coating tubes are highly labour intensive and spheroids produced are of variable size and quality. On the other hand, 96 well plate formats provide a more scalable alternative which could be combined with automation for consistent and timely spheroid formation. In efforts to address scalability a number of researchers have focused on using various substrates on which to culture DP cells with the aim of maintaining signature genes and promoting aggregation.

Prior to coating tubes, Poly(ethylene-co-vinyl alcohol) (EVAL) soft membranes were developed [81]. These membranes enhance cell motility and hence allowed greater self-aggregation of rat whisker DP cells. The cultured DP spheroids formed on the membrane were able to induce HF growth using the patch assay [81].

Other materials were then explored over the years, ranging from using platelet-rich plasma (PRP) membranes to soft self-assembling peptide scaffolds. Using a PRP gel, a fairly even distribution of dense aesthetically acceptable hair follicles can be induced. Here, murine DP co-cultured on 5% activated PRP infused gels with neonatal epidermal cells are transplanted into nude mice [82]. This method improves induction over the chamber graft utilised in the same study. However, co-culturing human DP with neonatal human keratinocytes proved non-inductive [82].

Similar to the EVAL membrane, the self-assembling peptide scaffold RAD16-I comprises the same principle of reduced adhesion to culture 3D DP cells [83]. DP cells show increased DP signatures when cultured on RAD16-I while adipogenic and osteogenic differentiation is also encouraged. A benefit to using biodegradable synthetic RAD16-I scaffolds is the control and reproducibility of the scaffold. Here synthetic scaffolds have an advantage compared to natural scaffolds such as collagen.

In a separate study, differing concentrations of poly-ethylene-glycol-diacrylate (PEGDA) was used for DP growth. It was found that the softer substrate of lower concentration allowed for more effective growth of inducible DP cells which was reflected by the upregulation of significant DP signalling pathway genes. This then presents a fresh perspective on how to optimise the substrate for which to culture 3D DP cells on since it seems that spatial organisation of the substrate plays a role as well. One slight drawback to the PEGDA and RAD-16 scaffold was that there was no in vivo testing done, which may limit the result's applicability to mice or even humans.

Recently, biomimetics approaches are starting to be incorporated and appear to be yielding positive results. By using 3D printed moulds, 4 mm indentations with a 500 μm diameter can be made within fibroblast containing collagen scaffolds [51]. This creates regularly spaced and repeating hair follicle like structures with this dermis equivalent. Cultured DP cells are subsequently seeded on the moulded collagen scaffold and aggregate at the bottom of the indentations. Aggregate size can be controlled by altering the width of the indentations, with aggregate diameter on average half the size of the indentation diameter. However, it is worth noting that variation in the aggregate diameter is quite high relative to the size of the aggregates themselves. Nevertheless, these aggregates show reduced αSMA expression in combination with increased expression of versican and alkaline phosphatase. When fibroblasts are seeded in the same manner aggregates still form, but they do not exhibit the same changes in molecular profile. This biomimetic scaffold is designed as part of an entire human skin equivalent and as such incorporates co-culturing of different cell types. Thus, providing us with an excellent segue into the final pillar of tissue engineering—Cells.

Cells

As alluded to earlier, when we look to bioengineer solutions for clinical issues, we need to take a combinatorial approach. With regards to tissue engineering, we focus on the three main pillars: Cells, Signals and Scaffolds. In this chapter we started with signals as the addition of supplements to culture providing the simplest way to bring about a phenotypic change. Scaffolds provide mechanical support and help to create a more in vivo-like environment. The choice of scaffold requires a multifaceted approach, taking into account: material, design, reproducibility, usage, and cost. Yet, progress here occurs at a fast pace as evidenced from the number of in-vitro scaffolds capable of supporting DP aggregation which have been developed over the past decade. Meanwhile cells, which we have yet to touch on, are possibly both the most important and the most challenging pillar of bioengineering. We have referred to signals in this chapter as extracellular additives for brevity, yet intra and inter cellular signalling which are fundamental to organ function are incredibly complex. This complexity is compounded when we take into account the number of cell types within specific tissues and then the heterogeneity within these cell types that single cell RNA sequencing has elucidated.

DP Cells

As we have highlighted throughout, DP cells possess an intrinsic ability to direct hair follicle morphogenesis and cycling. This is an ability that is not common among all mesenchymal cells. Closely related fibroblasts of the papillary dermis which share a common progenitor with DP cells are unable to direct hair follicle morphogenesis, neither in 2D culture nor in 3D culture [34, 79].

DP cells within the same papilla show distinct changes temporally and spatially depending on the stage of the hair cycle, as shown in the introduction. Furthermore, hair follicles vary in morphology; in mice pelage there are multiple hair types, including the whisker. Likewise, in humans, hair follicles vary. Cells within balding regions of the scalp show a higher expression of 5a reductase and androgen receptors than those adjacent in non-balding regions [24, 84]. Even more confounding is the contrasting response to androgens exhibited in hair follicles of pubescent regions such as the beard, armpit and pubis. At these loci hair follicles enlarge in response to androgen stimulation rather than miniaturise. Some studies have looked to understand androgen associated changes and highlighted genes like IGF1 as being upregulated in beard DP when stimulated with androgens [85]. In the same way that larger whisker DP prove more potent hair follicle inductors, beard DP cells are more inductive than DP cells from the scalp. This was linked to an increased expression of Wnt signalling mediator, secreted frizzled receptor protein 2 (SFRP2) where SFRP2 expression correlated to increased induction of de novo hair follicles. Furthermore, when SFRP2 was knocked down induction of de novo hair follicles was drastically reduced. It's worth noting however that while beard DP cells prove more inductive, unlike rat whisker DP, plate cultured beard DP are not inductive, demonstrating that 3D culture is also required for beard DP cells [86]. Nevertheless, the use of a more potent DP may overcome some of the limitations in using cultured human DP cells. To this end, transducing or transfecting DP cells in vitro may also prove beneficial. In an attempt to recapture the lost gene expression, factors involved in induction could be introduced to regain an inductive capacity. This could be done in an attempt to bypass 3D culture or with the idea of improving the quality of hair follicle induction. This latter attempt has been shown with the use of DP cells transfected with LEF1 overexpression vectors [51].

Hair Follicle Dermal Stem Cells

While the DP is the focus of this chapter, the DS, in particular the cells from the Dermal Sheath Cup (DSC), located at the below the DP, replenish the DP during cycling and are inductive on their own [21, 33, 87]. Therefore, these cells warrant consideration when devising bioengineered solutions to hair induction. With regards to improving inductivity in the DP, co-culture has been shown to restore inductivity in late passage rat DP cells [88]. By combining P60 DP cells (labelled with Dil) with early passage DS cells (labelled with EGFP), hair follicles were induced by performing the chamber assay in nude mice. The newly induced DP shows around a 50–50 split of EGFP and Dil labelled cells. Therefore, both low passage DS cells and aged DP cells contribute during hair induction. This means that DS cells can help restore an inductive potential to late passage DP cells, although it is possible that the inductive signal actually originates from the DS cells and that DP cells are simply incorporated into the induced follicle.

The ability of each cell type to contribute to the other also brings to light the plasticity of the two cell types. DP cells from DP induced hair follicles form a DS which surrounds the new follicle. Similarly, DS cells from DS induced hair follicles form the DP of the new follicle [87]. Hence, cells from each compartment can contribute to one another. This could be due to the presence of hfDSCs in both compartments and so it may be possible that induction is driven by these cells rather than the mature/committed cells of each compartment. We have already established that DP heterogeneity is present. Hence, it could be possible that in culture, inductive cell types such as hfDSCs are outgrown by more differentiated cells which proliferate in plate culture. Alternatively, stemness may be lost due to precocious proliferation. To evaluate this, single cell analysis of plate cultured cells would have to be carried out temporally.

Co-culture—Epidermal Cells

Similar to co-culture with the DS cells, culturing DP cells with keratinocyte—conditioned media has been shown to preserve the DP signature in culture [56]. Notably, rat whisker DP can maintain their inductive capacity up to 70 passages in keratinocyte conditioned media. However, direct co-culture of the two cell types is usually performed in order to stimulate the epidermal cells and induce a new hair follicle.

However, the choice of epidermal cells is also crucial. In almost all hair induction models, single cell suspensions derived from neonatal epidermis are used. Epidermal cells from neonatal sources respond to inductive signals better than adult keratinocytes, possibly due to an increased number of multipotent cells [89]. However, adult epidermal cells can also be inductive. Specifically, murine hair follicle stem cells of the bulge have been shown to be capable of reproducing all epidermal components and capable of producing fully functional hair follicles and sebaceous glands post dermal stimulation [90, 91]. However, the successful isolation of HFSCs has been challenging as cultured Outer Root Sheath (ORS) keratinocytes quickly lose their trichogenicity [92]. This loss of trichogenicity has been linked a downregulation in FOXA2 over prolonged culture [92, 93]. To date, induction using human HFSCs with human mesenchymal cells has not been shown.

Co-culture—Adipocyte-Derived Stem Cells

The bulb of the terminal hair follicle resides in the subcutaneous fat during anagen. As such many have linked adipocyte action to regulation of the DP niche [94]. Interestingly, when DP isolated from rat whiskers are co-cultured with adipocyte derived stem cells (ASCs) the results are variable [55]. When DP spheres are reconstituted with ASCs and injected subcutaneously with neonatal keratinocytes in the patch assay, they induce at similar levels as non-ASC patches. Alternatively, when ASC-DP spheres are constructed with the ASCs forming a shell around a DP core, the amount of hair follicles induced is doubled. However, if the same ASCs are incorporated inside the sphere, so that DP and ASCs cells form a mixed sphere, the inductivity of this structure is reduced. This shows how the engineering of a more representative organoid is beneficial. Here ASCs are positioned surrounding the DP similar to how they would surround the hair bulb in vivo. Remarkably though, the neonatal keratinocytes seem to more precociously respond to the inductive DP signal despite a physical shell of APCs surrounding the DP cells.

Reprogramming Cells

The source of DP cells may not need to necessarily come from microdissection of the hair follicle. Reprogramming of cells into DP has been shown to be possible and may provide an alternative source which requires less manual labour and can be more readily expandable than DP cells.

IPSCs are induced stem cells, which are well known for their pluripotent capacity and have been utilised to establish DP cells [89]. Here iPSCs are first converted from fibroblasts into induced Mesenchymal stem cells (iMSCs) and these cells are first treated with retinoic acid for 4 days and subsequently treated with DPAC medium (DMEM 10%FBS + BMP2, bFGF, and BIO) over a period of 5 days. These induced Dermal Papilla-substituting Cells (iDPSCs) showed DP-like gene expression and then were recombined with adult human keratinocytes in collagen gel drops, which were subsequently wrapped in human fibroblasts, and transplanted subcutaneously into SCID mice. Here they were able to induce hair morphogenesis at an efficiency of around 25–35%. This is a lower rate compared to human DP cells implanted in the same conformation (71%) but shows that DP like cells can also be sourced from non-hair follicle loci.

Another alternative source of cells could come from fibroblasts which can be isolated from skin biopsies and which can be easily expanded in culture. Fibroblasts provide an interesting alternative as they are a robust cell type that also proliferates faster than DP cells and may provide more timely expansion of source cells. However, as has been established by numerous studies, fibroblasts are unable to induce hair follicles in any conformation. Recently, both human neonatal and adult fibroblasts were reprogrammed into more DP-like cells by chemically supplementing with FGF2, BMP2, and BIO in low attachment culture conditions [95]. These cells show an increased expression in certain DP signature genes and are further able to induce hair follicle morphogenesis with a 65–70% efficiency in the patch assay when recombined with neonatal murine epidermal cells.

These two examples show early work in which alternative sources of DP cells may be utilised to provide more practical solutions when looking at tissue engineered solutions to hair loss.

The Trifecta!—Combining the Three Pillars

As we see in many studies related to induction, the level of induction and quality of induced hair follicles can vary significantly. In particular when using adult human cells, the hair follicles produced are substantially inferior to a healthy hair follicle. To address this shortcoming, we need to consider all three pillars of tissue engineering (Fig. 4.2). By developing mini organoids which (1) utilise several cells types, (2) take into account physiological structure and (3) allow for paracrine signals, we can achieve more clinically acceptable results.

To date a handful of studies have looked to actively combine all three pillars and we have already touched on each of them in the above sections where the most novel element of the study was demonstrated. However, it is important to emphasise how the three pillars can be combined to further focus on the future directions that should be considered.

In the first example, DP cells were co-cultured with keratinocytes on 3D scaffolds normally used to create skin equivalents [54]. These co-cultures were supplemented with 4 factors (RSPO1, MMP14, iPKC, and Wnt5a) each with specific timepoints for administration. Over the course of 6 days the skin equivalents begin to show aggregation and the formation of planar epidermis, demonstrating a complex self-regulation dependent on the interaction between the cells types, the in-vivo 3D environment and the supplemented growth factors. These constructs, which used murine adult cells, showed the capacity to induce hair follicle formation after implantation into full thickness wounds in nude mice. These newly induced hair follicles could in turn produce hair fibre, albeit unpigmented [54].

In the iPSC-based technique [89], the researchers formed spheroids by co-culturing reprogrammed induced DP-like cells with human keratinocytes with a drop of Matrigel. The drops were then subsequently coated with a second coating of Matrigel, which contained human fibroblasts. Here the choice of an alternative cell source results in the requirement to reprogramme iPSCs into iMCs and subsequently DP-like cells using a number of supplemented factors. Then through co-culturing

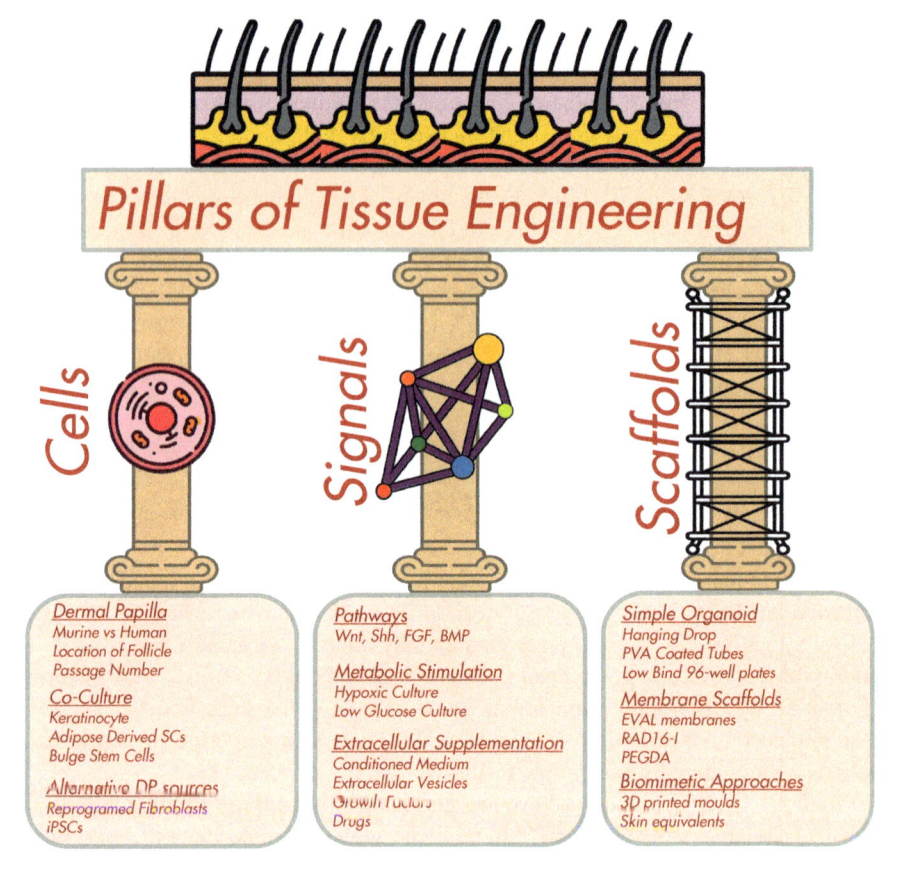

Fig. 4.2 Bioengineering Hair Follicles—The Pillars of Tissue Engineering. Cells—When selecting cells, we need to consider the differences between human and murine DP cells, hair follicle type and the amount of time spent within culture. In addition, can co-culture maintain a DP signature, or will certain epidermal cells respond more positively to DP induction? When selecting scaffolds, we need to consider how to aggregate cells, taking into account scalability, substrate stiffnesses, cost and bioactivity. Furthermore, can we create biomimetic scaffolds which better represent the in vivo environment? When looking at signals, we need to consider morphogenic signals which occur both in an autocrine and paracrine manner. Can additive factors increase or maintain inductivity?

the cells within a supportive gel scaffold to create a hair-germ like structure that allows for paracrine signalling between the cells, hair follicle induction is possible. This second technique looks to create a biomimetic structure based on early hair morphogenesis, and by doing so aiming to increase the inductive capacity [89].

In the final example, a fibroblast-infused collagen scaffold is moulded using a 3D printed stamp that creates hair follicle like invaginations within which dermal papilla cells are seeded [51]. The stamp replicates the vivo hair follicle distribution, providing spatial cues to the cells which can help them to self-regulate morphogenesis more effectively. Prior to seeding, the DP cells are transfected with a LEF1 overexpressing plasmid to increase their inductive capacity. Neonatal human keratinocytes were subsequently seeded on top of the dermal scaffold containing transfected DP aggregates. Remarkably, in vitro these skin equivalents are able to form hair follicles, capable of producing hair shafts. Furthermore, after seeding the fibroblast-infused collagen scaffold with human umbilical vein endothelial cells (HUVECs) at a high seeding density, spontaneous capillary generation occurred. The capillary network formed allowed the successful engraftment to nude mice where the skin equivalents induced the formation of human hair follicles. The hair follicles produced hair shafts which were unpigmented and in between the thickness of a vellus hair and a terminal hair. This approach stands out due to the level of complexity involved. The use of 4 cells types together which prior to implantation can form early hair follicle like structures on a natural biomimetic scaffold is truly ground-breaking and signals the next steps towards improving hair follicle induction.

The scope of this book chapter has been to summarise the work done in trying to bioengineer solutions that increase DP inductivity. Over the course of the manuscript, we have covered the early work demonstrating the capacity of the DP to induce de novo hair follicle morphogenesis and then proceeded to explain how DP cells lose their inductive potential when cultured. Maintenance of inductivity has been achieved using extracellular factors and conditioned medium while recovery of lost inductivity is possible through 3D culture. However, the scope of the book chapter is to improve DP inductivity through bioengineered methods, and as such we need to consider both inductive efficiency and hair follicle quality. To improve these, we cannot solely look at the DP cells but rather consider them as part of a complex system, while considering cell choice, extracellular factors, and supportive microenvironments that encourage organoid formation. By engineering micro-organoids that take into account compartmentalisation and temporal factors, we can assist inductive dermal cells to direct the proliferation and differentiation of multipotent epidermal cells into representative organs semi-autonomously. Like in all engineering solutions, we need to focus on the system as one. To increase the capabilities of DP cells means to increase the capabilities of its supporting niche.

References

1. Lin GG, Scott JG (2012) Novel tetra-peptide insertion in Gag-p6 ALIX-binding motif in HIV-1 subtype C associated with protease inhibitor failure. Semin Cell Dev Biol 100(2):130–134
2. Saxena N, Mok KW, Rendl M (2019) An updated classification of hair follicle morphogenesis. Exp Dermatol 28:332–344
3. Millar SE (2002) Molecular mechanisms regulating hair follicle development. J Invest Dermatol 118(2):216–225
4. Mok KW, Saxena N, Heitman N, Grisanti L, Srivastava D, Muraro MJ et al (2019) Dermal condensate niche fate specification occurs prior to formation and is placode progenitor dependent. Dev Cell 48(1):32–48.e5

5. Biggs LC, Mäkelä OJM, Myllymäki SM, Das Roy R, Närhi K, Pispa J et al (2018) Hair follicle dermal condensation forms via FGF20 primed cell cycle exit, cell motility, and aggregation. Elife 7:e36468
6. Ahtiainen L, Lefebvre S, Lindfors PH, Renvoisé E, Shirokova V, Vartiainen MK et al (2014) Directional cell migration, but not proliferation, drives hair placode morphogenesis. Dev Cell 28(5):588–602
7. Gupta K, Levinsohn J, Linderman G, Chen D, Sun TY, Dong D et al (2019) Single-cell analysis reveals a hair follicle dermal niche molecular differentiation trajectory that begins prior to morphogenesis. Dev Cell 48(1):17–31.e6
8. Paus R, Cotsarelis G (1999) The biology of hair follicles. N Engl J Med 341(7):491–497
9. Müller-Röver S, Handjiski B, Van Der Veen C, Eichmüller S, Foitzik K, McKay IA et al (2001) A comprehensive guide for the accurate classification of murine hair follicles in distinct hair cycle stages. J Invest Dermatol 117(1):3–15
10. Yang H, Adam RC, Ge Y, Hua ZL, Fuchs E (2017) Epithelial-mesenchymal micro-niches govern stem cell lineage choices. Cell 169(3):483–496.e13
11. Rompolas P, Deschene ER, Zito G, Gonzalez DG, Saotome I, Haberman AM et al (2012) Live imaging of stem cell and progeny behaviour in physiological hair-follicle regeneration. Nature 487(7408):496–499
12. Jacobson CM (1966) A comparative study of the mechanisms by which X-irradiation and genetic mutation cause loss of vibrissae in embryo mice. J Embryol Exp Morphol 16(2):369–379
13. van Scott EJ, Ekel TM (1958) Geometric relationships between the matrix of the hair bulb and its dermal papilla in normal and alopecic scalp. J Invest Dermatol 31(5):281–287
14. Elliott K, Stephenson TJ, Messenger AG (1999) Differences in hair follicle dermal papilla volume are due to extracellular matrix volume and cell number: Implications for the control of hair follicle size and androgen responses. J Invest Dermatol 113(6):873–877
15. Chi W, Wu E, Morgan BA (2013) Dermal papilla cell number specifies hair size, shape and cycling and its reduction causes follicular decline. Dev 140(8):1676–1683
16. Whiting DA (1998) Male pattern hair loss: current understanding. Int J Dermatol 37(8):561–566
17. Whiting DA (2001) Possible mechanisms of miniaturization during androgenetic alopecia or pattern hair loss. J Am Acad Dermatol 45(3 Suppl):81–86
18. Birch MP, Messenger JF, Messenger AG (2001) Hair density, hair diameter and the prevalence of female pattern hair loss. Br J Dermatol 144(2):297–304
19. Tobin DJ, Gunin A, Magerl M, Paus R (2003) Plasticity and cytokinetic dynamics of the hair follicle mesenchyme during the hair growth cycle: implications for growth control and hair follicle transformations. J Invest Dermatol Symp Proc 8(1):80–86
20. Stenn KS, Lawrence L, Veis D, Korsmeyer S, Seiberg M (1994) Expression of the bcl-2 protooncogene in the cycling adult mouse hair follicle. J Invest Dermatol 103(1):107–111
21. Rahmani W, Abbasi S, Hagner A, Raharjo E, Kumar R, Hotta A et al (2014) Hair follicle dermal stem cells regenerate the dermal sheath, repopulate the dermal papilla, and modulate hair type. Dev Cell 31(5):543–558
22. Agabalyan NA, Rosin NL, Rahmani W, Biernaskie J (2017) Hair follicle dermal stem cells and skin-derived precursor cells: exciting tools for endogenous and exogenous therapies. Exp Dermatol 26(6):505–509
23. Shin W, Rosin NL, Sparks H, Sinha S, Rahmani W, Sharma N et al (2020) Dysfunction of hair follicle mesenchymal progenitors contributes to age-associated hair loss. Dev Cell 53(2):185–198.e7
24. Sawaya ME, Price VH (1997) Different levels of 5α-reductase type I and II, aromatase, and androgen receptor in hair follicles of women and men with androgenetic alopecia. J Invest Dermatol 109(3):296–300
25. Oliver RF (1970) The induction of hair follicle formation in the adult hooded rat by vibrissa dermal papillae. J Embryol Exp Morphol 23(1):219–236
26. Cohen J (1961) The transplantation of individual rat and guineapig whisker papillae. J Embryol Exp Morphol 9(1):117–127

27. Lillie FR, Wang H (1941) Physiology of development of the feather V. Experimental morphogenesis. Physiol Zool 14(2):103–135
28. Dhouailly D (1973) Dermo epidermal interactions between birds and mammals: differentiation of cutaneous appendages. J Embryol Exp Morphol 30(3):587–603
29. Oliver RF (1966) Whisker growth after removal of the dermal papilla and lengths of follicle in the hooded rat. J Embryol Exp Morphol 15(3):331–347
30. Oliver RF (1966) Histological studies of whisker regeneration in the hooded rat. J Embryol Exp Morphol 16(2):231–244
31. Oliver RF (1967) The experimental induction of whisker growth in the hooded rat by implantation of dermal papillae. J Embryol Exp Morphol 18(1):43–51
32. Jahoda CAB (1992) Induction of follicle formation and hair growth by vibrissa dermal papillae implanted into rat ear wounds: vibrissa-type fibres are specified. Development 115(4):1103–1109
33. Reynolds AJ, Lawrence C, Cserhalmi-Friedman PB, Christiano AM, Jahoda CAB (1999) Trans-gender induction of hair follicles. Nature 402(6757):33–34
34. Higgins CA, Chen JC, Cerise JE, Jahoda CAB, Christiano AM (2013) Microenvironmental reprogramming by threedimensional culture enables dermal papilla cells to induce de novo human hair-follicle growth. Proc Natl Acad Sci U S A 110(49):19679–19688
35. Jahoda CAB, Oliver RF, Reynolds AJ, Forrester JC, Gillespie JW, Cserhalmi-Friedman PB et al (2001) Trans-species hair growth induction by human hair follicle dermal papillae. Exp Dermatol 10(4):229–237
36. Jahoda C, Oliver RF (1981) The growth of vibrissa dermal papilla cells in vitro. Br J Dermatol 105(6):623–627
37. Messenger AG (1984) The culture of dermal papilla cells from human hair follicles. Br J Dermatol 110(6):685–689
38. Jahoda CAB, Oliver RF (1984) Vibrissa dermal papilla cell aggregative behaviour in vivo and in vitro. J Embryol Exp Morphol 79:211–224
39. Jahoda CAB, Reynolds AJ (1993) Dermal-epidermal interactions-follicle-derived cell populations in the study of hair-growth mechanisms. J Invest Dermatol 101(1 Suppl)
40. Hillmer AM, Hanneken S, Ritzmann S, Becker T, Freudenberg J, Brockschmidt FF et al (2005) Genetic variation in the human androgen receptor gene is the major determinant of common early-onset androgenetic alopecia. Am J Hum Genet 77(1):140–148
41. Chew EGY, Tan JHJ, Bahta AW, Ho BSY, Liu X, Lim TC et al (2016) Differential expression between human dermal papilla cells from balding and non-balding scalps reveals new candidate genes for androgenetic alopecia. J Invest Dermatol 136(8):1559–1567
42. Randall VA, Thornton MJ, Hamada K, Redfern CPF, Nutbrown M, Ebling FJG et al (1991) Androgens and the hair follicle: cultured human dermal papilla cells as a model system. Ann N Y Acad Sci 642(1):355–375
43. Jahoda CAB, Reynolds AJ, Chaponnier C, Forester JC, Gabbiani G (1991) Smooth muscle α-actin is a marker for hair follicle dermis in vivo and in vitro. J Cell Sci 99(3):627–636
44. Messenger AG, Elliott K, Westgate GE, Gibson WT (1991) Distribution of extracellular matrix molecules in human hair follicles. Ann N Y Acad Sci 642(1):253–262
45. Jahoda CA, Horne KA, Oliver RF (1987) Vibrissa dermal papilla cells interact with epidermis to induce the formation of new hair-follicles producing vibrissa-type hairs. Br J Dermatol 422
46. Jahoda CAB, Horne KA, Oliver RF (1984) Induction of hair growth by implantation of cultured dermal papilla cells. Nature 311(5986):560–562
47. Horne KA, Jahoda CAB, Oliver RF (1986) Whisker growth induced by implantation of cultured vibrissa dermal papilla cells in the adult rat. J Embryol Exp Morphol 97:111–124
48. Topouzi H (2018) Morphological and behavioural characterisation of fibroblast sub-types found in human skin dermis. Imperial College London
49. Huang WY, Huang YC, Huang KS, Chan CC, Chiu HY, Tsai RY et al (2017) Stress-induced premature senescence of dermal papilla cells compromises hair follicle epithelial-mesenchymal interaction. J Dermatol Sci 86(2):114–122

50. Ohyama M, Kobayashi T, Sasaki T, Shimizu A, Amagai M (2012) Restoration of the intrinsic properties of human dermal papilla in vitro. J Cell Sci 125(17):4114–4125
51. Abaci HE, Coffman A, Doucet Y, Chen J, Jacków J, Wang E et al (2018) Tissue engineering of human hair follicles using a biomimetic developmental approach. Nat Commun 9(1)
52. Kwack MH, Seo CH, Gangadaran P, Ahn BC, Kim MK, Kim JC et al (2019) Exosomes derived from human dermal papilla cells promote hair growth in cultured human hair follicles and augment the hair-inductive capacity of cultured dermal papilla spheres. Exp Dermatol 28(7):854–857
53. Yoshida Y, Soma T, Matsuzaki T, Kishimoto J (2019) Wnt activator CHIR99021-stimulated human dermal papilla spheroids contribute to hair follicle formation and production of reconstituted follicle-enriched human skin. Biochem Biophys Res Commun 516(3):599–605
54. Weber EL, Lai YC, Lei M, Jiang TX, Chuong CM (2020) Human fetal scalp dermal papilla enriched genes and the role of R-Spondin-1 in the restoration of hair neogenesis in adult mouse cells. Front Cell Dev Biol 8:1454
55. Huang CF, Chang YJ, Hsueh YY, Huang CW, Wang DH, Huang TC et al (2016) Assembling composite dermal papilla spheres with adipose-derived stem cells to enhance hair follicle induction. Sci Rep 6:26436
56. Inamatsu M, Matsuzaki T, Iwanari H, Yoshizato K (1998) Establishment of rat derreal papilla cell lines that sustain the potency to induce hair follicles from afollicular skin. J Invest Dermatol 111(5):767–775
57. Chen Y, Huang J, Liu Z, Chen R, Fu D, Yang L et al (2020) miR-140-5p in small extracellular vesicles from human papilla cells stimulates hair growth by promoting proliferation of outer root sheath and hair matrix cells. Front Cell Dev Biol 8:593638
58. Chen Y, Huang J, Chen R, Yang L, Wang J, Liu B et al (2020) Sustained release of dermal papilla-derived extracellular vesicles from injectable microgel promotes hair growth. Theranostics 10(3):1454–1478
59. Rajendran RL, Gangadaran P, Seo CH, Kwack MH, Oh JM, Lee HW et al (2020) Macrophage-derived extracellular vesicle promotes hair growth. Cells 9(4):856
60. Rajendran RL, Gangadaran P, Bak SS, Oh JM, Kalimuthu S, Lee HW et al (2017) Extracellular vesicles derived from MSCs activates dermal papilla cell in vitro and promotes hair follicle conversion from telogen to anagen in mice. Sci Rep 7(1):1–12
61. le Riche A, Aberdam E, Marchand L, Frank E, Jahoda C, Petit I et al (2019) Extracellular vesicles from activated dermal fibroblasts stimulate hair follicle growth through dermal papilla-secreted norrin. Stem Cells 37(9):1166–1175
62. Hwang KA, Hwang YL, Lee MH, Kim NR, Roh SS, Lee Y et al (2012) Adenosine stimulates growth of dermal papilla and lengthens the anagen phase by increasing the cysteine level via fibroblast growth factors 2 and 7 in an organ culture of mouse vibrissae hair follicles. Int J Mol Med 29(2):195–201
63. Rendl M, Polak L, Fuchs E (2008) BMP signaling in dermal papilla cells is required for their hair follicle-inductive properties. Genes Dev 22(4):543–557
64. Osada A, Iwabuchi T, Kishimoto J, Hamazaki TS, Okochi H (2007) Long-term culture of mouse vibrissal dermal papilla cells and De Novo hair follicle induction. Tissue Eng 13(5):975–982
65. Choi M, Choi YM, Choi SY, An IS, Bae S, An S et al (2020) Glucose metabolism regulates expression of hair-inductive genes of dermal papilla spheres via histone acetylation. Sci Rep 10(1)
66. Zheng M, Jang Y, Choi N, Kim DY, Han TW, Yeo JH et al (2019) Hypoxia improves hair inductivity of dermal papilla cells via nuclear NADPH oxidase 4-mediated reactive oxygen species generation'. Br J Dermatol 181(3):523–534
67. Yin RH, Zhao SJ, Wang ZY, Zhu YB, Yin RL, Bai M et al (2020) LncRNA-599547 contributes the inductive property of dermal papilla cells in cashmere goat through miR-15b-5p/Wnt10b axis. Anim Biotechnol 18:1–15
68. Lachgar S, Charveron M, Gall Y, Bonafe JL (1998) Minoxidil upregulates the expression of vascular endothelial growth factor in human hair dermal papilla cells. Br J Dermatol 138(3):407–411

69. Miao Y, Sun Y Bin, Sun XJ, Du BJ, Jiang JD, Hu ZQ (2013) Promotional effect of platelet-rich plasma on hair follicle reconstitution in vivo. Dermatol Surg 39(12):1868–1876
70. Kishimoto J, Burgeson RE, Morgan BA (2000) WNT signaling maintains the hair-inducing activity of the dermal papilla. Genes Dev 14(10):1181–1185
71. Harel S, Higgins CA, Cerise JE, Dai Z, Chen JC, Clynes R et al (2015) Clinical medicine: pharmacologic inhibition of JAK-STAT signaling promotes hair growth. Sci Adv 1(9):e1500973–e1500973
72. Couchman JR, McCarthy KJ, Woods A (1991) Proteoglycans and glycoproteins in hair follicle development and cycling. Ann N Y Acad Sci 642(1):243–251
73. Couchman JR (1986) Rat hair follicle dermal papillae have an extracellular matrix containing basement membrane components. J Invest Dermatol 87(6):762–767
74. Messenger AG, Elliott K, Temple A, Randall VA (1991) Expression of basement membrane proteins and interstitial collagens in dermal papillae of human hair follicles. J Invest Dermatol 96(1):93–97
75. Weber L, Kirsch E, Muller P, Krieg T (1984) Collagen type distribution and macromolecular organization of connective tissue in different layers of human skin. J Invest Dermatol 82(2):156–160
76. Harrison RG (1910) The outgrowth of the nerve fiber as a mode of protoplasmic movement. J Exp Zool 9(4):787–846
77. Higgins CA, Richardson GD, Ferdinando D, Westgate GE, Jahoda CAB (2010) Modelling the hair follicle dermal papilla using spheroid cell cultures. Exp Dermatol 19(6):546–548
78. Huang YC, Chan CC, Lin WT, Chiu HY, Tsai RY, Tsai TH et al (2013) Scalable production of controllable dermal papilla spheroids on PVA surfaces and the effects of spheroid size on hair follicle regeneration. Biomaterials 34(2):442–451
79. Kang BM, Kwack MH, Kim MK, Kim JC, Sung YK (2012) Sphere formation increases the ability of cultured human dermal papilla cells to induce hair follicles from mouse epidermal cells in a reconstitution assay. J Invest Dermatol 132(1):237–239
80. Reynolds AJ, Jahoda CAB (1992) Cultured dermal papilla cells induce follicle formation and hair growth by transdifferentiation of an adult epidermis. Development 115(2):587–593
81. Young TH, Lee CY, Chiu HC, Hsu CJ, Lin SJ (2008) Self assembly of dermal papilla cells into inductive spheroidal microtissues on poly(ethylene-co-vinyl alcohol) membranes for hair follicle regeneration. Biomaterials 29(26):3521–3530
82. Xiao SE, Miao Y, Wang J, Jiang W, Fan ZX, Liu XM et al (2017) As a carrier-Transporter for hair follicle reconstitution, platelet-rich plasma promotes proliferation and induction of mouse dermal papilla cells. Sci Rep 7(1):1–11
83. Betriu N, Jarrosson-Moral C, Semino CE (2020) Culture and differentiation of human hair follicle dermal papilla cells in a soft 3D self-assembling peptide scaffold. Biomolecules 10(5)
84. Hibberts NA, Howell AE, Randall VA (1998) Balding hair follicle dermal papilla cells contain higher levels of androgen receptors than those from non-balding scalp. J Endocrinol 156(1):59–65
85. Randall VA (2008) Androgens and hair growth. Dermatol Ther 21(5):314–328
86. Kwack MH, Ahn JS, Jang JH, Kim JC, Sung YK, Kim MK (2016) SFRP2 augments Wnt/β-catenin signalling in cultured dermal papilla cells. Exp Dermatol 25:813–815
87. McElwee KJ, Kissling S, Wenzel E, Huth A, Hoffmann R (2003) Cultured peribulbar dermal sheath cells can induce hair follicle development and contribute to the dermal sheath and dermal papilla. J Invest Dermatol 121(6):1267–1275
88. Yamao M, Inamatsu M, Ogawa Y, Toki H, Okada T, Toyoshima KE et al (2010) Contact between dermal papilla cells and dermal sheath cells enhances the ability of DPCs to induce hair growth. J Invest Dermatol 130(12):2707–2718
89. Veraitch O, Mabuchi Y, Matsuzaki Y, Sasaki T, Okuno H, Tsukashima A et al (2017) Induction of hair follicle dermal papilla cell properties in human induced pluripotent stem cell-derived multipotent LNGFR(+)THY-1(+) mesenchymal cells. Sci Rep 7
90. Blanpain C, Lowry WE, Geoghegan A, Polak L, Fuchs E (2004) Self-renewal, multipotency, and the existence of two cell populations within an epithelial stem cell niche. Cell 118(5):635–648

91. Morris RJ, Liu Y, Marles L, Yang Z, Trempus C, Li S et al (2004) Capturing and profiling adult hair follicle stem cells. Nat Biotechnol 22(4):411–417
92. Bak SS, Kwack MH, Shin HS, Kim JC, Kim MK, Sung YK (2018) Restoration of hair-inductive activity of cultured human follicular keratinocytes by co-culturing with dermal papilla cells. Biochem Biophys Res Commun 505(2):360–364
93. Bak SS, Park JM, Oh JW, Kim JC, Kim MK, Sung YK (2020) Knockdown of FOXA2 impairs hair-inductive activity of cultured human follicular keratinocytes. Front Cell Dev Biol 8
94. Festa E, Fretz J, Berry R, Schmidt B, Rodeheffer M, Horowitz M et al (2011) Adipocyte lineage cells contribute to the skin stem cell niche to drive hair cycling. Cell 146(5):761–771
95. Zhao Q, Li N, Zhang H, Lei X, Cao Y, Xia G et al (2019) Chemically induced transformation of human dermal fibroblasts to hair-inducing dermal papilla-like cells. Cell Prolif 52(5):e12652

Chapter 5
Dermal Sheath Cells and Hair Follicle Regeneration

Yuzo Yoshida, Ryoji Tsuboi, and Jiro Kishimoto

Abstract The dermal sheath (DS) cells are present in a sheath-like connective tissue on the outermost side of the hair follicle; DS connects to the dermal papilla (DP) in the lower part of the hair bulb. The ability of DP cells to induce and regenerate hair follicles has been demonstrated in a great number of studies as discussed in other chapters. In this chapter, the characteristics and functions of DS cells in hair follicle regeneration have been reviewed. The potential of a clinical application using DS cells for male and female pattern loss is also discussed in this chapter. The PubMed database was used for the selection of papers with keywords including 'dermal sheath', 'connective tissue sheath', and 'hair'. Through a comprehensive reading of the abstracts, papers on hair follicle regeneration were further selected and cited. Several papers have shown that DS cells can induce growth of new hair follicles. Recent studies reported the presence of hair follicle dermal stem cells in the DS, suggesting these may be responsible for hair follicle regeneration. Based on these properties, clinical studies have been conducted on male and female pattern hair loss using lower DS cells. The initial results from these studies show safety and some efficacy of lower DS cells in hair follicle regeneration. In conclusion, several rodent studies have shown that the DS and cells derived from it have both hair regenerative and skin regenerative potential. Further studies characterizing human DS cells and identifying if there are human hair follicle dermal stem cells need to precede further understanding of human hair follicle regeneration, for the development of an efficient clinical application for male and female pattern hair loss.

Keywords Hair follicles · Dermal sheath · Connective tissue sheath · Androgenetic alopecia · Cell-based therapy

---

Y. Yoshida (✉) · J. Kishimoto
Shiseido FS Innovation Center, Regenerative Medicine Research & Business Development Section, Yokohama, Japan
e-mail: yuzo.yoshida1@shiseido.com

R. Tsuboi
Department of Dermatology, Tokyo Medical University Hospital, Tokyo, Japan

J. Kishimoto
Hayashibara Co., Ltd. Personal Healthcare Division, Okayama, Japan

© The Author(s), under exclusive license to Springer Nature Switzerland AG 2022
F. Jimenez and C. Higgins (eds.), *Hair Follicle Regeneration*, Stem Cell Biology and Regenerative Medicine 72, https://doi.org/10.1007/978-3-030-98331-4_5

Summary

Various methods have been proposed as scientific approaches to hair follicle regeneration. However, regardless of the methods selected, interactions between the different cell types in the epithelium and mesenchyme appear essential. Research on mesenchymal cells in terms of hair-inducing properties has overwhelmingly focused on dermal papilla (DP) cells, and many research reports and reviews have been published on their hair induction ability. While another chapter in this book has focused on the clinical use of DP cells for hair follicle regeneration, in the present chapter we will review the functional role of the dermal sheath (DS) cells in hair follicle regeneration compared with DP cells. We discuss factors which affect the regenerative potential of the hair follicle; the stem cell characteristics of DS cells and clinical trials using DS cells as a possible therapy for hair loss disorders.

Key Points

- DP cells and DS cells can induce de novo hair follicles, but interfollicular dermal fibroblast cells themselves do not have the ability to induce hair follicles even in inductive culture conditions.
- Recent findings indicate the presence of hair follicle dermal stem cells in the DS, suggesting that DS cells could be a key element for hair follicle regeneration.
- Clinical studies using DS cells injected into the scalp of for male and female pattern hair loss show some positive results, suggesting their potential use for hair follicle regeneration in the clinical setting.

Dermal Fibroblast Cells

The mesenchymal cells within the dermal skin tissue, namely dermal fibroblasts (DF) are broadly heterogeneous [1, 2], however in the context of the hair follicle they can be divided into two types: interfollicular and follicular DF cells. Although 4 sub-types of interfollicular DF have been described based on transcriptomic profiling results [1], interfollicular DF can also be subdivided by their location in the skin; papillary DF are present in the upper papillary dermis while reticular DF are present in the lower dermis [3]. The follicular DF can also be further subdivided into dermal papilla (DP) cells and dermal sheath (DS) cells, again based on location within the follicle. Work by Driskell and colleagues showed an intrinsic developmental connection between interfollicular and follicular DF, the latter of which are derived from the same developmental progenitors as papillary, but not reticular fibroblasts [4]. In neonatal, but not adult skin, papillary DF retain embryonic plasticity, and are able to switch to hair follicle DF to instruct hair growth [4].

Despite this, the conclusion from many experimental observations utilizing assay systems to validate hair follicle induction ability in animal models is that adult interfollicular DF cells themselves do not have the ability to induce hair follicles. Although some findings have suggested an auxiliary role for DF cells as bedding cells, no reports to date have clearly demonstrated that adult interfollicular DF cells have hair follicle-inducing potential even through the addition of growth factors such as FGF, EGF, or spheroid-formed culture condition.

That adult interfollicular DF can turn into hair follicle DF though is not unassailable. Collins et al. [5] also demonstrated by the gene profiling analysis that murine dermis can be reprogrammed to a neonatal state with ectopic hair follicle induced by sustained epidermal activation of β-catenin. This suggests that reprogramming interfollicular adult DF to an embryonic state (capable of turning into DP and DS) could lead to the achievement of hair induction, although the adult DFs themselves are not inductive.

Dermal Sheath Cells

Anatomy, Biological Function and Hair Regenerative Potential of Dermal Sheath Cells

The DS is a sheath-like connective tissue that encases the entire hair follicle; it is located on the outermost side of the hair follicle, and physically connects to and provides a source of stem cells that fuels the DP (Fig. 5.1). In mouse and rat whiskers and coat hair, the DS is a thin layer composed of connective tissue outside the basement membrane that forms at the border of the epithelial tissue, comprising the outer hair root sheath. In human hair follicles, the DS is formed by multiple layers of connective tissue; an inner layer composed by longitudinally oriented collagen fibers to the hair axis, a middle layer in which collagen fibers are oriented in a transverse direction harboring a few blood vessels, and an outer layer in which collagen fibers are oriented randomly and blood vessels are present. [6, 7] (Fig. 5.1). DS cells are observed in the middle and outer layer of DS, but there are few cells in the inner layer (Fig. 5.1). In rat and human hair follicles, alpha-smooth muscle actin (αSMA) has been reported to be expressed in DS, but not DP, suggesting αSMA could be a DS marker [8]. It has recently been demonstrated that the DS express the molecular machinery of smooth muscles and its contraction throughout catagen seems to be essential for the regression of the follicle and niche relocation [9]. Lineage tracing analyses for the identification of developmental origins of DS cells revealed that they were derived from the dermal condensation that occurs early during hair follicle formation; the same condensation from which DP cells are derived [10]. As well as having the same lineage as DP cells, it is well established that DS cells have hair follicle regenerative potential, similar to DP cells.

In 1966, Oliver et al. [11] suggested that both DS and DP may be able to induce hair follicles, on the basis of amputation experiments of rat whisker hair follicles. In these experiments, it was also found that hair follicles did not regenerate when more than one-third of the lower hair follicle was removed. Subsequent experiments showed that implantation of lower DS cup against the exposed epithelial surface of follicles inactivated by amputation of their lower half, could rescue inactivated

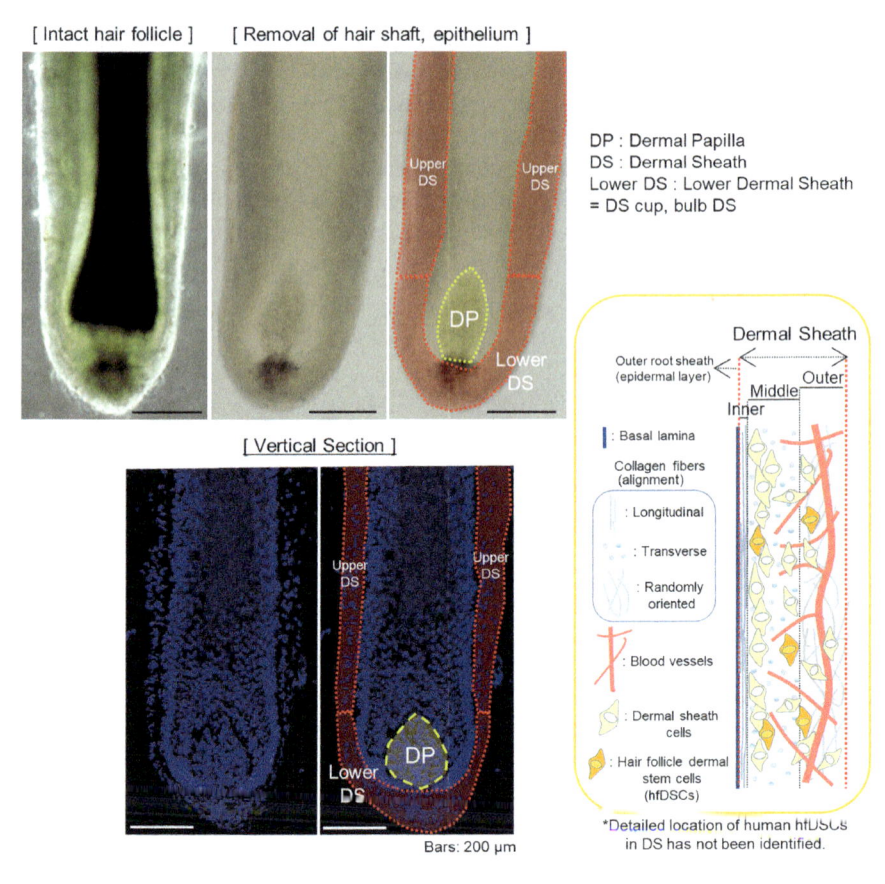

Fig. 5.1 Location and schematic view of dermal sheath cells in human hair follicle

follicles, demonstrating that DS has the capacity to instruct hair fibre growth [12]. Analyses on human hair follicles has also shown the hair follicle regenerative potential of DS. When human hair follicles with amputation of their hair bulb area, i.e., hair follicles with the DP and lower DS removed, were implanted beneath mouse skin, the entire hair bulb containing a small DP, was regenerated from the remaining upper DS [13]. Intradermal implantation of lower DS, harvested from human male terminal hair follicles, into the upper arm of a female recipient was also shown to regenerate an entire hair follicle [14]. These observations allow us to understand that the hair follicle regenerative potential of DS is not restricted to autologous hair follicle transplantation or to specific types of hair follicles, but is a universal feature.

The cells constituting both DP and DS can generally be harvested by an outgrowth/explant method. While nearly all reports, especially with rodent DP, have confirmed their inductivity, the results with DS have had conflicting results. Transplantation experiments in the wounded rat ear, a model that drives new hair follicle induction and hair follicle regeneration, showed that cultured rat whisker DP cells were able to induce enlarged (whisker like) hair follicles, however cultured DS cells had no such effect [15]. However, similar experiments demonstrated that cultured rat whisker DS cells, in combination with hair matrix germinative epidermal cells, had the ability to induce whisker like hair follicles, similar to the case for DP cells [16]. Contrastingly, transplantation of cultured mouse whisker DS cup cells into the foot pad resulted in the generation of new hair follicles [17]. This study also showed that transplanted lower DS cup cells into ear wounds were incorporated into the DP and DS tissue of existing hair follicles, where they prolonged the growth phase of in situ follicles [17]. A publication in the same year (2003) also showed that transplanted lower DS cells from rat whiskers could incorporate into existing hair follicles in rat ear and body skin wounds [18]. On the other hand, neither upper follicle DS cells or DF were capable of instructing hair follicle regeneration [17] nor were they observed incorporated into hair follicles in skin wounds [17, 18]. Despite this, transplantation of upper DS into follicles inactivated by amputation could promote hair shaft growth, although the size was small [19]. Collectively, the role of DS in the regeneration of hair follicles seems to differ depending on if upper or lower (hair bulb) DS are used. Going forward, studies need to be careful to specify whether they are using DS from the upper or the lower follicle. Nevertheless, given the previous literature [17, 18] we can take away the understanding that DS in the hair bulb area, in other words, the lower DS or DS cup, have the ability to regenerate hair follicles. This regenerative potential both aligns well with and reflects the location of recently discovered dermal (hair follicle) stem cells, which are found in the lower dermal sheath in mouse hair follicles [20] and are discussed in more detail in the following section.

Mechanisms for Hair Follicle Regeneration by Dermal Sheath Cells and Factors Affecting Their Hair Follicle Regenerative Potential

DS and DP of adult hair follicles share the developmental dermal condensate as their progenitor [10], suggesting a functional similarity between DS cells and DP cells. Furthermore, our recent work on cultured human DS and DP cells showed more similarities in terms of gene expression profiles between both cell types when compared with cultured DF cells [21]. Addition of thrombin has been reported to induce the differentiation of DP cells into myofibroblast-like cells in vitro, which is a phenotype characteristically observed in DS cells [22]. Studies monitoring the hair cycle in mice have also suggested plasticity and interchangeability of DP and DS cells [23]. In mouse pelage hair, cells constituting the DP gradually increases in number during the initial growth phase, yet their proliferation rate is markedly lower than that of adjacent DS cells [23]. This observation led to the suggestion that DS cells are supplied to the DP during the early growth phase [23], while genetic lineage tracing in mice directly revealed the entity of cell supply [20, 24]. Firstly, using the *Corin* gene locus (*Corin-cre*; *ROSA*), to mark DP cells during the hair cycle, researchers found that at the start of a new growth phase (anagen), DP-constituting cells were also supplied from locations other than DP tissue [24]. In a parallel set of experiments by another research group, the lineage of DS cells in the hair cycle was tracked using the alpha-smooth muscle actin (αSMA) gene locus (αSMA-*cre*; *ROSA*), a marker of DS cells; the results revealed that a subset of DS cells contributed to the DP at the start of a new anagen [20]. The authors used the name hair follicle dermal stem cells (hfDSCs), referring to cells located in lower DS cup in mouse hair follicles that supplied DP cells in early anagen [20], while later work from the same group further characterized these hfDSCs showing them be responsive to DP-secreted R-spondins [25]. The ability of hfDSC to contribute to the DP is affected by several factors, for example, the numbers of cells supplied by hfDSCs to DP tissue is known to increase after skin wounding [26], yet the number of hfDSCs decreases with ageing [27].

Further reports have suggested that not only can cells move between DP and DS tissues, but that secreted factors and/or adhesins expressed by DP and DS cells can synergistically enhance the inductivity of hair follicle and hair follicle-regenerating potential [28, 29]. Although DP cells lose their ability to induce hair follicles after serial passages in vitro, coculture of DP cells with DS cells can prevent this reduction in hair inductivity [28]. Transplantation of rat DS and DP cells mixed together led to their more frequent incorporation into regenerating hair follicles and further accelerated the hair follicle regeneration compared with transplantation of DS cells or DP cells alone [29].

Various signalling pathways have been implicated in the regulation of the DS, and hfDSC. Firstly, the Wnt signaling pathway has been implicated as a factor affecting the hair follicle inductivity and regenerative potential of DS cells. When αSMA was used as a DS cell marker to activate Wnt signaling specifically in DS cells, ectopic hair follicles were formed derived from the lower region including the bulb of pre-existing hair follicles [30]. When collagen type I alpha 2 chain (*Col1α2*) was used to activate Wnt signaling specifically in other DS cells, ectopic hair follicles were formed in a manner derived from the bulge region of hair follicles [31]. Secondly, PDGF signaling has been reported to play a role in maintaining hfDSCs in DS, for the supply of DP cells [32]. Using αSMA as a DS cell marker, it was shown that when expression of PDGFRα (a receptor for PDGF) was abolished in DS cells containing hfDSCs, the number of DS cells contributing to the DP at the start of anagen was reduced [32].

The Plasticity and Stemness of Dermal Sheath Cells

The plasticity of DS cells, which can become DP cells, has been described previously, but several reports have suggested that DS cells can differentiate into cells other than hair follicle cells. SKP (skin-derived precursor) cells, obtained by dispersing skin samples, were reported to differentiate into cells of the nervous system, including neurons, and cells of the mesodermal lineage, including smooth muscle cells and adipocytes, indicating that they are pluripotent in nature [33]. Further analyses showed that SKP cells had similar properties to nascent neural crest cells, and neural crest-derived cells were observed in the DP of hair follicles [34]. These results and common characteristics of marker genes between DP and SKP cells suggested that DP cells in hair follicles may act as a source of SKP cells in adults [34]. Subsequently, cells positive for Sox2 in the DP and lower DS were shown to be the origin of SKP cells [35]. These Sox2-positive DP and lower DS cells were demonstrated to have hair follicle-inducing ability and to become localized to the DP and DS of hair follicles after transplantation into skin, with some differentiating into DF cells [35]. In humans, SKPs can be acquired in higher numbers from skin tissues with higher hair follicle density suggesting an intrinsic link between SKPs and hair follicle mesenchyme [36]. Several reports have also indicated that cultured DP cells and some lower DS cells have properties associated with SKP cells, as described above [34, 35]. Cultured DS cells can differentiate into muscle-type cells with myotubes, fat cells with oil droplets, and osteocytes, alluding to their multipotency [37]. Although these experiments were performed using mouse or rat DS cells, in vitro studies with human DS cells also showed that isolated human DS cells were capable of differentiating into adipocytes, osteocytes, and chondrocytes, in a similar manner to mesenchymal stem cells [38]. Taking these reports together, it is understood that the DP and/or lower DS contains cells with high potency and can be considered as a dermal stem cell niche for maintenance of stemness. Moreover, as previously described, some cells in the lower DS are hfDSCs that fuel the DP cells at the start of a new anagen [20]. It is highly likely therefore that these hfDSCs are the origin of SKP cells [39]. The multipotency of DP cells has been shown [40], and it is quite reasonable that they have this potency due to their source being the hfDSC.

Dermal Sheath Cells in Wound Healing and Skin Regeneration

In 2001, Jahoda et al. [41] pointed out that DS cells may emerge outside of the hair follicle at the time of skin wound healing and act as myofibroblasts. It is well known that myofibroblasts with the properties of smooth muscle cells and fibroblasts emerge at the time of wound healing and are responsible for tissue repair either as cellular material or by secreting repair factors [42]. Based on the expression of αSMA (a marker of myofibroblasts) in DS cells, the high abundance of αSMA-positive cells in tissues containing densely packed hair follicles and the self-renewal capacity of hair follicles, the potential functions of DS cells during skin wound healing have been alluded [41]. It has been suggested that the origins of myofibroblasts may be DS cells and vascular pericytes, on the basis of their commonality of expression markers such as αSMA [43]. As previously described, DP cells stimulated with thrombin, which become activated upon wound healing, will differentiate into myofibroblasts and DS-like cells [22], while culture supernatant from human DS cells can enhance skin wound healing in diabetic excisional mouse wound model [38]. Taken together, DS cells may be the central players in not only maintenance of skin (dermal) homeostasis as dermal stem cells, but also skin wound healing as myofibroblasts. In three-dimensional skin models, it was demonstrated that compared with DP cells and DF cells, DS cells were able to form enriched basement membrane components between epithelial layers [44], suggesting that they may be multifunctional cells with involvement in skin regeneration in addition to hair regeneration.

Attempts to Treat Alopecia Using Human Dermal Sheath Cells as Cell-Based Therapy

The diameter of the DP, and thus the number of DP cells, has been reported to be correlated with the thickness of the hair fiber produced by a follicle. One study directly showed that loss of some DP cells led to smaller hair fibers in mouse coat hair [45], while another study showed that human hair thickness was correlated with DP size and DP cell number [46]. In addition, the size of the DP is reduced in miniaturized hair follicles characteristically observed in androgenetic alopecia [47]. In the 1990's another hypothesis was raised, suggesting that androgenetic alopecia arises due to a disruption of the traffic balance between lower DS cells and DP cells in the hair cycle, reducing the number of DP cells and resulting in miniaturisation [48]. Based on these findings, the idea of injecting DS cells as a treatment for alopecia is a reasonable concept. Given the consequences of some DS cells supplying DP cells via the hair cycle [20], it has been implied that transplanted lower DS cells, incorporated into DS tissue, may supply DP cells as the hair cycle progresses, resulting in restoration of DP and recovery from miniaturization of hair follicles. Under this concept, a clinical study on cell-based therapy using autologous lower DS cells, DS cup cells in other words, was recently conducted and demonstrated efficacy against androgenetic alopecia (otherwise known as male and female pattern hair loss) [49]. Lower DS cells were dissected from an approximately 6 mm diameter occipital scalp tissue biopsy and cultured in a cell processing facility for expansion. Six and nine months after injections of cultured lower DS cells (1 ml of 3.0×10^5 cells/ml in total, to the hair loss area of approx. 2cm^2), total hair density and cumulative hair diameter at the injection site was significantly increased compared with the placebo [49] (Fig. 5.2). However, twelve months after injection the hair density started to decrease, suggesting that a repeated injection protocol would be beneficial in subsequent trials.

To further determine the behavior of transplanted human lower DS cells within human skin, injection of human lower DS cells into a human reconstituted skin with hair follicles, which was de novo generated from the combination of Wnt signalling activated human DP spheroids, keratinocytes and fibroblasts, revealed their incorporation within DS tissue of hair follicles [50]. DS cells were shown to be relatively less responsive to pro-inflammatory challenge than DF cells [51], supporting their clinical application in cell-based therapy. The earlier finding that DS had immune privilege [14] further supports their potential clinical application for allogeneic transplantation.

5 Dermal Sheath Cells and Hair Follicle Regeneration

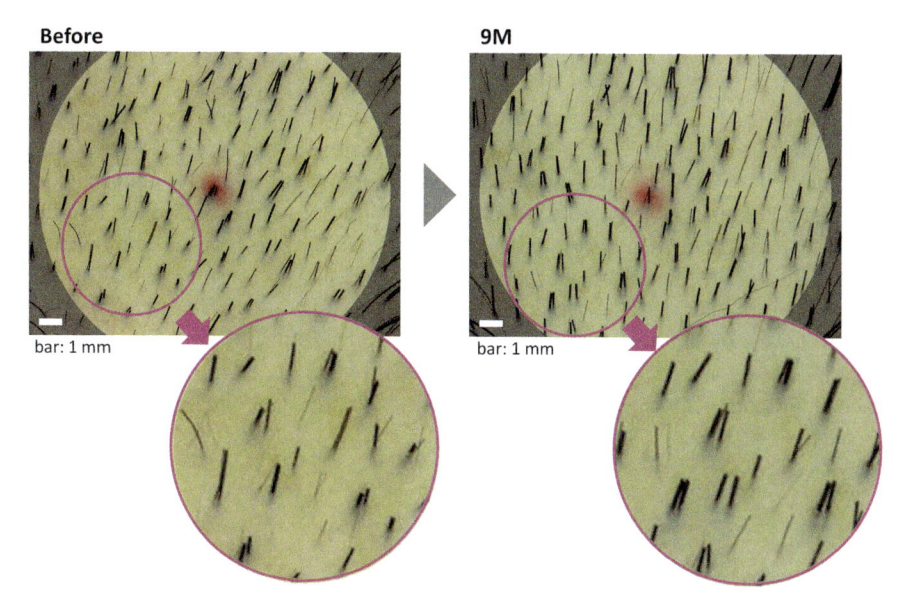

Fig. 5.2 Phototrichogram image of female subject before the treatment and 9 month after the injection of lower DS cells

The number of DP cells in the human scalp hair follicle is approximately 1200 cells [46], in contrast, the number of DP cells in the mouse pelage hair follicle (Awl hair, relatively large hair) is approximately 50 cells as mean value [45]. The number of hfDSC in Mouse pelage hair follicles is believed to be between 3 and 6 cells, so 10% of the total number of DP [20, 27]. If a similar ratio of DP to hfDSCs governs the supply of DP cells in humans, there could be approximately 100 hfDSCs present in the DS of a human scalp hair follicle. In human hair follicles, αSMA is expressed in DS cells within the middle layer, while DS cells in the other outer layer are αSMA negative except those around the vascular network [7], suggesting that human DS are composed of heterogeneous populations of cells with different properties. For clinical application, it is assumed that the culture process will be carried out to achieve the target cell yield. During this culturing process, a change in the proportion of cell population could occur and the proportion of cells with prominent hair follicle regeneration potential, like hfDSCs, could change among DS cells. We have recently proceeded with gene expression profiling for human DS tissue and DS cells [21, 52], and attempted to identify the specific properties of human DS cells. Gene expression profiling of cultured human DS cells showed that vasculature-related genes are highly expressed in cultured DS cells [21]. Signature genes for human lower DS include extracellular matrix components and bone morphogenetic protein-binding molecules, as well as transforming growth factor beta 1 as an upstream regulator [52].

Concluding Remarks

The characteristics and functions of DS cells have been reviewed with citation of papers in this chapter. Selected papers with functional studies on DS (cells) are summarized in Table 5.1. In the table, the papers are listed per experimental model, species, and hair type as source for DS and key findings focusing on the function of DS. Overall, transplantation studies in rat unveiled the hair regenerative potential of the DS, while lineage tracing studies in mouse showed the entity of DS cells which have regenerative potential and factors that affect this potential. This work was followed by studies in human including clinical application of DS as a cell-based therapy for alopecia. For further details, the experimental models employed in each of the studies should be considered to correctly understand the findings obtained.

Table 5.1 Summary of functional studies on dermal sheath (DS) cells

	Experimental models	Source Species of DS (cells)	Tissue or Cells	Key findings, focusing on function of DS (cells)
Oliver [11]	Regenerated whisker hair follicle (rat) after amputation of DP or hair bulb	Rat	Remaining DS tissue in amputated whisker hair follicle	Hair follicles were regenerated (implying regeneration induced by DS tissue)
Horne and Jahoda [12]	Regenerated whisker hair follicle (rat) after amputation of lower half	Rat	Lower DS tissue of whisker hair follicle, attached to bottom end of amputated hair follicle	Hair follicles were regenerated, induced by lower DS tissue
Jahoda et al. [15]	Transplantation into wounded ear skin (rat)	Rat	Lower DS cells from whisker hair follicle	Lower DS cells didn't induce large hair follicles (DP cells induced large hair follicles)
Reynolds and Jahoda [16]	Transplantation into wounded ear skin (rat)	Rat	Lower DS cells from whisker hair follicle, transplanted with hair germinative epidermal cells	Large hair follicles were induced by lower DS cells
Matsuzaki et al. [19]	Regenerated whisker hair follicle (rat) after amputation of lower half, implanted into kidney capsule (rat)	Rat	Remaining upper DS tissue in amputated whisker hair follicle	Small hair follicles were regenerated (implying regeneration induced by upper DS tissue)

(continued)

5 Dermal Sheath Cells and Hair Follicle Regeneration

Table 5.1 (continued)

	Experimental models	Source Species of DS (cells)	Tissue or Cells	Key findings, focusing on function of DS (cells)
Gharzi et al. [18]	Transplantation into wounded ear and back skin (rat)	Rat	Lower DS cells and upper DS cells from whisker hair follicle	Only lower DS cells were incorporated into existing hair follicles
Yamao et al. [28]	Hair follicle reconstitution assay (chamber- graft assay) on back skin (mouse)	Rat	DS cells from whisker hair follicle, transplanted with DP cells and newborn epidermal cells	DS cells promoted hair follicle generation induced by co-transplanted DP cells, restored hair inductivity of DP cells by DS cells
Yamao et al. [29]	Transplantation into wounded back skin where hair follicles are amputated (rat)	Rat	DS cells from whisker hair follicle, transplanted with DP cells	DS cells improved hair follicle regeneration
McElwee et al. [17]	Transplantation into wounded ear skin, or into foot pad (mouse)	Mouse	Lower DS cells, upper DS cells from whisker hair follicle	In wounded ear, growth phase of hair were prolonged through incorporation of transplanted lower DS cells into hair follicles. In foot pad, lower DS cells generated new hair follicles. Upper DS cells didn't show such effects
Chi et al. [24]	Genetic lineage tracing study during hair cycle (mouse)	Mouse	(Cells including DS cells)	DP-constituting cells were also supplied from locations other than DP tissue
Rahmani et al. [20]	Genetic lineage tracing study during hair cycle (mouse)	Mouse	hfDSCs (hair follicle dermal stem cells) in DS tissue	hfDSC progeny were recruited into DP in early anagen, indicating hfDSCs supply DP-constituting cells

(continued)

Table 5.1 (continued)

	Experimental models	Source Species of DS (cells)	Tissue or Cells	Key findings, focusing on function of DS (cells)
Zhou et al. [31]	Gene function study (mouse)	Mouse	DS cells (Wnt signaling activated)	Ectopic hair follicles were formed from bulge region of existing hair follicles
Tao et al. [30]	Gene function study (mouse)	Mouse	hfDSCs in DS tissue (Wnt signaling activated)	Ectopic hair follicles were formed from lower region of existing hair follicles
Gonzalez et al. [32]	Gene function study (mouse)	Mouse	hfDSCs in DS tissue (PDGFRa knockout)	Number of hfDSC progeny recruited into DP was reduced
Abbasi and Biernaskie [26]	Genetic lineage tracing study in wounded back skin (mouse)	Mouse	hfDSCs in DS tissue	Number of hfDSC progeny recruited into DP was increased in wounded skin
Jahoda et al. [13]	Regenerated hair follicle (human) after amputation of hair bulb, implanted beneath skin (mouse)	Human	Remaining DS tissue in amputated hair follicle	Hair follicles were regenerated (implying regeneration induced by DS tissue)
Reynolds et al. [14]	Implantation into skin (human)	Human	Lower DS tissue of scalp hair follicle	Lower DS tissues generated new hair follicles
Yoshida et al. [50]	Transplantation into humanized skin with hair follicles, generated on back skin (mouse)	Human	Lower DS cells from scalp hair follicle	Transplanted lower DS cells were incorporated into existing hair follicles
Tsuboi et al. [49]	Clinical study	Human	Lower DS cells from scalp hair follicle	Lower DS cells increase hair density and cumulative hair diameter

In conclusion, the DS and cells derived from it have hair regenerative potential and also show a potential role for skin regeneration. Especially, αSMA-positive hfDSCs located in the DS of rodent hair follicles may facilitate the hair regeneration, while Wnt and PDGF signaling can affect hair regeneration observed during the normal hair cycle. However, the entity of human hfDSCs and their precise location within the DS remains unclear. Further studies for characterization of human DS cells and identification of the human hfDSCs need to precede further understanding of human hair follicle regeneration for development of an efficient clinical application for male and female pattern hair loss.

Acknowledgements We are grateful to Prof. Takashi Matsuzaki (Faculty of Life and Environmental Science, Shimane University), Dr. Francisco Jimenez (University Fernando Pessoa Canarias, Las Palmas de Gran Canaria), and Dr. Claire Higgins (Department of Bioengineering, Imperial College London) for their helpful discussion and critical reading of the manuscript.

Conflicts of Interest The authors have no conflict of interest to declare.

References

1. Philippeos C, Telerman SB, Oules B, Pisco AO, Shaw TJ, Elgueta R, Lombardi G, Driskell RR, Soldin M, Lynch MD, Watt FM (2018) Spatial and single-cell transcriptional profiling identifies functionally distinct human dermal fibroblast subpopulations. J Invest Dermatol 138:811–825. https://doi.org/10.1016/j.jid.2018.01.016
2. Tabib T, Morse C, Wang T, Chen W, Lafyatis R (2018) SFRP2/DPP4 and FMO1/LSP1 define major fibroblast populations in human skin. J Invest Dermatol 138:802–810. https://doi.org/10.1016/j.jid.2017.09.045
3. Sorrell JM, Caplan AI (2004) Fibroblast heterogeneity: more than skin deep. J Cell Sci 117:667–675. https://doi.org/10.1242/jcs.01005
4. Driskell RR, Lichtenberger BM, Hoste E, Kretzschmar K, Simons BD, Charalambous M, Ferron SR, Herault Y, Pavlovic G, Ferguson-Smith AC, Watt FM (2013) Distinct fibroblast lineages determine dermal architecture in skin development and repair. Nature 504:277–281. https://doi.org/10.1038/nature12783
5. Collins CA, Kretzschmar K, Watt FM (2011) Reprogramming adult dermis to a neonatal state through epidermal activation of beta catenin. Development 138:5189–5199. https://doi.org/10.1242/dev.064592
6. Ito M, Sato Y (1990) Dynamic ultrastructural changes of the connective tissue sheath of human hair follicles during hair cycle. Arch Dermatol Res 282:434–441. https://doi.org/10.1007/BF00402618
7. Urabe A, Furumura M, Imayama S, Nakayama J, Hori Y (1992) Identification of a cell layer containing alpha-smooth muscle actin in the connective tissue sheath of human anagen hair. Arch Dermatol Res 284:246–249. https://doi.org/10.1007/BF00375803
8. Jahoda CA, Reynolds AJ, Chaponnier C, Forester JC, Gabbiani G (1991) Smooth muscle alpha-actin is a marker for hair follicle dermis in vivo and in vitro. J Cell Sci 99(Pt 3):627–636
9. Heitman N, Sennett R, Mok KW, Saxena N, Srivastava D, Martino P, Grisanti L, Wang Z, Ma'ayan A, Rompolas P, Rendl M (2020) Dermal sheath contraction powers stem cell niche relocation during hair cycle regression. Science 367:161–166. https://doi.org/10.1126/science.aax9131
10. Grisanti L, Clavel C, Cai X, Rezza A, Tsai SY, Sennett R, Mumau M, Cai CL, Rendl M (2013) Tbx18 targets dermal condensates for labeling, isolation, and gene ablation during embryonic hair follicle formation. J Invest Dermatol 133:344–353. https://doi.org/10.1038/jid.2012.329
11. Oliver RF (1966) Whisker growth after removal of the dermal papilla and lengths of follicle in the hooded rat. J Embryol Exp Morphol 15:331–347
12. Horne KA, Jahoda CA (1992) Restoration of hair growth by surgical implantation of follicular dermal sheath. Development 116:563–571
13. Jahoda CA, Oliver RF, Reynolds AJ, Forrester JC, Horne KA (1996) Human hair follicle regeneration following amputation and grafting into the nude mouse. J Invest Dermatol 107:804–807. https://doi.org/10.1111/1523-1747.ep12330565

14. Reynolds AJ, Lawrence C, Cserhalmi-Friedman PB, Christiano AM, Jahoda CA (1999) Transgender induction of hair follicles. Nature 402:33–34. https://doi.org/10.1038/46938
15. Jahoda CA, Reynolds AJ, Oliver RF (1993) Induction of hair growth in ear wounds by cultured dermal papilla cells. J Invest Dermatol 101:584–590
16. Reynolds AJ, Jahoda CA (1996) Hair matrix germinative epidermal cells confer follicle-inducing capabilities on dermal sheath and high passage papilla cells. Development 122:3085–3094
17. McElwee KJ, Kissling S, Wenzel E, Huth A, Hoffmann R (2003) Cultured peribulbar dermal sheath cells can induce hair follicle development and contribute to the dermal sheath and dermal papilla. J Invest Dermatol 121:1267–1275. https://doi.org/10.1111/j.1523-1747.2003.12568.x
18. Gharzi A, Reynolds AJ, Jahoda CA (2003) Plasticity of hair follicle dermal cells in wound healing and induction. Exp Dermatol 12:126–136. https://doi.org/10.1034/j.1600-0625.2003.00106.x
19. Matsuzaki T, Inamatsu M, Yoshizato K (1996) The upper dermal sheath has a potential to regenerate the hair in the rat follicular epidermis. Differentiation 60:287–297. https://doi.org/10.1046/j.1432-0436.1996.6050287.x
20. Rahmani W, Abbasi S, Hagner A, Raharjo E, Kumar R, Hotta A, Magness S, Metzger D, Biernaskie J (2014) Hair follicle dermal stem cells regenerate the dermal sheath, repopulate the dermal papilla, and modulate hair type. Dev Cell 31:543–558. https://doi.org/10.1016/j.devcel.2014.10.022
21. Yoshida Y, Soma T, Kishimoto J (2019) Characterization of human dermal sheath cells reveals CD36-expressing perivascular cells associated with capillary blood vessel formation in hair follicles. Biochem Biophys Res Commun 516:945–950. https://doi.org/10.1016/j.bbrc.2019.06.146
22. Feutz AC, Barrandon Y, Monard D (2008) Control of thrombin signaling through PI3K is a mechanism underlying plasticity between hair follicle dermal sheath and papilla cells. J Cell Sci 121:1435–1443. https://doi.org/10.1242/jcs.018689
23. Tobin DJ, Gunin A, Magerl M, Handijski B, Paus R (2003) Plasticity and cytokinetic dynamics of the hair follicle mesenchyme: implications for hair growth control. J Invest Dermatol 120:895–904. https://doi.org/10.1046/j.1523-1747.2003.12237.x
24. Chi WY, Enshell-Seijffers D, Morgan BA (2010) De novo production of dermal papilla cells during the anagen phase of the hair cycle. J Invest Dermatol 130:2664–2666. https://doi.org/10.1038/jid.2010.176
25. Hagner A, Shin W, Sinha S, Alpaugh W, Workentine M, Abbasi S, Rahmani W, Agabalyan N, Sharma N, Sparks H, Yoon J, Labit E, Cobb J, Dobrinski I, Biernaskie J (2020) Transcriptional profiling of the adult hair follicle mesenchyme reveals R-spondin as a novel regulator of dermal progenitor function. iScience 23:101019. https://doi.org/10.1016/j.isci.2020.101019
26. Abbasi S, Biernaskie J (2019) Injury modifies the fate of hair follicle dermal stem cell progeny in a hair cycle-dependent manner. Exp Dermatol 28:419–424. https://doi.org/10.1111/exd.13924
27. Shin W, Rosin NL, Sparks H, Sinha S, Rahmani W, Sharma N, Workentine M, Abbasi S, Labit E, Stratton JA, Biernaskie J(2020) Dysfunction of hair follicle mesenchymal progenitors contributes to age-associated hair loss. Dev Cell 53:185–198 e187. https://doi.org/10.1016/j.devcel.2020.03.019
28. Yamao M, Inamatsu M, Ogawa Y, Toki H, Okada T, Toyoshima KE, Yoshizato K (2010) Contact between dermal papilla cells and dermal sheath cells enhances the ability of DPCs to induce hair growth. J Invest Dermatol 130:2707–2718. https://doi.org/10.1038/jid.2010.241
29. Yamao M, Inamatsu M, Okada T, Ogawa Y, Tateno C, Yoshizato K (2017) Enhanced restoration of in situ-damaged hairs by intradermal transplantation of trichogenous dermal cells. J Tissue Eng Regen Med 11:977–988. https://doi.org/10.1002/term.1997
30. Tao Y, Yang Q, Wang L, Zhang J, Zhu X, Sun Q, Han Y, Luo Q, Wang Y, Guo X, Wu J, Li B, Yang X, He L, Ma G (2019) beta-catenin activation in hair follicle dermal stem cells induces ectopic hair outgrowth and skin fibrosis. J Mol Cell Biol 11:26–38. https://doi.org/10.1093/jmcb/mjy032

31. Zhou L, Yang K, Wickett RR, Andl T, Zhang Y (2016) Dermal sheath cells contribute to postnatal hair follicle growth and cycling. J Dermatol Sci 82:129–131. https://doi.org/10.1016/j.jdermsci.2016.02.002
32. Gonzalez R, Moffatt G, Hagner A, Sinha S, Shin W, Rahmani W, Chojnacki A, Biernaskie J (2017) Platelet-derived growth factor signaling modulates adult hair follicle dermal stem cell maintenance and self-renewal. NPJ Regen Med 2:11. https://doi.org/10.1038/s41536-017-0013-4
33. Toma JG, Akhavan M, Fernandes KJ, Barnabe-Heider F, Sadikot A, Kaplan DR, Miller FD (2001) Isolation of multipotent adult stem cells from the dermis of mammalian skin. Nat Cell Biol 3:778–784. https://doi.org/10.1038/ncb0901-778
34. Fernandes KJ, McKenzie IA, Mill P, Smith KM, Akhavan M, Barnabe-Heider F, Biernaskie J, Junek A, Kobayashi NR, Toma JG, Kaplan DR, Labosky PA, Rafuse V, Hui CC, Miller FD (2004) A dermal niche for multipotent adult skin-derived precursor cells. Nat Cell Biol 6:1082–1093. https://doi.org/10.1038/ncb1181
35. Biernaskie J, Paris M, Morozova O, Fagan BM, Marra M, Pevny L, Miller FD (2009) SKPs derive from hair follicle precursors and exhibit properties of adult dermal stem cells. Cell Stem Cell 5:610–623. https://doi.org/10.1016/j.stem.2009.10.019
36. Hill RP, Gledhill K, Gardner A, Higgins CA, Crawford H, Lawrence C, Hutchison CJ, Owens WA, Kara B, James SE, Jahoda CA (2012) Generation and characterization of multipotent stem cells from established dermal cultures. PLoS One 7:e50742. https://doi.org/10.1371/journal.pone.0050742
37. Jahoda CA, Whitehouse J, Reynolds AJ, Hole N (2003) Hair follicle dermal cells differentiate into adipogenic and osteogenic lineages. Exp Dermatol 12:849–859. https://doi.org/10.1111/j.0906-6705.2003.00161.x
38. Ma D, Kua JE, Lim WK, Lee ST, Chua AW (2015) In vitro characterization of human hair follicle dermal sheath mesenchymal stromal cells and their potential in enhancing diabetic wound healing. Cytotherapy 17:1036–1051. https://doi.org/10.1016/j.jcyt.2015.04.001
39. Agabalyan NA, Rosin NL, Rahmani W, Biernaskie J (2017) Hair follicle dermal stem cells and skin-derived precursor cells: Exciting tools for endogenous and exogenous therapies. Exp Dermatol 26:505–509. https://doi.org/10.1111/exd.13359
40. Driskell RR, Clavel C, Rendl M, Watt FM (2011) Hair follicle dermal papilla cells at a glance. J Cell Sci 124:1179–1182. https://doi.org/10.1242/jcs.082446
41. Jahoda CA, Reynolds AJ (2001) Hair follicle dermal sheath cells: unsung participants in wound healing. Lancet 358:1445–1448. https://doi.org/10.1016/S0140-6736(01)06532-1
42. Sarrazy V, Billet F, Micallef L, Coulomb B, Desmouliere A (2011) Mechanisms of pathological scarring: role of myofibroblasts and current developments. Wound Repair Regen 19(Suppl 1):s10-15. https://doi.org/10.1111/j.1524-475X.2011.00708.x
43. Juniantito V, Izawa T, Yuasa T, Ichikawa C, Yamamoto E, Kuwamura M, Yamate J (2012) Immunophenotypical analyses of myofibroblasts in rat excisional wound healing: possible transdifferentiation of blood vessel pericytes and perifollicular dermal sheath cells into myofibroblasts. Histol Histopathol 27:515–527. https://doi.org/10.14670/HH-27.515
44. Higgins CA, Roger MF, Hill RP, Ali-Khan AS, Garlick JA, Christiano AM, Jahoda CAB (2017) Multifaceted role of hair follicle dermal cells in bioengineered skins. Br J Dermatol 176:1259–1269. https://doi.org/10.1111/bjd.15087
45. Chi W, Wu E, Morgan BA (2013) Dermal papilla cell number specifies hair size, shape and cycling and its reduction causes follicular decline. Development 140:1676–1683. https://doi.org/10.1242/dev.090662
46. Elliott K, Stephenson TJ, Messenger AG (1999) Differences in hair follicle dermal papilla volume are due to extracellular matrix volume and cell number: implications for the control of hair follicle size and androgen responses. J Invest Dermatol 113:873–877. https://doi.org/10.1046/j.1523-1747.1999.00797.x
47. Alcaraz MV, Villena A, Perez de Vargas I (1993) Quantitative study of the human hair follicle in normal scalp and androgenetic alopecia. J Cutan Pathol 20:344–349. https://doi.org/10.1111/j.1600-0560.1993.tb01273.x

48. Jahoda CA (1998) Cellular and developmental aspects of androgenetic alopecia. Exp Dermatol 7:235–248
49. Tsuboi R, Niiyama S, Irisawa R, Harada K, Nakazawa Y, Kishimoto J (2020) Autologous cell-based therapy for male and female pattern hair loss using dermal sheath cup cells: a randomized placebo-controlled double-blinded dose-finding clinical study. J Am Acad Dermatol 83:109–116. https://doi.org/10.1016/j.jaad.2020.02.033
50. Yoshida Y, Soma T, Matsuzaki T, Kishimoto J (2019) Wnt activator CHIR99021-stimulated human dermal papilla spheroids contribute to hair follicle formation and production of reconstituted follicle-enriched human skin. Biochem Biophys Res Commun 516:599–605. https://doi.org/10.1016/j.bbrc.2019.06.038
51. Hill RP, Haycock JW, Jahoda CA (2012) Human hair follicle dermal cells and skin fibroblasts show differential activation of NF-kappaB in response to pro-inflammatory challenge. Exp Dermatol 21:158–160. https://doi.org/10.1111/j.1600-0625.2011.01401.x
52. Niiyama S, Ishimatsu-Tsuji Y, Nakazawa Y, Yoshida Y, Soma T, Ideta R, Mukai H, Kishimoto J (2018) Gene expression profiling of the intact dermal sheath cup of human hair follicles. Acta Derm Venereol 98:694–698. https://doi.org/10.2340/00015555-2949

Chapter 6
Epithelial-Mesenchymal Interactions Between Hair Follicles and Dermal Adipose Tissue

Raul Ramos and Maksim V. Plikus

Abstract The skin forms the outer layer of the body and provides animals with an array of functions vital for their survival and adaptation to external environment. Hair follicles and dermal adipose tissue (dWAT) are two major components of mammalian skin, each with its distinct anatomy and cellular lineage. Yet, in late embryogenesis they come in direct contact and become physiologically coupled in adulthood into a remodelling cycle. Indeed, in species, such as Mus musculus, dWAT prominently and reversibly expands in volume when adjacent hair follicles transition from resting to active growth phase. Recent studies reveal that such coupling is supported by direct epithelial-mesenchymal interactions between hair follicle and dWAT cells via several paracrine signaling pathways. Here, we review classic and emerging literature regarding epithelial-mesenchymal interactions between hair follicles and dWAT and its implications for mammalian skin physiology under thermoregulatory, immuno-protective, and regenerative contexts. We conclude that reciprocal molecular interactions between hair follicles and dWAT are essential for normal physiology of mammalian skin. Paracrine signals between closely associated hair follicles and dWAT profoundly affect each other's cyclic behavior and tissue volume. Upon excisional skin wounding, newly regenerated hair follicles in the center of the scar secrete signaling factors that induce reprograming of the cellular fate of wound fibroblasts into new adipocytes. Further, upon bacterial infection, both dermal adipocytes and, likely, hair follicles install a rapid innate immune response to combat pathogenic invasion. In summary, cyclic epithelial-mesenchymal interactions between hair follicles and dWAT are essential for optimizing their joint thermoregulatory, immune-protective and regenerative functions. Further study of their communications will not only provide answers to standing questions in basic skin biology, but can also inform

R. Ramos · M. V. Plikus (✉)
Department of Developmental and Cell Biology, University of California, Irvine, CA 92697, USA
e-mail: plikus@uci.edu

Sue and Bill Gross Stem Cell Research Center, University of California, Irvine, CA 92697, USA

Center for Complex Biological Systems, University of California, Irvine, CA 92697, USA

NSF-Simons Center for Multiscale Cell Fate Research, University of California, Irvine, CA 92697, USA

© The Author(s), under exclusive license to Springer Nature Switzerland AG 2022
F. Jimenez and C. Higgins (eds.), *Hair Follicle Regeneration*, Stem Cell Biology and Regenerative Medicine 72, https://doi.org/10.1007/978-3-030-98331-4_6

new solutions to unmet clinical needs, including therapies for alopecia, lipodystrophy and skin scarring.

Keywords Adipocyte · Hair follicle · Regeneration · Signaling · Thermoregulation

Key Points

- Hair follicles (HFs) and dermal white adipose (dWAT) tissue are essential thermoregulatory components of the mammalian skin.
- HFs and dWAT can communicate and synchronize their activities, particularly across the hair cycle.
- In close cooperation with adjacent HFs, dWAT has evolved a range of immunoregulatory and regenerative functions specialized to skin.

Summary

In this chapter, we will review the current knowledge on why and how HFs and dWAT may have evolved such a close physiological connection and how other skin appendages, such as sweat and mammary glands, may also be connected with dWAT.

Introduction: What Do Hairs and Fat Have in Common?

One of the key features of mammalian metabolism is the ability to autonomously generate body heat and maintain near-constant body temperature [1, 2]. In rodents, the skin has evolved as a key thermoregulatory organ, mainly with the aid of two distinct tissue elements: dermal white adipose tissue (dWAT), which consists of lipid-laden adipocytes that form the deepest skin layer; and pelage, formed by hair fibers on the outer surface of the skin. Because of their low thermal conductivity, lipid droplets in dWAT function as insulators, preventing heat loss [3–5]. Hair fibers trap warm air at the epidermal surface of the skin, further retarding heat loss [6–8], and also shield it from infrared radiation to prevent overheating [9]. In terms of their development and anatomy, dWAT and hair follicles (HFs) are distinct and yet, they form an intimate physiological relationship. They jointly change their morphologies throughout the animal's life: expanding during the active hair growth phase, called *anagen* and regressing during the hair rest phase, called *telogen* [10–13]. Considering their shared role as thermal insulators, synchronization of HF and dWAT growth would appear logical; however, their vastly different cellular composition and stem cell lineage dynamics put into question both the need for this synchronized growth as well as the shared control mechanisms.

Hair Follicles are Ectodermal Organs

Mammalian skin has evolved a diverse array of specialized appendages that allow it to perform its thermoregulatory functions, including sudoriferous (sweat) glands, which secrete sweat onto the epidermis to cool it down [14, 15]; HFs, which form hair fibers to cover the skin from direct heat and to retain body temperature; and HF-associated sebaceous glands, which secrete oil-rich sebum onto hair fibers and surrounding epidermis to make them water-repellent [16, 17]. HFs are particularly interesting in this regard, as they have the ability to cyclically renew throughout the organism's lifetime in what is known as the hair growth cycle [10–13]. Like other ectodermal skin appendages, HFs form as derivatives of embryonic epidermis via a morphogenetic process mediated by reciprocal molecular interactions with underlying dermal fibroblast precursors [18–21]. In mice, HF development starts in utero and completes during the early postnatal period (Fig. 6.1), occurring in three consecutive waves of morphogenesis [22–24]. The first wave initiates at around embryonic day (E) 14.5, and it produces the so-called primary HFs that make the longest hair type, called guard hair. The second wave starts at around E16.5 and generates many more HFs that grow smaller hairs with bends called awl and auchene. The third and last morphogenetic wave generates the rest of the HFs that predominantly grow the

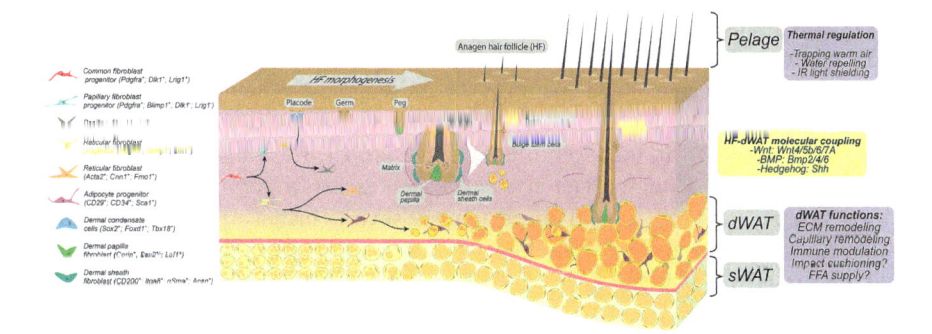

Fig. 6.1 Hair follicle and dWAT morphogenesis couple late during embryonic development. At around embryonic day (E) 12.5 in mice, common fibroblast progenitors populate what will become the future dermal compartment of the skin. By E14.5, these fibroblast progenitors undergo a series of lineage restrictions to give rise to papillary progenitors in the upper dermis, that in turn give rise to specialized dermal condensate cells of the developing hair follicle (HF). Dermal condensate cells will further differentiate toward dermal papilla and dermal sheath cells. E14.5 progenitors in the lower dermis give rise to reticular progenitors, a portion of which will then specify toward adipogenic precursors of developing dWAT by E16.5. At this time, only a very small proportion of developing HFs have attained maturity. By E17.5–18.5, fully developed HFs activate multiple pro-adipogenic signaling pathways (listed in the yellow box). In response to this, adipose progenitors differentiate and small multilocular adipocytes begin to appear around the HF, rapidly maturing into large, unilocular adipocytes. Distinct skin structures and cell types are color-coded and annotated. Gene markers of various embryonic mouse skin fibroblast progenitors and their progenies are listed on the left. Key properties of pelage and dWAT that support skin thermoregulation are listed on the right in the purple boxes

smallest hair type, called zigzag [25, 26] due to its numerous bends. In rodents like mice and rats, follicles mainly exist as simple HFs, i.e. HFs are not grouped in follicular units and each active follicle contains only one growing hair shaft. However, many other mammals including humans [27–29], dogs [30], sheep [31], and rabbits [32] have compound HFs, where each follicular unit contains "bundles" of multiple individual follicles. In human scalp, most follicular units are composed of two to four terminal hairs, in contrast with the vellous hair follicular units that populate the rest of the body [33]. In dogs and rabbits, follicular units with three or more growing shafts are common [34]. In combination, HF density, their distribution pattern, solitary versus compound follicular unit type and hair shaft morphotype all have direct impact on pelage density and represent a *developmental mechanism of hair density control.*

At the tissue level, HFs are mainly comprised by an epithelial lineage that includes long-termed self-renewing stem cells in the so-called *bulge* [35–37]—located at the depth of the follicular isthmus, in the upper reticular dermis, beneath the sebaceous gland [38, 39], as well as a dermal lineage comprised by the specialized fibroblasts of *dermal papilla (DP)* [40–42] and *dermal sheath (DS)* [43–45]. The former resides at the base of the follicle, while the latter envelop it from the outside and is physically continuous with the DP (Fig. 6.1). During anagen, HFs extend through the reticular dermis and into dWAT. Within its base portion, called the *bulb,* coordinated proliferative and differentiation activities by the so-called *hair matrix* cells give rise to the differentiated products: the hair shaft, which extrudes over the skin surface and the surrounding cylinder of the *inner root sheath,* which eventually disintegrates at the level of sebaceous gland opening [46]. Skin in most mammals contains dense arrays of thousands of HFs, leading to dense pelages and effective thermal insulation (Fig. 6.1, right). In addition to the follicle proper, HF units can also contain additional functionally important structures such as the *arrector pili* muscle [47–49], which physically raises and lowers the hair shaft to increase or decrease pelage volume, respectively, to further modulate heat insulation, and in the case of inguinal and axillary HFs in humans, an apocrine gland that secretes sweat [50].

In addition to body heat retention, other important auxiliary functions of the HF have emerged across the mammalian clade. For example, specialized follicles in hedgehogs [51] and echidnas [52] produce hollow spines to dampen impact from falls and to deter predators. African crested rats chew on the bark of the poison arrow tree (*Acokanthera schimperi*) and then apply the toxin-laden spit mixture onto a strip of specialized, sponge-like hairs that retain it for self-defense purposes [53, 54]. In sloths, specialized grooved hairs can harbor algae which provide them with camouflage and additional dietary sources [55]. In many mammalian species, whisker hairs (*aka* vibrissae) develop in isolation from other HFs and in close association with nerves, enabling increased tactile sensory perception [56]. Hair shafts also aid in skin-wide sebum and sweat dispersion from the glands, and camouflage via their variable pigmentation [39]. The HFs' ability to cyclically self-renew adds another level of control over hair morphology [57] and pigmentation [58], allowing for more versatile pelage adaptations to seasonal and other environmental challenges. Self-renewing ability of adult HFs represents a *regenerative mechanism of hair density control*.

The Role of Hair Cycle in Modulating Pelage Density

Growth cycle of adult HFs can be divided into three consecutive phases of active growth *(anagen)*, regression *(catagen)*, and relative quiescence *(telogen)* [10, 11] (Fig. 6.2). Duration of anagen phase is the primary parameter that determines the final length of the hair shaft, with longer anagen leading to proportionately longer hairs. In mice, as in other species, HFs are able to switch the hair morphotype they produce between consecutive cycles. For example, follicles that originally produced zigzag hair will often produce awl or auchene hair in the following anagen [57]. Hair growth is terminated during the catagen phase, and the HF regresses back to its resting state. HF regression is driven by coordinated apoptosis, phagocytosis and terminal

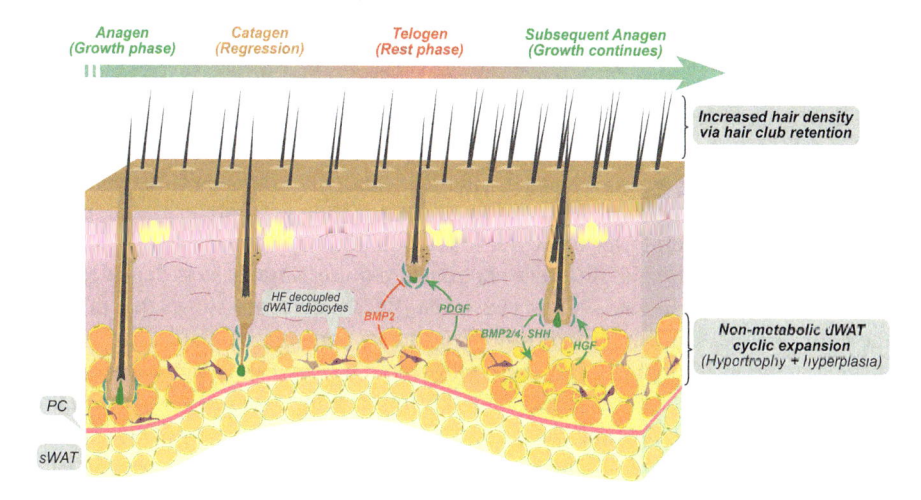

Fig. 6.2 dWAT cycling is tightly coupled to the hair growth cycle. During mature anagen phase of the hair growth cycle (on the left), the lower portion of the HF becomes embedded within the dWAT compartment. Upon HF growth regression (catagen phase), dWAT rapidly shrinks, leaving a reduced, yet persistent population of adipocytes. These adipocytes secrete BMP signals that help to maintain early telogen HFs in a highly quiescent state, a property known as telogen refractivity. Eventually, dWAT-generated BMP signals reduce, and PDGF ligands secreted by adipose progenitors, signal to now competent, late telogen HFs to stimulate their anagen reentry. Upon anagen initiation, expanding HFs produce several paracrine pro-adipogenic signals, including BMP2/4 and SHH, that contribute to rapid anagen-coupled dWAT expansion. At the cellular level, anagen-coupled dWAT expansion is driven both by hyperplasia and adipocyte hypertrophy. dWAT likely reciprocally contributes to the maintenance of the anagen phase, such as by secreting HGF ligands. Expanding dWAT likely modulates several skin properties, that are listed in the box on the right bottom. PC—*panniculus carnosus*

differentiation of many of its epithelial progenies [59–62]. These cellular events are survived by the DP, which begins to migrate upwardly towards the HF bulge along with other surviving epithelial cells that reorganize to form the so-called secondary hair germ, a quiescent but primed epithelial progenitor population situated between bulge and DP of the resting telogen follicle [63, 64]. Recent evidence shows that this upward migration of the DP is facilitated by the smooth muscle-like contraction of DS cells [65]. After some time, telogen HFs exit quiescence and begin new growth, which results in either the replacement of the old fiber (*aka* club hair), or its retention within the follicle along with the new, growing hair shaft. In this regard, many mammals, including mice, are able to retain hair shafts from several previous cycles and accumulate them, essentially leading to hair "bundles" within each solitary HF (Fig. 6.2, top). This increases pelage density without the need for more HFs. The process of hair fiber shedding, known as *exogen* has a complex molecular mechanism [66, 67] and can be coordinated over large skin areas in what is known as molt [68]. When coordinated with new anagen, exogen contributes both to thinning of the overall pelage and to its morphotype and pigmentation change. Thus, pelage density is a direct outcome of *three main factors*—HF density, which is determined during embryogenesis; exogen frequency, which determines the speed at which old hair fibers are shed; and hair cycle speed, which determines the rate at which new hair shafts are produced. In combination, these three factors represent *a multifactorial mechanism of hair density control*.

HF density is defined during skin morphogenesis and cannot further increase after embryonic development. It only decreases in the postnatal period as skin surface continues to expand and as some HFs permanently degenerate with advanced age [69]. However, hair cycle speed (i.e., how quickly a follicle reenters anagen) can be modulated to be made faster or slower to attain a thicker or thinner pelage, respectively. In many mammals, different body sites require variable hair densities, and regional differences in hair cycle speeds exist. Recent evidence shows that in mice, the entire haired skin behaves as a heterogeneous regenerative field composed of distinct anatomical domains (i.e., ventrum, chin, dorsum, ears, etc.), each with its unique hair growth cycling dynamics [70]. Distinct cycling dynamics allow each skin region to achieve variable hair densities optimized for their function. For example, shorter, lower density fur around the eyes and ears can allow for unimpeded visual and auditory perception, respectively. Similarly, short hairs in the ventral skin of low gaited animals can prevent them from becoming entangled with the ground. Interestingly, jerboa species of rodents grow a prominent tuft of hair at the tip of the tail, which may function as a lure to distract potential predators during escape [71]. It is emerging that complex molecular interactions between HFs and other distinct tissue elements within skin, such as dWAT [72–74], cartilage [70], nerves [75], and blood vessels [76], can determine regionally-variable pelage renewal dynamics and characteristics.

Modes for Regulating Hair Cycle Speed

Many of the molecular mechanisms that drive cyclic hair growth are intrinsic to the HF's *micro*-environment [77]. For example, when individual micro-dissected whisker HFs are transplanted under the immune privileged kidney capsule, they remain viable and continue cyclic growth [78, 79]. WNT (Wingless-type mouse mammary tumor virus integration site) [80], BMP (Bone morphogenetic protein) [42], FGF (Fibroblast growth factor) [81, 82], and Hedgehog [83] are now well-recognized among other pathways for their critical roles as *micro*-environmental regulators of cyclic hair growth. Specific to HF stem cell activation, that is critical for initiating new anagen, BMP6 and FGF18, secreted by bulge cells, signal to maintain the state of quiescence during early telogen [84]. During late telogen, DP fibroblasts secrete TGFβ2, Noggin as well as FGF7/10 to trigger adjacent hair germ cell activation and anagen reentry [64, 85].

The above mentioned and many other molecular events occur cyclically and allow each follicle to grow multiple rounds of new hair. Intriguingly, however, many key molecular players of the HF stem cell *micro*-environment also play important roles in the development and growth of other skin tissue elements; because of this, HFs and non-hair cell types can cross-regulate. Indeed, HFs have been shown to molecularly interact with other HFs [74], immune cells [86], nerve endings [87], and, prominently, dermal adipocytes [73], that collectively constitute the signaling *macro*-environment [74]. Further, anatomically distinct skin regions can interact with one another across the boundary, driving skin-wide pelage growth dynamics [88]. As such, modulation of autonomous HF cycle by various extra-follicular tissue elements enables both the synchronization of hair growth phases over long distances and their regionalization. This mechanism has the following physiological impacts:

First, it partitions important energetic and protein resources required to grow a healthy pelage. Rather than growing all of the body's hair at the same time, small patches of growing hairs initiate discretely across the body, activating adjacent HFs and advancing across the skin in wave-like patterns [89], using fewer resources at once. Importantly, combined signaling effects of *micro*- [82, 90] and *macro*-environmental signaling [70, 74] during early telogen phase set the follicle into the so-called refractory state, during which is it largely "disabled" from responding to anagen-inducing signals, like those originating from neighboring anagen HFs. Such refractory signaling prevents HFs from entering new hair cycle prematurely.

Second, it allows for the maintenance of regionally distinct fur densities across the body. For instance, the mouse external ear *macro*-environment contains highly inhibitory signals, largely preventing hair growth on the ear which can potentially negatively affect sound perception. In the ventrum, locally available activating signals induce fast cycling of hairs, leading to denser pelage as compared to dorsum. This can be essential to better protect ventral skin from abrasion in low-gaited animals like mice [70].

Third, it allows for a local response to stimulus. Hairs can efficiently and precociously grow around wounds [91, 92], which may provide injured skin with additional protection from further insult, direct UV exposure, and, potentially, infection. The mechanism for wound-induced hair growth likely involves active signaling by tissue-resident immune cells. This mechanism also activates upon relatively minor skin insults, such as hair plucking. Indeed, when hairs are purposefully plucked in a proper arrangement, tissue-resident myeloid cells signal, in part via TNFα, to induce new hair growth of plucked as well as non-plucked neighboring HFs [86]. Intriguingly, when the number of plucked hairs is below a certain minimal threshold, all HFs in the region, including plucked HFs, remain in a quiescent state, saving on tissue energy resources. Similar immune-mediated and tensile-sensitive mechanism for hair growth activates in response to long-lasting skin stretching [93]. These types of localized but collective, all-or-nothing decisions by HF populations are possible to achieve with the help of locally available, long-distance *macro*-environmental signaling mechanisms. Recent studies have established dWAT as a critical effector of macro-environmental signaling in the skin by (*a*) directly influencing HF stem cell fate decisions via multiple secreted growth factors and cytokines, (*b*) contributing to a robust wound healing response via molecular synergy with HFs and different skin-resident cell types, and *c)* modulating local immune responses upon infection and injury. Below, we will discuss how dWAT associates with HFs and other cells in the skin to support its thermoregulatory, regenerative, and immune properties.

dWAT Is a Specialized White Adipose Depot

White adipose tissue (WAT) has evolved as a key metabolic hub and principal energy storage [94, 95]. WAT is largely composed by clusters of lipid-laden adipocytes embedded in a network of fine, web-like extracellular matrix (ECM). If nutrients are available, adipocytes produce neutral lipids via lipogenesis and store them in form of stable neutral lipid droplets. However, upon energy-demanding exercise or starvation, they can rapidly mobilize them through lipolysis [96, 97]. WAT depots are distributed throughout the body, where they exert critical metabolic roles beyond energy storage and mobilization. These include secretion of systemic hormones, primarily adipokines such as Adiponectin, Resistin, Leptin, and Chemerin [98] that regulate glucose homeostasis, satiety sensing and inflammation [99–101]; as well as an impact-dampening role, as exemplified by intra-articular adipose tissue found in skeletal joints [102]. In many mammals, including humans, major WAT depots exist in the subcutaneous (sWAT) and visceral (vWAT) spaces, allowing for rapid volumetric alterations brought on by lipogenesis and lipolysis without affecting vital organ function. Despite a similar general appearance, significant heterogeneity exists between sWAT and vWAT depots in terms of embryonic origin, adipokine secretion, and turnover rates [103, 104]. A distinct adipocyte depot called dWAT exists in the skin, where it comes in close physical and physiological association with ectodermal skin organs, including HFs and sweat glands (Fig. 6.1).

Developmental Origin of dWAT

Anatomically diverse WAT depots have distinct embryonic origins [104]. With regards to dWAT, at around embryonic day (E) 12.5 in mice, $Pdgfra^+/Dlk1^+/Lrig1^+$ common fibroblast progenitors populate the future dermis [105] (Fig. 6.1, left). By E14.5, these fibroblast progenitors undergo a series of lineage restrictions such that progenitor cells in the upper dermis give rise to $Pdgfra^+/Blimp1^+/Dlk1^-/Lrig1^-$ fibroblast progenitors, which then further differentiate into papillary dermal fibroblasts, HF-associated DP and DS cells, and contractile cells of the *arrector pili* muscle. In contrast, E14.5 progenitors in the lower dermis give rise to $Pdgfra^+/Blimp1^-/Dlk1^+$ fibroblast progenitors, which then give rise to reticular and dWAT fibroblasts. The latter become adipogenic and differentiate into mature, lipid-filled dermal adipocytes [105]. Despite their close proximity in the developing skin, the early stages of HF and dWAT development are largely uncoupled from one another. By E16.5 dWAT precursors first appear in the lower reticular dermis, when only a very small proportion of HFs have attained maturity [34]. This means that adipose progenitors do not require signaling input from HFs to initiate and to mature. Indeed, prominent dWAT exists in the skin of largely hairless mammals, such as dolphins [106]. That early HF and dWAT development is uncoupled is further supported by the evidence that dWAT development occurs in mouse models with genetically suppressed HF morphogenesis [107]. It is not until later during HF morphogenesis that adipose progenitors become coupled with and dependent on HF signaling (Fig. 6.1, right). Early-stage multilocular adipocytes first appear in mouse skin between E17.5–18.5 around early anagen HFs [108, 109]. Their size and number then expand quickly until approximately postnatal day (P) 12, when HFs are in mature anagen [109].

Unique Microanatomy of dWAT

Unlike vWAT and sWAT depots, which grow into large, often amorphous masses, dWAT has a very fine and organized microanatomy. In the mouse dorsum, for example, dWAT exists as a continuous or semi-continuous thin layer [32], isolated from the underlying sWAT by the *panniculus carnosus* muscle. Importantly, dWAT is in contact with distinct structures above (i.e. reticular dermis, HFs) and below (i.e. skeletal muscle), which likely creates anatomically different signaling environment. Unique to dWAT is their close interaction with periodically cycling HFs, which grow down and embed themselves into the dWAT layer with every new anagen and then retract away during telogen. Thus, with each hair growth cycle, a large number of distinct cell types (i.e. DP and DS fibroblasts, HF matrix and *outer root sheath* keratinocytes) periodically come in contact with dWAT and bring in high levels of distinct signaling ligands and antagonists. These in turn generate periodic signaling environment for the dWAT. In humans and other mammals which largely lack *panniculus carnosus,* dWAT and sWAT come into direct contact. Yet, they maintain unique cellular compositions and metabolic activities [110], further highlighting the unique status of dWAT as a skin-associated tissue. In pigs, three distinct layers of skin-associated WAT exist separated by thin layers of muscle, each with a unique metabolic, lipidomic, and enzymatic profile [111, 112]. In rabbits, dWAT is organized as a discontinuous layer, where adipocytes form small clusters specifically around follicular units [32]. Thus, despite its overall similarities with other adipose depots, its development, anatomical location and fine organization make *dWAT a unique WAT compartment.*

Metabolic and Non-Metabolic dWAT Expansion

One of the characteristic features of dWAT is its ability to undergo significant morphological remodeling in coordination with the hair growth cycle [113, 114] (Fig. 6.2). During anagen, dWAT prominently thickens and eventually envelops the mature HF at its base. Subsequently, dWAT significantly thins out when HFs transition into telogen. It is widely thought that such hair growth phase-coupled dWAT cycling comes as a result of adipogenesis-modulating *macro-*environmental signaling by HFs, that involves one or several pathways critical in adipose biology. Importantly, however, dWAT expansion is not exclusively coupled to hair growth, and like other WAT depots it also prominently responds to systemic metabolic stimuli. Indeed, alterations to systemic metabolism override local HF-dWAT coupling. For instance, upon long-term high fat diet (HFD), healthy obese mice expand their dWAT compartment by more than 1mm even when HFs are in telogen [115]. Recent evidence shows that metabolic stress from HFD-induced obesity leads to hair thinning and accelerated hair loss in mice through various stem cell-centric mechanisms, including damage from reactive oxygen species [116]. Further, inflammatory cytokines associated with pathological obesity exacerbate oxidative damage in the skin, leading to a rapid deficiency in hair regeneration [116]. Although the extent to which dWAT drives hair loss in these obesity models requires further investigation, it is possible that it plays a paracrine role by secreting obesity-associated inflammatory effectors such as IL-6, a known adipose-derived factor [117]. Mirroring these observations of obesity-associated hair loss, long-term caloric restriction in mice leads to a stimulus in hair growth rates via HF stem cell population expansion as well as an increase in hair fiber retention, despite a significant reduction to dWAT volume [118]. Also, responding to a drop in external temperature, dWAT significantly expands independent of hair cycle stage, whereas sWAT normally responds by lipolysis [119]. As such, dWAT represents a unique adipose tissue, capable of performing classic metabolic functions normally attributed to other depots while also maintaining a synergistic relationship with HFs and possibly other cutaneous elements to help regulate mammalian skin homeostasis.

Why Do HFs and dWAT Associate?

Despite clear examples of HF-dWAT association, the reasons why they form such an intimate coupling remain unclear. Possible reasons include:

Cushioning of anagen HFs. Physical properties of lipid droplets make WAT an ideal impact-dampener. For example, WAT depots exist in the intra-articular spaces of skeletal joints, like the acetabular fossa of the hip joint and the infrapatellar space of the knee joint, where they are thought to cushion the joint and reduce friction between adjacent skeletal elements [102]. In elephants, the largest extant land mammal, the foot skeleton rests in a thick pad of plantar WAT, which cushions it during walking [120]. Similarly, plantar skin in humans contains adipose pads that decrease impact and redistribute pressures to protect the skeleton [121, 122]. In the skin, during anagen, the bulb part of the follicle harbors a population of rapidly dividing epithelial matrix cells that differentiate into distinct epithelial structures, including the hair shaft. During this stage, the bulb becomes encased in dWAT. Considering impact-dampening role of fat pads and adipocytes in other body locations, it is possible that dWAT serves a cushioning function to protect actively growing HFs from impact damage or shear stress from skin stretching. dWAT is also observed to envelop human eccrine sweat glands, where it may also provide an impact- or sheer stress-dampening role.

ECM remodeling. The expansion of different adipose depots, including dWAT, is associated with rapid ECM remodeling [123]. It is possible that dWAT serves a tissue remodeling role in the skin by biochemically altering its ECM and that of adjacent reticular dermis, such as via secretion of ECM-remodeling enzymes, that could aid in better accommodating anagen HFs within the skin during their rapid anagen enlargement. Moreover, fatty acids released by dWAT adipocytes during telogen-coupled lipogenesis have been shown to directly affect the ECM-making activity by fibroblasts in the reticular dermis [124].

Local source of nutrients. In the mammary gland, stroma is mainly composed of mature WAT [125]. During lactation, stromal adipocytes mobilize their lipid depots and delipidate to provide the secretory epithelial cells with a source of lipid for milk fat production. During weaning, these adipocytes expand again and repopulate the mammary stroma [125]. In the heart, clusters of mature adipocytes are frequently adjacent to the subepicardial myocardium of the right ventricle [126] and may act as readily available, direct sources of free fatty acids (FFAs) for cardiomyocytes [127]. Considering that active hair growth is an energy-demanding process, it is plausible that dWAT provides anagen HFs with FFAs to satisfy energetic demands.

Capillary network remodeling. WAT is heavily vascularized, and metabolic homeostasis is dependent on effective crosstalk between adipocytes and blood vessels [128, 129]. In the heart, epicardial WAT has been shown to secrete vasoactive compounds that signal to remodel epicardial vascular network, effectively modulating the availability of nutrients and hormones to the epicardium [130]. In the skin, both dWAT and HFs are also extensively vascularized. This includes lymphatic vessels. In WAT depots, lymphatics are necessary for immune cell circulation and interstitial fluid collection [131], and changes to WAT, such as upon high fat diet, have been shown in mice to disrupt lymphatic networks and impair lymph and dendritic cell dynamics [132, 133]. Lymphatics also serve as signaling *macro*-environment for HF stem cells, and their disruption in transgenic mice results in hair cycle defects [134]. It is plausible that lymphatics may serve as essential conduits in mediating HF-dWAT interactions.

Thermoregulatory synergy. At the epidermal level, hair fibers aid in thermoregulation by trapping warm air [135], providing partial impermeability to water [136], and shielding from infrared radiation [9]. Below the dermis, dWAT insulates the organism from heat loss due to the low thermal conductance of its lipid contents [3]. Dynamic changes to pelage from the hair cycle are accompanied by a response in dWAT, which enlarges in synchrony with anagen. It is plausible that anagen HFs and dWAT synergize to enhance thermoregulatory properties of the skin.

How Do dWAT and HF Associate?

Crosstalk between closely positioned tissue elements is commonly mediated by soluble signaling factors. Indeed, close anatomical association between ectodermal appendages and dWAT, as well as shared themes in their signaling biology can create permissive conditions for efficient cross-tissue regulation even despite their vastly different anatomical organization and cellular lineage composition. During active growth, anagen HFs secrete ligands and antagonists for multiple signaling pathways with known roles in adipogenesis, including BMP, WNT, Platelet-derived growth factor (PDGF), FGF, and Hedgehog pathways among others. Similarly, adipose tissue is a rich source of soluble ligands such as BMPs [137–139], adipokines like Leptin and Adiponectin [140–142], and inflammatory cytokines [143], that can reciprocally influence HF stem cell fate decisions.

BMP signaling. BMPs form part of the Transforming Growth Factor β (TGFβ) superfamily, and many developmental processes including cell growth, apoptosis, and differentiation rely on BMP signaling. Briefly, BMP ligands signal through two distinct types of transmembrane serine/threonine kinase receptors and this results in phosphorylation-mediated activation of SMAD1/5/8 effector proteins, which modulate gene expression in the nucleus [144, 145]. During WAT development, BMP ligands are essential for the commitment of mesenchymal precursors toward an adipogenic fate, mainly though the activation PPARγ via the transcriptional regulator ZFP423 [138]. In a similar way, BMP-dependent activation of PPARγ is necessary for maturation of dWAT under normal conditions and in response to wounding. Expression of multiple BMP ligands by anagen HFs [146] raises the possibility of BMP-mediated crosstalk between HFs and dWAT. Indeed, genetic ablation of *Zfp423* or BMP signaling in normal skin causes a significant reduction of dWAT size and results in smaller adipocytes [147]. In the context of wound-induced neogenesis in adult mouse skin, epithelial overexpression of BMP antagonist Noggin largely precludes regeneration of new dWAT around new HFs [147]. In normal mouse skin, dWAT can reciprocally signal to HFs via BMP2 during early telogen phase, helping to maintain their refractory state [74] (Fig. 6.2).

WNT signaling. Another important regulator of dWAT adipogenesis is canonical WNT signaling. WNT ligands regulate multiple cellular processes, including cell fate determination [148] and cell polarity [149]. WNT ligand binding to the Frizzled-LRP5/6 receptor complex results in cytoplasmic stabilization of β-catenin, leading to changes in target gene expression [150]. In WAT, including dWAT, canonical WNT signaling has antiadipogenic role [151]. Indeed, overactivation of canonical WNT through overexpression of stabilized β-catenin in dermal cells leads to a prominent reduction to the dWAT compartment [107]. On the other hand, telogen and anagen HFs have been shown to respond to and modulate WNT signals respectively, through the expression of both agonist and antagonist for the pathway [80, 152, 153]. Intriguingly, however, dWAT expansion is largely coupled to anagen phase, when WNT ligand secretion and activity peak in HFs. This suggests that anagen HFs can either balance their WNT signaling outputs into the surrounding skin via secreted antagonists and/or override inhibitory WNT activities on dWAT via other signaling channels. In line with embryonic adipogenesis, the anti-adipogenic role of canonical WNT signaling has been shown in a regenerative context, where neogenic dWAT adipocytes fail to form around neogenic HFs in large wounds in mice that overexpress canonical WNT ligand *Wnt7a* [147].

PDGF signaling. PDGF pathway signaling is mediated by two receptor genes PDGFR-α and -β and their ligands, PDGF-A/B/C/D. Recent data shows that the maintenance of dermal adipose progenitors is under the control of an autocrine PDGF-A/PI3K-AKT signaling pathway axis [154] (Fig. 6.2). In the HF, PDGF signaling is required for anagen induction and maintenance [155, 156], as well as HF dermal stem cell maintenance and self-renewal [157]. Considering the robust expression and secretion of PDGF-A by adipose progenitors during late telogen, it is possible that dWAT-derived PDGF signaling might have an anagen-inducing effect on adjacent resting follicles (Fig. 6.2).

Hedgehog signaling. Hedgehog signaling in adipose depots, such as vWAT, has been shown to have a strong anti-adipogenic effect [158], potentially through the inhibition of adipose fate in progenitor cells to favor other mesenchymal fates [159]. Interestingly, in mouse skin, anagen HF-sourced Sonic hedgehog (SHH) ligands stimulate proliferation of adipocyte precursors and growth of dWAT pre-adipocytes, partly through PPARγ activation [160] (Fig. 6.2). Genetic ablation of *Shh* in HF epithelium stops dWAT expansion. Similarly, genetic ablation of the SHH receptor *Smo* in dWAT, which effectively renders them Hedgehog-insensitive, also prevents dWAT from anagen-coupled expansion [160]. Furthermore, skin-specific overexpression of SHH leads to a nearly two-fold increase in dWAT size. Such distinct Hedgehog responsiveness by dWAT relative to other WAT depots might have evolved to make it "compatible" with anagen HF signaling.

HF-dWAT Connection in Human Scalp, a Unique Scenario

In human scalp, HFs can remain in continuous anagen phase for many years, with relatively short catagen and telogen phases [161–164], forming a connection with dWAT that is essentially permanent. Compared to mouse skin, where a thorough remodeling of dWAT is possible thanks to relatively similar lengths of anagen and telogen, human scalp dWAT may not have an opportunity to remodel as much and as such it may not contribute to the telogen-anagen transition as seen in mice. Despite the fact that dWAT has recently emerged as an important modulator of HF regeneration, wound healing, and immune activity in mice, little is known about its biology in humans, particularly in scalp skin. Recently, however, morphological changes associated with scalp HF cycling have been reported. Compared to adipocytes adjacent to anagen scalp HFs, dWAT adipocytes adjacent to catagen scalp HFs display large numbers of autophagosome-like vacuoles within their lipid droplets as well as elevated LC3 protein—a marker of autophagosome membranes [165]. This raises the possibility that dWAT lipophagy plays an important role in the maintenance and turnover of dWAT during human scalp HF cycling. Recently, human scalp HF-dWAT co-culture experiments have shown that dWAT adipocytes and associated pericytes secrete Hepatocyte growth factor (HGF) that acts upon its receptor, c-Met, in the adjacent HF matrix and DP, upregulating canonical WNT signaling and resulting in extended anagen duration in vitro [166] (Fig. 6.2). Human dWAT has also been documented to closely interact with other skin appendages like sweat glands [167], which do not cycle like HFs, and likely provide it with a continuous source of signals. These examples underscore the importance of the signaling *macro*-environment in a human context, where other HF-dWAT signaling axes may exist, offering exciting therapeutic possibilities to target different hair disorders.

HF-dWAT Synergy During Wound Response

Recent studies show that a robust wound healing response in the mammalian skin requires synergy between HFs, mature adipocytes, adipocyte progenitors, and other skin-resident cell types. An inducible mouse model of fat ablation [168] shows that mature dWAT adipocytes are required for proper establishment of the wound bed during the early wound healing response (Fig. 6.3b). Namely, lack of dWAT adipocytes leads to a significant thinning of the wound bed, improper keratinocyte migration over it, as well as a reduction in collagen deposition and overall changes to ECM organization in the healing wound [115]. During the catagen phase of normal hair growth cycle, dWAT undergoes a dramatic reduction in size, largely due to adipocyte lipolysis and FFA release [73]. Recent studies show that these FFAs directly act upon dermal fibroblasts and alter their metabolism and ECM output [124]. In line with these observations, in mouse model of wound healing dWAT lipolysis and FFA release have been identified as important drivers for inflammatory macrophage

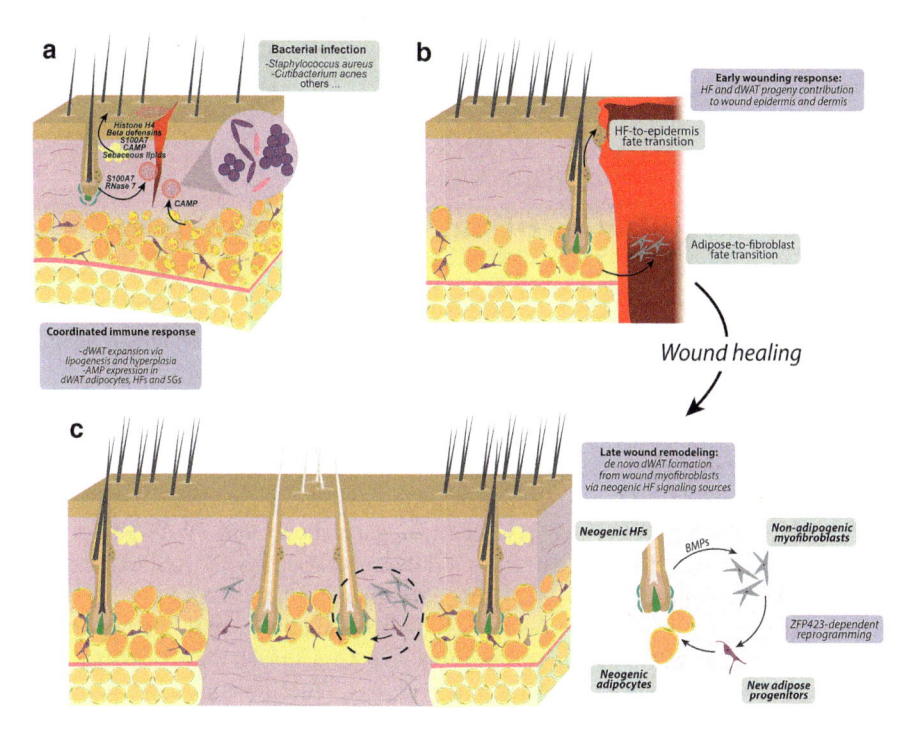

Fig. 6.3 Coordinated HF-dWAT responses to infection and wounding. **a** HFs and dWAT respond to bacterial infection by producing antimicrobial molecules. Upon intradermal infection by *Staphylococcus aureus*, dWAT undergoes rapid expansion via hypertrophy and hyperplasia, that are not coupled to the HF growth cycle. At the same time, dWAT adipocytes secrete Cathelicidin antimicrobial peptide (CAMP) onto the surrounding infected dermis to combat infection. Similarly, an antibacterial response is triggered across the HF upon skin infection by *Cutibacterium acnes*. **b** During wound healing, HF-derived keratinocytes migrate into the interfollicular epidermis to contribute with differentiated progeny to the wound. Similarly, dWAT progenitors migrate into the wound bed to stimulate recruitment of fibroblasts and contribute to scar formation. **c** During the late wound healing response, neogenic adipocytes are formed surrounding neogenic HFs from non-adipogenic dermal myofibroblasts (dashed outline). Mechanistically, this phenomenon of dWAT neogenesis is regulated by BMP ligands secreted from neogenic HFs, which signal to induce ZFP423-based adipogenic reprogramming of myofibroblasts

infiltration into the wound [169]. In mice, large skin excisions often lead to newly regenerating HFs at the center of the healed wound via molecular mechanisms that partially recapitulate embryonic HF development, a process termed wound-induced hair neogenesis (WIHN) [170, 171]. Along with de novo HFs, new adipocytes that are functionally and morphologically indistinguishable from normal dWAT adipocytes appear exclusively in the center of the wound, in close association with neogenic HFs [147]. Lineage tracing experiments show that neogenic adipocytes originate from a myofibroblast progenitors that undergo lineage reprogramming through the BMP-ZFP423 axis, the core transcriptional mechanism of preadipocyte commitment in embryonic development [138, 147] (Fig. 6.3c). Disruption of this axis, either through ablation of BMP receptors in myofibroblasts or by overexpression of BMP antagonist Noggin in skin epithelium leads to lack of new adipocytes, despite normal hair neogenesis.

Possible Synergy Between HFs and dWAT During Microbial Infection

One of the most important functions of the skin is to protect the organism against infections, and the skin achieves this via expression of antimicrobial peptides (Fig. 6.3a). Indeed, the human antimicrobial peptide Dermicidin is constitutively produced and secreted by sweat glands [172]. Several genes encoding for antimicrobial peptides, including Cathelicidin antimicrobial peptide (*Camp*) and Chemokine (C–C motif) ligand 4 (*Ccl4*) have been shown to be expressed in multiple WAT depots [115, 173], including dWAT [174]. Recent evidence shows that dWAT adipocytes are critical for skin protection against bacterial infections [174, 175]. Skin infection by *Staphylococcys aureus* leads to a significant expansion of the dWAT compartment via adaptive lipogenesis along with expression upregulation of multiple antimicrobial peptide genes, including *Camp*, by adipocytes. Conversely, mice with impaired adipogenesis show a reduced innate immune response. Additionally, upon microbial invasion, HFs have been shown to express classic antimicrobial peptides such as S100 family of proteins and Psoriasin, further adding onto the antimicrobial defense mechanisms of skin [176]. Similarly, infection by *Cutibacterium acnes* has been implicated in an increased production of antimicrobial sebum and long chain fatty acids by the sebaceous gland [177–179]. It is not known whether dWAT and HFs act in synergy to produce an antimicrobial defense battery of molecules but considering the close association and constant molecular cross-feed between them, it would not be illogical to think this is the case. Future work should elucidate any possible antimicrobial "domino effect" between HFs and dWAT.

Are HF-dWAT Themes Unique or Do They Occur Elsewhere?

The evolution of WAT brought along not only the ability to store energy via fatty acids, but also metabolic homeostasis through specialized endocrine functions and paracrine interactions with other tissue elements. In mammals, WAT supports endothermy, high metabolic rates, and lactation; all required for survival and reproduction [125, 180, 181]. Considering the functional overlaps between specialized dWAT and other non-skin WAT depots, it is reasonable to think that the molecular crosstalk and interaction themes between HFs and dWAT can occur elsewhere. For example, epicardial adipose tissue (EAT) is located on the surface of the heart in close association with large coronary arteries, supplying the heart with energy-rich FFAs for sustained myocardial function [130]. Like dWAT and other fat depots, EAT stores lipids in accordance with systemic nutrient availability and secretes many classic adipokines [127, 182]. Yet, similar to dWAT, EAT appears to have acquired additional specialized functions to better support adjacent heart tissue requirements. Compared to other WAT depots, EAT has an increased capacity for FFA uptake and mobilization as well as lower rates of glucose utilization [183], protecting the heart from cytotoxic FFAs and providing quick energy to the adjacent myocardium.

In the mammary gland, lactation leads to delipidation of associated WAT, which supports the production of milk by secreting FFAs into the gland [125]. Mammary gland-associated adipocytes have also been shown to de-differentiate into pre-adipocyte-like precursors during lactation, proliferate and then re-differentiate back into adipocytes after weaning [184], mirroring the de-differentiation/re-differentiation phenomenon seen in dWAT during the hair cycle [115]. It is currently unknown whether this phenomenon is also seen under physiological conditions in other WAT depots or if it is exclusive to skin-associated WATs and if so, why this is the case. It is possible that dWAT and mammary gland-associated WAT take advantage of this strategy in order to "protect" themselves from lipolysis-based depletion.

Translating HF-dWAT Interactions into Clinical Practice

Most of our understanding of HF-dWAT interactions is based on studies done in mice and there is a large information gap that needs to be filled with regards to the relationship between HFs and dWAT in human skin. For example, in mice, dWAT and sWAT compartments develop and function independently from each other and are separated by the *panniculus carnosus* muscle continuously across the body [108, 185]. In contrast, humans largely lack *panniculus carnosus* [186], and our skin-associated WAT compartments—traditionally called *superficial* and *deep* WAT [187], are not physically demarcated [188]. Despite this, metabolic studies have shown a stronger association with insulin resistance and obesity-associated complications in deep WAT relative to its superficial counterpart [110, 189]. Other physiological differences between skin-associated WAT compartments are possible, including in their interactions with HFs. Further studies on human dWAT may warrant novel therapeutic strategies to combat hair loss disorders, including androgenetic alopecia [190]. Indeed, autologous dWAT-derived cell preparation treatments show promise as a hair growth-promoting strategy [191, 192]. Considering the immunomodulatory role of dWAT in mouse skin, dWAT-based strategies may emerge for treating alopecia areata, an autoimmune disorder characterized by the transient, non-scarring loss of scalp and body hair [193].

A significant loss of dWAT has been reported in patients suffering from systemic scleroderma, characterized by fibrotic tissue buildup in the skin and other organs [194, 195]. Though the extent at which HFs contribute to the onset of the disease is still unexplored, recent evidence implicates canonical WNT ligands, well-known

HF-derived signals and anti-adipogenic factors, in the pathogenesis of systemic scleroderma [196]. Pharmacological inhibition of WNT-β-catenin signaling, which can prove to be pro-adipogenic could also potentially delay the full onset of this condition.

In humans, complete regeneration of hair- and fat-bearing skin rather than scarification is a clinically-desired outcome. Hypertrophic scarring in human skin has been correlated with areas with a high density of "dermal cones"—adipocyte-rich structures that directly connect the lower portion of the follicle with the rest of the WAT beneath them [197]. The extent at which human dWAT modulates scarification, if at all, in the skin is still not fully known. Because a human model of skin wound healing remains, at best, a challenge, innovative experimental strategies including humanized rodent models as well as xenograft studies are still very much needed to address this clinical need.

Future Perspectives

Despite a large body of evidence for molecular communication and growth coupling between HFs and dWAT, several aspects of the associated mechanism and the functional significance of such coupling remain unexplored. For instance, *what is the underlying mechanism of rapid adipocyte shrinkage during catagen?* Recent data shows that in addition to lipolysis, human dWAT adipocytes can reduce their volume via the autophagy pathway [165], which typically activates in other WATs under pathological conditions [198]. Shrinkage of dWAT adipocytes during catagen proceeds all the way to lipid-free pre-adipocyte state, and this process is reversible, as such pre-adipocytes can proliferate and re-differentiate into new mature adipocytes upon new anagen re-entry [115]. Catagen HF-derived paracrine drivers of dWAT lipolysis and autophagy are currently unknown.

Intriguingly, a significant population of dWAT adipocytes remains in telogen skin, where it likely maintains non-hair cycle-dependent functions. *What makes telogen adipocytes "immune" to catagen pro-lipolysis signaling?* One possibility is that additional signals from *panniculus carnosus* muscle may support telogen adipocytes. Hypothetical molecular interactions between dWAT and *panniculus carnosus* could support the thermogenic capabilities of the skin. Additionally, *panniculus carnosus* could also serve as a physical barrier to decouple dWAT adipocytes from metabolic influence from the underlying sWAT. dWAT displays contradictory response to certain signals compared to other WAT counterparts. For example, anagen HF-secreted SHH induces proliferation and differentiation of dWAT progenitors [160], while in non-dWAT contexts SHH represses adipocyte differentiation through C/EBPα and PPARγ downregulation [199]. Similarly, WNT ligands have an anti-adipogenic effect on adipocyte precursors [151]; however, dWAT significantly expands during anagen, when HF-derived WNT ligands are the highest. These signaling response differences by dWAT may have rose as a mechanism to partially uncouple it from common embryonic mechanisms of adipose differentiation to better synergize with HF functions. Further studies of the epigenetic landscape of adipose progenitors from distinct origins could clarify how these depot-specific mechanisms are enabled.

dWAT adipocytes respond to systemic metabolic changes and contribute to the organism's energetic status [200, 201]; however, the effect of metabolic disorders such as Type 2 diabetes and pathological obesity on dWAT remains largely unexplored. Future studies of dWAT upon metabolic stress could clarify its metabolic relevance and expand our knowledge on the role of the skin as a metabolically responsive organ. As a whole, the mammalian skin works as a formidable barrier organ largely due to the synergy created by epithelial-mesenchymal interactions, including those between its appendages, like HFs and glands, and dWAT. While dWAT maintains essential energetic functions and remains responsive to the organism's metabolic demands, it has also adapted novel functions that separate it from other WAT depots, including a unique dynamic coupling with surrounding HFs, fast innate immune responses to infection, modulation of skin ECM architecture, and cell progeny contributions during wound healing.

References

1. Tan CL et al (2016) Warm-sensitive neurons that control body temperature. Cell 167(1):47–59 e15
2. Bartfai T, Conti B (2012) Molecules affecting hypothalamic control of core body temperature in response to calorie intake. Front Genet 3:184
3. Alexander CM et al (2015) Dermal white adipose tissue: a new component of the thermogenic response. J Lipid Res 56(11):2061–2069
4. Kasza I et al (2022) Contrasting recruitment of skin-associated adipose depots during cold challenge of mouse and human. J Physiol 600(4):847–868
5. Kasza I et al (2016) Thermogenic profiling using magnetic resonance imaging of dermal and other adipose tissues. JCI Insight 1(13):e87146
6. Knight FM (1987) The development of pelage insulation and its relationship to homeothermic ability in an altricial rodent, peromyscus leucopus. Physiol Zool 60(2):181–190
7. Kauffman AS, Cabrera A, Zucker I (2001) Energy intake and fur in summer- and winter-acclimated Siberian hamsters (Phodopus sungorus). Am J Physiol Regul Integr Comp Physiol 281(2):R519–R527
8. Batavia M et al (2010) Influence of pelage insulation and ambient temperature on energy intake and growth of juvenile Siberian hamsters. Physiol Behav 101(3):376–380
9. Hofmeyr MD (1985) Thermal properties of the pelages of selected African ungulates. S Afr J Zool 20(4):179–189
10. Muller-Rover S et al (2001) A comprehensive guide for the accurate classification of murine hair follicles in distinct hair cycle stages. J Invest Dermatol 117(1):3–15
11. Oh JW et al (2016) A guide to studying human hair follicle cycling in vivo. J Invest Dermatol 136(1):34–44
12. Paus R, Foitzik K (2004) In search of the "hair cycle clock": a guided tour. Differentiation 72(9–10):489–511
13. Alonso L, Fuchs E (2006) The hair cycle. J Cell Sci 119(Pt 3):391–393
14. Machado-Moreira CA et al (2015) Temporal and thermal variations in site-specific thermoregulatory sudomotor thresholds: precursor versus discharged sweat production. Psychophysiol 52(1):117–123

15. Wenger CB (1972) Heat of evaporation of sweat: thermodynamic considerations. J Appl Physiol 32(4):456–459
16. Butcher EO, Coonin A (1949) The physical properties of human sebum. J Invest Dermatol 12(4):249–254
17. Porter AM (2001) Why do we have apocrine and sebaceous glands? J R Soc Med 94(5):236–237
18. Saxena N, Mok KW, Rendl M (2019) An updated classification of hair follicle morphogenesis. Exp Dermatol 28(4):332–344
19. Paus R et al (1999) A comprehensive guide for the recognition and classification of distinct stages of hair follicle morphogenesis. J Invest Dermatol 113(4):523–532
20. Chuong CM et al (2013) Module-based complexity formation: periodic patterning in feathers and hairs. Wiley Interdiscip Rev Dev Biol 2(1):97–112
21. Millar SE (2002) Molecular mechanisms regulating hair follicle development. J Invest Dermatol 118(2):216–225
22. Tsai SY et al (2014) Wnt/beta-catenin signaling in dermal condensates is required for hair follicle formation. Dev Biol 385(2):179–188
23. Grisanti L et al (2013) Tbx18 targets dermal condensates for labeling, isolation, and gene ablation during embryonic hair follicle formation. J Invest Dermatol 133(2):344–353
24. Sennett R, Rendl M (2012) Mesenchymal-epithelial interactions during hair follicle morphogenesis and cycling. Semin Cell Dev Biol 23(8):917–927
25. Villani RM et al (2020) Murine dorsal hair type is genetically determined by polymorphisms in candidate genes that influence BMP and WNT signalling. Exp Dermatol 29(5):450–461
26. Chi W, Wu E, Morgan BA (2015) Earlier-born secondary hair follicles exhibit phenotypic plasticity. Exp Dermatol 24(4):265–268
27. Loewenthal LJ (1947) Compound and grouped hairs of the human scalp; their possible connection with follicular infections. J Invest Dermatol 8(5):263–273
28. Jimenez F and Ruifernandez JM (1999) Distribution of human hair in follicular units. A mathematical model for estimating the donor size in follicular unit transplantation. Dermatol Surg 25(4):294–8
29. Headington JT (1984) Transverse microscopic anatomy of the human scalp. A basis for a morphometric approach to disorders of the hair follicle. Arch Dermatol 120(4):449–56
30. Welle MM, Wiener DJ (2016) The hair follicle: a comparative review of canine hair follicle anatomy and physiology. Toxicol Pathol 44(4):564–574
31. Mobini B (2012) Histology of the skin in an Iranian native breed of sheep at different ages. J VetY Adv 2:226–231
32. Plikus MV et al (2011) Self-organizing and stochastic behaviors during the regeneration of hair stem cells. Science 332(6029):586–589
33. Pinkus H (1951) Multiple hairs (Flemming-Giovannini; report of two cases of pili multigemini and discussion of some other anomalies of the pilary complex. J Invest Dermatol 17(5):291–301
34. Guerrero-Juarez CF, Plikus MV (2018) Emerging nonmetabolic functions of skin fat. Nat Rev Endocrinol 14(3):163–173
35. Blanpain C et al (2004) Self-renewal, multipotency, and the existence of two cell populations within an epithelial stem cell niche. Cell 118(5):635–648
36. Morris RJ et al (2004) Capturing and profiling adult hair follicle stem cells. Nat Biotechnol 22(4):411–417
37. Cotsarelis G, Sun TT, Lavker RM (1990) Label-retaining cells reside in the bulge area of pilosebaceous unit: implications for follicular stem cells, hair cycle, and skin carcinogenesis. Cell 61(7):1329–1337
38. Jimenez F, Izeta A, Poblet E (2011) Morphometric analysis of the human scalp hair follicle: practical implications for the hair transplant surgeon and hair regeneration studies. Dermatol Surg 37(1):58–64
39. Schneider MR, Schmidt-Ullrich R, Paus R (2009) The hair follicle as a dynamic miniorgan. Curr Biol 19(3):R132–R142

40. Morgan BA (2014) The dermal papilla: an instructive niche for epithelial stem and progenitor cells in development and regeneration of the hair follicle. Cold Spring Harb Perspect Med 4(7): a015180
41. Driskell RR et al (2011) Hair follicle dermal papilla cells at a glance. J Cell Sci 124(Pt 8):1179–1182
42. Rendl M, Polak L, Fuchs E (2008) BMP signaling in dermal papilla cells is required for their hair follicle-inductive properties. Genes Dev 22(4):543–557
43. Martino PA, Heitman N, Rendl M (2021) The dermal sheath: An emerging component of the hair follicle stem cell niche. Exp Dermatol 30(4):512–521
44. Abbasi S, Biernaskie J (2019) Injury modifies the fate of hair follicle dermal stem cell progeny in a hair cycle-dependent manner. Exp Dermatol 28(4):419–424
45. Rahmani W et al (2014) Hair follicle dermal stem cells regenerate the dermal sheath, repopulate the dermal papilla, and modulate hair type. Dev Cell 31(5):543–558
46. Kiani MT, Higgins CA, Almquist BD (2018) The hair follicle: an underutilized source of cells and materials for regenerative medicine. ACS Biomater Sci Eng 4(4):1193–1207
47. Torkamani N et al (2017) The arrector pili muscle, the bridge between the follicular stem cell niche and the interfollicular epidermis. Anat Sci Int 92(1):151–158
48. Fujiwara H et al (2011) The basement membrane of hair follicle stem cells is a muscle cell niche. Cell 144(4):577–589
49. Poblet E, Ortega F, Jimenez F (2002) The arrector pili muscle and the follicular unit of the scalp: a microscopic anatomy study. Dermatol Surg 28(9):800–803
50. Nawrocki S, Cha J (2019) The etiology, diagnosis, and management of hyperhidrosis: a comprehensive review: etiology and clinical work-up. J Am Acad Dermatol 81(3):657–666
51. Vincent JFV, Owers P (1986) Mechanical design of hedgehog spines and porcupine quills. J Zool 210(1):55–75
52. Goncalves GL et al (2018) Divergent genetic mechanism leads to spiny hair in rodents. PLoS One 13(8):e0202219
53. Kingdon J et al (2012) A poisonous surprise under the coat of the African crested rat. Proc Biol Sci 279(1729):675–680
54. Plikus MV, Astrowski AA (2014) Deadly hairs, lethal feathers: convergent evolution of poisonous integument in mammals and birds. Exp Dermatol 23(7):466–468
55. Pauli JN et al (2014) A syndrome of mutualism reinforces the lifestyle of a sloth. Proc Biol Sci 281(1778):20133006
56. Adibi M (2019) Whisker-mediated touch system in rodents: from neuron to behavior. Front Syst Neurosci 13:40
57. Chi W, Wu E, Morgan BA (2013) Dermal papilla cell number specifies hair size, shape and cycling and its reduction causes follicular decline. Development 140(8):1676–1683
58. Enshell-Seijffers D et al (2010) Beta-catenin activity in the dermal papilla of the hair follicle regulates pigment-type switching. Proc Natl Acad Sci U S A 107(50):21564–21569
59. Foitzik K et al (2000) Control of murine hair follicle regression (catagen) by TGF-beta1 in vivo. FASEB J 14(5):752–760
60. Mesa KR et al (2015) Niche-induced cell death and epithelial phagocytosis regulate hair follicle stem cell pool. Nature 522(7554):94–97
61. Botchkarev VA et al (2003) p75 Neurotrophin receptor antagonist retards apoptosis-driven hair follicle involution (catagen). J Invest Dermatol 120(1):168–169
62. Lindner G et al (1997) Analysis of apoptosis during hair follicle regression (catagen). Am J Pathol 151(6):1601–1617
63. Panteleyev AA (2018) Functional anatomy of the hair follicle: the secondary hair germ. Exp Dermatol 27(7):701–720
64. Greco V et al (2009) A two-step mechanism for stem cell activation during hair regeneration. Cell Stem Cell 4(2):155–169
65. Heitman N et al (2020) Dermal sheath contraction powers stem cell niche relocation during hair cycle regression. Science 367(6474):161–166

66. Higgins CA et al (2009) Exogen involves gradual release of the hair club fibre in the vibrissa follicle model. Exp Dermatol 18(9):793–795
67. Higgins CA, Westgate GE, Jahoda CA (2009) From telogen to exogen: mechanisms underlying formation and subsequent loss of the hair club fiber. J Invest Dermatol 129(9):2100–2108
68. Milner Y et al (2002) Exogen, shedding phase of the hair growth cycle: characterization of a mouse model. J Invest Dermatol 119(3):639–644
69. Matsumura H et al (2016) Hair follicle aging is driven by transepidermal elimination of stem cells via COL17A1 proteolysis. Sci 351(6273):aad4395
70. Wang Q et al (2017) A multi-scale model for hair follicles reveals heterogeneous domains driving rapid spatiotemporal hair growth patterning. Elife 6
71. Caro T (2009) Contrasting coloration in terrestrial mammals. Philos Trans R Soc Lond B Biol Sci 364(1516):537–548
72. Kruglikov IL, Zhang Z, Scherer PE (2019) The role of immature and mature adipocytes in hair cycling. Trends Endocrinol Metab 30(2):93–105
73. Festa E et al (2011) Adipocyte lineage cells contribute to the skin stem cell niche to drive hair cycling. Cell 146(5):761–771
74. Plikus MV et al (2008) Cyclic dermal BMP signalling regulates stem cell activation during hair regeneration. Nature 451(7176):340–344
75. Zhang B et al (2020) Hyperactivation of sympathetic nerves drives depletion of melanocyte stem cells. Nature 577(7792):676–681
76. Li KN et al (2019) Skin vasculature and hair follicle cross-talking associated with stem cell activation and tissue homeostasis. Elife 8
77. Rompolas P, Greco V (2014) Stem cell dynamics in the hair follicle niche. Semin Cell Dev Biol 25–26:34–42
78. Kobayashi K, Nishimura E (1989) Ectopic growth of mouse whiskers from implanted lengths of plucked vibrissa follicles. J Invest Dermatol 92(2):278–282
79. Jahoda CA et al (1996) Human hair follicle regeneration following amputation and grafting into the nude mouse. J Invest Dermatol 107(6):804–807
80. Choi YS et al (2013) Distinct functions for Wnt/beta-catenin in hair follicle stem cell proliferation and survival and interfollicular epidermal homeostasis. Cell Stem Cell 13(6):720–733
81. Harshuk-Shabso S et al (2020) Fgf and Wnt signaling interaction in the mesenchymal niche regulates the murine hair cycle clock. Nat Commun 11(1):5114
82. Kimura-Ueki M et al (2012) Hair cycle resting phase is regulated by cyclic epithelial FGF18 signaling. J Invest Dermatol 132(5):1338–1345
83. Oro AE, Higgins K (2003) Hair cycle regulation of Hedgehog signal reception. Dev Biol 255(2):238–248
84. Hsu YC, Pasolli HA, Fuchs E (2011) Dynamics between stem cells, niche, and progeny in the hair follicle. Cell 144(1):92–105
85. Oshimori N, Fuchs E (2012) Paracrine TGF-beta signaling counterbalances BMP-mediated repression in hair follicle stem cell activation. Cell Stem Cell 10(1):63–75
86. Chen CC et al (2015) Organ-level quorum sensing directs regeneration in hair stem cell populations. Cell 161(2):277–290
87. Peters EM, Arck PC, Paus R (2006) Hair growth inhibition by psychoemotional stress: a mouse model for neural mechanisms in hair growth control. Exp Dermatol 15(1):1–13
88. Plikus MV and Chuong CM (2014) Macroenvironmental regulation of hair cycling and collective regenerative behavior. Cold Spring Harb Perspect Med 4(1): a015198
89. Plikus MV et al (2009) Analyses of regenerative wave patterns in adult hair follicle populations reveal macro-environmental regulation of stem cell activity. Int J Dev Biol 53(5–6):857–868
90. Wu P et al (2019) The balance of Bmp6 and Wnt10b regulates the telogen-anagen transition of hair follicles. Cell Commun Signal 17(1):16
91. Jiang S et al (2010) Small cutaneous wounds induce telogen to anagen transition of murine hair follicle stem cells. J Dermatol Sci 60(3):143–150
92. Sun ZY et al (2009) A very rare complication: new hair growth around healing wounds. J Int Med Res 37(2):583–586

93. Chu SY et al (2019) Mechanical stretch induces hair regeneration through the alternative activation of macrophages. Nat Commun 10(1):1524
94. Berry DC et al (2013) The developmental origins of adipose tissue. Development 140(19):3939–3949
95. Ottaviani E, Malagoli D, Franceschi C (2011) The evolution of the adipose tissue: a neglected enigma. Gen Comp Endocrinol 174(1):1–4
96. Lafontan M, Langin D (2009) Lipolysis and lipid mobilization in human adipose tissue. Prog Lipid Res 48(5):275–297
97. Wolfe RR (1998) Fat metabolism in exercise. Adv Exp Med Biol 441:147–156
98. Luo L, Liu M (2016) Adipose tissue in control of metabolism. J Endocrinol 231(3):R77–R99
99. Rosen ED, Spiegelman BM (2006) Adipocytes as regulators of energy balance and glucose homeostasis. Nature 444(7121):847–853
100. Ouchi N et al (2011) Adipokines in inflammation and metabolic disease. Nat Rev Immunol 11(2):85–97
101. Mancuso P (2016) The role of adipokines in chronic inflammation. Immunotargets Ther 5:47–56
102. Draghi F et al (2016) Hoffa's fat pad abnormalities, knee pain and magnetic resonance imaging in daily practice. Insights Imaging 7(3):373–383
103. Kwok KH, Lam KS and Xu A (2016) Heterogeneity of white adipose tissue: molecular basis and clinical implications. Exp Mol Med 48: e215
104. Lee KY et al (2019) Developmental and functional heterogeneity of white adipocytes within a single fat depot. EMBO J 38(3)
105. Driskell RR et al (2013) Distinct fibroblast lineages determine dermal architecture in skin development and repair. Nature 504(7479):277–281
106. Springer MS et al (2021) Genomic and anatomical comparisons of skin support independent adaptation to life in water by cetaceans and hippos. Curr Biol 31(10): 2124–2139 e3
107. Donati G et al (2014) Epidermal Wnt/beta-catenin signaling regulates adipocyte differentiation via secretion of adipogenic factors. Proc Natl Acad Sci U S A 111(15):E1501–E1509
108. Wojciechowicz K et al (2013) Development of the mouse dermal adipose layer occurs independently of subcutaneous adipose tissue and is marked by restricted early expression of PPARγ. PLoS One 8(3): e59811
109. Wojciechowicz K, Markiewicz E, Jahoda CA (2008) C/EBPalpha identifies differentiating preadipocytes around hair follicles in foetal and neonatal rat and mouse skin. Exp Dermatol 17(8):675–680
110. Walker GE et al (2007) Deep subcutaneous adipose tissue: a distinct abdominal adipose depot. Obesity (Silver Spring) 15(8):1933–1943
111. Monziols M et al (2007) Comparison of the lipid content and fatty acid composition of intermuscular and subcutaneous adipose tissues in pig carcasses. Meat Sci 76(1):54–60
112. Anderson DB, Kauffman RG, Kastenschmidt LL (1972) Lipogenic enzyme activities and cellularity of porcine adipose tissue from various anatomical locations. J Lipid Res 13(5):593–599
113. Hansen LS et al (1984) The influence of the hair cycle on the thickness of mouse skin. Anat Rec 210(4):569–573
114. Chase HB, Montagna W, Malone JD (1953) Changes in the skin in relation to the hair growth cycle. Anat Rec 116(1):75–81
115. Zhang Z et al (2019) Dermal adipose tissue has high plasticity and undergoes reversible dedifferentiation in mice. J Clin Invest 129(12):5327–5342
116. Morinaga H et al (2021) Obesity accelerates hair thinning by stem cell-centric converging mechanisms. Nature 595;266–271
117. Kim DW et al (2014) Adipose-derived stem cells inhibit epidermal melanocytes through an interleukin-6-mediated mechanism. Plast Reconstr Surg 134(3):470–480
118. Forni MF et al (2017) Caloric restriction promotes structural and metabolic changes in the skin. Cell Rep 20(11):2678–2692

119. Kruglikov IL, Scherer PE (2016) Dermal adipocytes: from irrelevance to metabolic targets? Trends Endocrinol Metab 27(1):1–10
120. Weissengruber GE et al (2006) The structure of the cushions in the feet of African elephants (Loxodonta africana). J Anat 209(6):781–792
121. Boyle CJ et al (2019) Morphology and composition play distinct and complementary roles in the tolerance of plantar skin to mechanical load. Sci Adv 5(10): eaay0244
122. Fontanella CG et al (2016) Biomechanical behavior of plantar fat pad in healthy and degenerative foot conditions. Med Biol Eng Comput 54(4):653–661
123. Seo BR et al (2015) Obesity-dependent changes in interstitial ECM mechanics promote breast tumorigenesis. Sci Transl Med 7(301): 301ra130
124. Zhang Z et al (2021) Dermal adipocytes contribute to the metabolic regulation of dermal fibroblasts. Exp Dermatol 30(1):102–111
125. Zwick RK et al (2018) Adipocyte hypertrophy and lipid dynamics underlie mammary gland remodeling after lactation. Nat Commun 9(1):3592
126. Iacobellis G, Corradi D, Sharma AM (2005) Epicardial adipose tissue: anatomic, biomolecular and clinical relationships with the heart. Nat Clin Pract Cardiovasc Med 2(10):536–543
127. Iacobellis G, Bianco AC (2011) Epicardial adipose tissue: emerging physiological, pathophysiological and clinical features. Trends Endocrinol Metab 22(11):450–457
128. Herold J and Kalucka J (2020) Angiogenesis in adipose tissue: the interplay between adipose and endothelial cells. Front Physiol 11: 624903
129. Loesch A, Dashwood MR (2018) Nerve-perivascular fat communication as a potential influence on the performance of blood vessels used as coronary artery bypass grafts. J Cell Commun Signal 12(1):181–191
130. Rabkin SW (2007) Epicardial fat: properties, function and relationship to obesity. Obes Rev 8(3):253–261
131. Redondo PAG et al (2020) Lymphatic vessels in human adipose tissue. Cell Tissue Res 379(3):511–520
132. Blum KS et al (2014) Chronic high-fat diet impairs collecting lymphatic vessel function in mice. PLoS One 9(4); e94713
133. Aschen S et al (2012) Regulation of adipogenesis by lymphatic fluid stasis: part II. Expression of adipose differentiation genes. Plast Reconstr Surg 129(4): 838–847
134. Gur-Cohen S et al (2019) Stem cell-driven lymphatic remodeling coordinates tissue regeneration. Science 366(6470):1218–1225
135. Webb DR, McClure PA (1988) Insulation development in an altricial rodent: neotoma floridana Thomas. Funct Ecol 2(2):237–248
136. Nankey P et al. Under pressure: effects of instrumentation methods on fur seal pelt function. Mar Mammal Sci
137. Qian S, Tang Y, and Tang QQ (2021) Adipose tissue plasticity and the pleiotropic roles of BMP signaling. J Biol Chem 100678
138. Gupta RK et al (2010) Transcriptional control of preadipocyte determination by Zfp423. Nature 464(7288):619–623
139. Huang H et al (2009) BMP signaling pathway is required for commitment of C3H10T1/2 pluripotent stem cells to the adipocyte lineage. Proc Natl Acad Sci U S A 106(31):12670–12675
140. Musovic S, Olofsson CS (2019) Adrenergic stimulation of adiponectin secretion in visceral mouse adipocytes is blunted in high-fat diet induced obesity. Sci Rep 9(1):10680
141. Xie L et al (2008) Adiponectin and leptin are secreted through distinct trafficking pathways in adipocytes. Biochim Biophys Acta 1782(2):99–108
142. Cammisotto PG, Bukowiecki LJ (2002) Mechanisms of leptin secretion from white adipocytes. Am J Physiol Cell Physiol 283(1):C244–C250
143. Makki K, Froguel P and Wolowczuk I (2013) Adipose tissue in obesity-related inflammation and insulin resistance: cells, cytokines, and chemokines. ISRN Inflamm 2013: 139239
144. Miyazawa K and Miyazono K (2017) Regulation of TGF-beta family signaling by inhibitory smads. Cold Spring Harb Perspect Biol 9(3)

145. Wagner DO et al (2010) BMPs: from bone to body morphogenetic proteins. Sci Signal 3(107):mr1
146. Kobielak K et al (2003) Defining BMP functions in the hair follicle by conditional ablation of BMP receptor IA. J Cell Biol 163(3):609–623
147. Plikus MV et al (2017) Regeneration of fat cells from myofibroblasts during wound healing. Science 355(6326):748–752
148. Van Camp JK et al (2014) Wnt signaling and the control of human stem cell fate. Stem Cell Rev Rep 10(2):207–229
149. Chen J, Chuong CM (2012) Patterning skin by planar cell polarity: the multi-talented hair designer. Exp Dermatol 21(2):81–85
150. MacDonald BT and He X (2012) Frizzled and LRP5/6 receptors for Wnt/beta-catenin signaling. Cold Spring Harb Perspect Biol 4(12)
151. Ross SE et al (2000) Inhibition of adipogenesis by Wnt signaling. Science 289(5481):950–953
152. Lowry WE et al (2005) Defining the impact of beta-catenin/Tcf transactivation on epithelial stem cells. Genes Dev 19(13):1596–1611
153. Kwack MH et al (2012) Dickkopf 1 promotes regression of hair follicles. J Invest Dermatol 132(6):1554–1560
154. Rivera-Gonzalez GC et al (2016) Skin adipocyte stem cell self-renewal is regulated by a PDGFA/AKT-Signaling axis. Cell Stem Cell 19(6):738–751
155. Rezza A et al (2016) Signaling networks among stem cell precursors, transit-amplifying progenitors, and their niche in developing hair follicles. Cell Rep 14(12):3001–3018
156. Tomita Y, Akiyama M, Shimizu H (2006) PDGF isoforms induce and maintain anagen phase of murine hair follicles. J Dermatol Sci 43(2):105–115
157. Gonzalez R et al (2017) Platelet-derived growth factor signaling modulates adult hair follicle dermal stem cell maintenance and self-renewal. NPJ Regen Med 2:11
158. Shi Y and Long F (2017) Hedgehog signaling via Gli2 prevents obesity induced by high-fat diet in adult mice. Elife 6
159. James AW et al (2010) Sonic Hedgehog influences the balance of osteogenesis and adipogenesis in mouse adipose-derived stromal cells. Tissue Eng Part A 16(8):2605–2616
160. Zhang B et al (2016) Hair follicles' transit-amplifying cells govern concurrent dermal adipocyte production through Sonic Hedgehog. Genes Dev 30(20):2325–2338
161. Halloy J et al (2000) Modeling the dynamics of human hair cycles by a follicular automaton. Proc Natl Acad Sci U S A 97(15):8328–8333
162. Paus R, Cotsarelis G (1999) The biology of hair follicles. N Engl J Med 341(7):491–497
163. Ebling FJ (1988) The hair cycle and its regulation. Clin Dermatol 6(4):67–73
164. Kligman AM (1961) Pathologic dynamics of human hair loss. I. Telogen effluvium. Arch Dermatol 83:175–198
165. Nicu C et al (2019) Do human dermal adipocytes switch from lipogenesis in anagen to lipophagy and lipolysis during catagen in the human hair cycle? Exp Dermatol 28(4):432–435
166. Nicu C et al (2021) Dermal adipose tissue secretes HGF to promote human hair growth and pigmentation. J Invest Dermatol 141:1633–45.
167. Poblet E et al (2018) Eccrine sweat glands associate with the human hair follicle within a defined compartment of dermal white adipose tissue. Br J Dermatol 178(5):1163–1172
168. Pajvani UB et al (2005) Fat apoptosis through targeted activation of caspase 8: a new mouse model of inducible and reversible lipoatrophy. Nat Med 11(7):797–803
169. Shook BA et al (2020) Dermal adipocyte lipolysis and myofibroblast conversion are required for efficient skin repair. Cell Stem Cell 26(6): 880–895 e6
170. Gay D et al (2013) Fgf9 from dermal gammadelta T cells induces hair follicle neogenesis after wounding. Nat Med 19(7):916–923
171. Ito M et al (2007) Wnt-dependent de novo hair follicle regeneration in adult mouse skin after wounding. Nature 447(7142):316–320
172. Schittek B et al (2001) Dermcidin: a novel human antibiotic peptide secreted by sweat glands. Nat Immunol 2(12):1133–1137

173. Hochberg A et al (2021) Serum levels and adipose tissue gene expression of cathelicidin antimicrobial peptide (CAMP) in obesity and during weight loss. Horm Metab Res 53(3):169–177
174. Zhang LJ et al (2015) Innate immunity. Dermal adipocytes protect against invasive Staphylococcus aureus skin infection. Sci 347(6217): 67–71
175. Zhang LJ et al (2019) Age-related loss of innate immune antimicrobial function of dermal fat is mediated by transforming growth factor beta. Immun 50(1): 121–136e5
176. Reithmayer K et al (2009) Human hair follicle epithelium has an antimicrobial defence system that includes the inducible antimicrobial peptide psoriasin (S100A7) and RNase 7. Br J Dermatol 161(1):78–89
177. Iinuma K et al (2009) Involvement of Propionibacterium acnes in the augmentation of lipogenesis in hamster sebaceous glands in vivo and in vitro. J Invest Dermatol 129(9):2113–2119
178. Alestas T et al (2006) Enzymes involved in the biosynthesis of leukotriene B4 and prostaglandin E2 are active in sebaceous glands. J Mol Med (Berl) 84(1):75–87
179. Georgel P et al (2005) A toll-like receptor 2-responsive lipid effector pathway protects mammals against skin infections with gram-positive bacteria. Infect Immun 73(8):4512–4521
180. Choe SS et al (2016) Adipose tissue remodeling: its role in energy metabolism and metabolic disorders. Front Endocrinol (Lausanne) 7:30
181. Jastroch M, Seebacher F (2020) Importance of adipocyte browning in the evolution of endothermy. Philos Trans R Soc Lond B Biol Sci 37(1793):20190134
182. Marchington JM, Pond CM (1990) Site-specific properties of pericardial and epicardial adipose tissue: the effects of insulin and high-fat feeding on lipogenesis and the incorporation of fatty acids in vitro. Int J Obes 14(12):1013–1022
183. Marchington JM, Mattacks CA, Pond CM (1989) Adipose tissue in the mammalian heart and pericardium: structure, foetal development and biochemical properties. Comp Biochem Physiol B 94(2):225–232
184. Wang QA et al (2018) Reversible de-differentiation of mature white adipocytes into preadipocyte-like precursors during lactation. Cell Metab 28(2):282–288e3
185. Naldaiz-Gastesi N et al (2018) The panniculus carnosus muscle: an evolutionary enigma at the intersection of distinct research fields. J Anat 233 275–288
186. Bahri OA et al (2019) The panniculus carnosus muscle: a novel model of striated muscle regeneration that exhibits sex differences in the mdx mouse. Sci Rep 9(1):15964
187. Schneider MR (2014) Coming home at last: dermal white adipose tissue. Exp Dermatol 23(9):634–635
188. Chen SX, Zhang LJ, Gallo RL (2019) Dermal white adipose tissue: a newly recognized layer of skin innate defense. J Invest Dermatol 139(5):1002–1009
189. Kelley DE et al (2000) Subdivisions of subcutaneous abdominal adipose tissue and insulin resistance. Am J Physiol Endocrinol Metab 278(5):E941–E948
190. Piraccini BM, Alessandrini A (2014) Androgenetic alopecia. G Ital Dermatol Venereol 149(1):15–24
191. Kuka G et al (2020) Cell enriched autologous fat grafts to follicular niche improves hair regrowth in early androgenetic alopecia. Aesthet Surg J 40(6):NP328–NP339
192. Tak YJ et al (2020) A randomized, double-blind, vehicle-controlled clinical study of hair regeneration using adipose-derived stem cell constituent extract in androgenetic alopecia. Stem Cells Transl Med 9(8):839–849
193. Pratt CH et al (2017) Alopecia areata. Nat Rev Dis Primers 3:17011
194. Sobolewski P et al (2019) Systemic sclerosis—multidisciplinary disease: clinical features and treatment. Reumatologia 57(4):221–233
195. Varga J, Marangoni RG (2017) Systemic sclerosis in 2016: dermal white adipose tissue implicated in SSc pathogenesis. Nat Rev Rheumatol 13(2):71–72
196. Mastrogiannaki M et al (2016) beta-Catenin Stabilization in Skin Fibroblasts Causes Fibrotic Lesions by Preventing Adipocyte Differentiation of the Reticular Dermis. J Invest Dermatol 136(6):1130–1142

197. Matsumura H et al (2001) Cones of skin occur where hypertrophic scar occurs. Wound Repair Regen 9(4):269–277
198. Kosacka J et al (2015) Autophagy in adipose tissue of patients with obesity and type 2 diabetes. Mol Cell Endocrinol 409:21–32
199. Spinella-Jaegle S et al (2001) Sonic hedgehog increases the commitment of pluripotent mesenchymal cells into the osteoblastic lineage and abolishes adipocytic differentiation. J Cell Sci 114(Pt 11):2085–2094
200. Kasza I et al (2014) Syndecan-1 is required to maintain intradermal fat and prevent cold stress. PLoS Genet 10(8):e1004514
201. Khan T et al (2009) Metabolic dysregulation and adipose tissue fibrosis: role of collagen VI. Mol Cell Biol 29(6):1575–1591

Chapter 7
Lymphatic Vasculature and Hair Follicle Regeneration

Anna Cazzola and Mirna Perez-Moreno

Abstract *Introduction*: When fluids build up in the skin, it is the function of lymphatic vessels (LV) that allows their proper drainage. LV exert vital roles in macromolecules transport, immune cells' trafficking, and immune response regulation. Recent findings have exposed a novel role for LV in skin regeneration and hair follicle (HF) growth through their connections with hair follicle stem cells (HFSC) and the dermal papilla (DP). This chapter will introduce the readers to the LV organization and function, to later define their distribution in skin and association with HFSC as new components of the HFSC niche. The emerging models of the HFSC-LV crosstalk in the modulation of HF growth will be discussed as well as their potential implications in regenerative therapies and disease. *Methods*: Revision of peer-reviewed published literature using different databases including Pubmed. *Results*: HFSC recruit LV to their niche, and, in turn, LV surround the HF bulge and interconnect neighboring HF across the tissue. Interestingly, the HFSC regulation of LV seems to be dependent on the HF stage. Changes in LV flow and tissue drainage associate with HFSC quiescent and active stages, along with LV caliber and interconnection of neighboring HF rearrangements. Gene expression changes in both HFSC and LV associate with changes in genes involved in LV reorganization, and thorough analyses exposed the contribution of switches of HFSC-secreted angiopoietin-like proteins. Other findings also revealed a DP-dependent modulation of LV and anagen HF growth. *Conclusions*: LV are far from being passive channels and show a complex role in different tissues. Although their activity in the modulation of HF regeneration and cycle in the skin has been recently revealed, many questions remain unanswered. Therefore, future studies in both skin homeostasis and disease will be critical for the further characterization of the LV-HFSC crosstalk and identifying potential therapies for a wide range of pathological conditions.

A. Cazzola (✉) · M. Perez-Moreno
Section of Cell Biology and Physiology, Department of Biology, University of Copenhagen, 2100 Copenhagen, Denmark
e-mail: anna.cazzola@bio.ku.dk

M. Perez-Moreno
e-mail: mirna.pmoreno@bio.ku.dk

Keywords Hair follicle · Lymphatic vessels · Lymphatic endothelial cells · Hair follicle regeneration · Hair follicle stem cells · Hair follicle cycle

Summary

Our understanding of the lymphatic vessels (LV) function has remarkably improved in recent years. Indeed, it has been shown that LV are far from being passive channels, and play complex roles in different tissues, in both health and disease. In the skin, LV are closely associated with the hair follicles (HF). Lymphatic capillaries are critical hair follicle stem cells-niche components that control HF development, cycling, and organization.Although the LV activity in modulating HF regeneration and cycling in the skin has been recently revealed, many questions remain unanswered. Therefore, future studies will be critical to further characterize the signalling events of the LV-HFSC crosstalk in both skin homeostasis and disease and identify potential therapies for a wide range of pathological conditions.

Keypoints

1. Hair follicle stem cells drive the polarized anterior association of lymph vessels to the hair follicle stem cell niche, which in turn interconnect hair follicles across the skin.
2. Lymph vessels are critical regulators of the activation of hair follicle stem cells and hair follicle growth.
3. Lymph vessel changes in caliber and permeability associate to the activation status of hair follicle stem cells at different stages of hair follicle cycle.

Lymphatic Vessels Organization and Functions

The discovery of lymphatic vessels (LV) occurred concurrently with the description of the blood circulation by William Harvey [1], dating to the seventeenth century, when Gaspare Aselli identified veins with a milky appearance in the mesentery of a well-fed dog. He called *them lacteis veins* "milky veins" [2]. Later studies described the lymphatic system's anatomy, and their role started to be elucidated only at the beginning of the twentieth century [3].

LV are traditionally described as an extensive vessel network located throughout the body that drains excess tissue fluid coming from blood capillaries. The lymphatic system has a hierarchically branched structure that consists of blind-ended capillaries, also called initial lymphatic vessels, pre-collecting, collecting vessels, and lymph nodes [4]. Lymphatic transport begins with lymphatic capillaries converging into pre-collecting and collecting vessels that flow to LV afferent vessels [5, 6]. At intervals, the afferent LV connect to small masses of lymph tissue, the so-called lymph nodes, which receive the unfiltered lymph (fluid, macromolecules, metabolites, and immune cells) from tissues and remove pollutants and foreign substances such as infectious microorganisms by filtering out lymph before emptying into efferent vessels [7, 8].

While acting as barriers, filtering noxious elements to prevent their backflow to blood circulation, these nodes, containing immune cells, act as secondary lymphoid organs that initiate and expand immune responses [7, 8].

Lymph nodes also monitor the fluid flow to and from the heart. LV carry lymph at low pressure, and the distinct morphological characteristics of the hierarchical LV branched structure favor its transport [7, 9, 10]. The thin-walled initial lymphatic capillaries consist of a single layer of endothelial cells connected by discontinuous button-like junctions [9, 11]. They also lack a continuous basement membrane and perivascular mural cells. They are highly permeable, facilitating the interstitial fluid uptake and entry of tissue-resident immune cells. Lymphatic capillaries connect to pre-collecting vessels and collecting vessels. The structure of collecting vessels differs from that of lymphatic capillaries. They display zipper-like endothelial junctions that connect more elongated endothelial cells [9, 11], contain valves to control the lymph's unidirectional flow, and are surrounded by pericytes, smooth muscle cells, which provide contractile activity, and adventitia that anchors LV to the surrounding tissue. Overall, the LV structural characteristics allow the largest LV equipped with valves to prevent lymph reflux under the influence of gravity [9].

The skeletal muscles' contraction compresses the LV, increases the pressure inside, and ensures a progressive upward movement of the lymph. The speed at which lymph is pushed through a lymphatic vessel is also influenced by other factors, such as the vessel's size, body location, and shape. Interestingly, the tissues in which LV reside influence LV functions through their exposure to tissue-specific environments. For instance, in the small intestine, lymphatic capillaries known as lacteals carry out dietary fats' absorption [7, 12].

Nowadays, the lymphatic vasculature is still intensively studied, and new insights into their role in other organs [13, 14] and stem cell niches are emerging. Indeed, recent discoveries have shed light on the role of LV in the regulation of hair follicle (HF) growth and skin regeneration [15–17], as we will describe in a subsequent section.

Identification of Lymphatic Vessels Markers: Insights into Their Origins and Functional Regulation

After years of scientific advances, the study and understanding of the lymphatic system have been revolutionized by discovering several proteins expressed in lymphatic endothelial cells (LEC). These proteins serve as major LV markers, and using them in combination improves the truthfulness of detection, allowing the specific discrimination of LV from blood vessels or other cell types. These markers include (1) the Prospero-related homeobox 1 (PROX1) protein, (2) the lymphatic vessel endothelial hyaluronan receptor 1 (LYVE1), and (3) the transmembrane O-glycoprotein podoplanin (PDPN).

PROX1 is a transcription factor essential for the specification and maintenance of the LEC identity [18]. It induces, among others, the expression of the vascular endothelial growth factor receptor 3 (VEGFR3), stimulating the exit of LEC from cardinal veins during lymphangiogenesis and the migration of LEC towards the VEGFR3 ligands [19–22]. Although PROX1 function is not limited to lymphatics [23–25], it is instrumental in discriminating LV from blood vessels.

LYVE1 is a homolog of the CD44 receptor. It is involved in the transport of hyaluronan from the tissues to the lymph and has been described as restricted to lymphatics in several normal tissues [26, 27], but it can also be expressed by certain macrophages subsets [28–30]. LYVE1 and hyaluronan interactions might stimulate leukocyte migration through LV and tumor cell adhesion to LV. It is mainly expressed by lymphatic capillaries rather than collecting vessels and does not harbor a specific developmental or regulatory role [31, 32].

PDPN is a transmembrane glycoprotein expressed in multiple tissues [33]. It has critical roles during development and adulthood, as shown by an abnormal lymphatic vessel formation in podoplanin-deficient mice [34, 35].

LV can also be detected by the expression of the VEGF-receptor VEGFR3 and neuropilin-2 (NRP2) [36, 37], although they are markers of blood vasculature as well. The VEGF family regulates the growth of both lymphatic and blood vessels. Whereas all VEGF family members seem to influence blood endothelial cell proliferation and migration, VEGF-A, -C, -D stimulates LEC growth, being the VEGFR3/VEGF-C the main signaling axis that drives lymphangiogenesis. VEGFR3 is mainly expressed in LEC but has been detected in blood vessels during angiogenesis and tumor-associated vessels [38–40]. NRP2, a receptor for Class II semaphorins and VEGFR co-receptor, is a LV marker that is observed to a lesser extent in veins [41]. It binds VEGF-C and is involved in lymphatic sprouting [37, 42].

Most of the described LEC markers harbor a critical role in lymphatic vasculature development. Although the lymphatic vasculature displays peculiar traits that ensure its physiological functions, it is closely related to the blood vasculature. They show some common features regarding markers, growth factor sensitivity, cell proliferation/new vessel formation, structure, and also embryonic origin.

Indeed, the embryonic origin of LV has been a matter of debate for over one century [43, 44]. In 1902, Sabin proposed that mammalian LV originate from venous endothelial cells during early development [45]. A few years later, Huntington and McClure postulated that LV derive from mesenchymal precursors and that the connections with the veins are defined in later stages of development [46]. Recent studies have shed light on this controversy, and it is now clear that both mechanisms may contribute to the formation of the lymphatic vasculature [47, 48]. In addition, alternative non-venous precursors seem to contribute to the origin of dermal LV and LV of the mesentery and heart [49–52]. These studies now open several questions, including the extent of LV ontogeny and heterogeneity involvement in the diverse LV functional roles in tissues, such as the skin, which offers a perfect, easily accessible system for dissecting molecular events regulated by LV during morphogenesis, homeostasis, and skin disorders.

Dermal Blood and Lymphatic Vasculature

Dermal blood vessels provide oxygen, nutrients, and dispose waste. They are organized into two horizontal plexuses: the lower and the upper plexus, which connect through a network of extending capillaries [53]. Angiogenesis, blood vessel recruitment, and blood vasculature remodeling have been described to occur at the HF level and influence HF growth [54–56]. Additionally, a permanent vascular structure, known as the upper venule annulus, is located in the upper region of the HF starting from morphogenesis, although its role is still not understood [57].

Dermal LV constitute the main route of transport of immune cells, inflammatory mediators, antigens, and pathogens that enter the body through the skin from the periphery to lymph nodes, where the immune responses are initiated [58]. They are also involved in the reverse transport of cholesterol from peripheral tissues to blood circulation [59, 60]. Dermal LV are organized into two plexuses: a superficial plexus located near the blood vessels and a subcutaneous one. The lymphatic capillary network resides in the upper dermis; on the contrary, larger collecting LV locate in the lower dermis and the uppermost area of the subcutaneous tissue [3]. Branches of lymphatic capillaries have been described to extend to the HF dermal papilla (DP), a specialized mesenchymal condensate associated with the HF [61–64], and drain into collecting vessels to the DP base [65]. A more detailed organization of LV in the dermis (Fig. 7.1), specifically in the HF vicinity, has been recently elucidated together with the identification of a novel role of LV in the coordination of the HF cycling [15–17], which will be discussed in the following sections.

Novel Connections Between Hair Follicles and Lymphatic Vessels

Mature HF are mini organs structurally formed by concentric, distinct, epithelial layers [66]. HF undergo a unique cyclic regeneration as multicellular systems, where the HF cycle through phases of growth (anagen), degeneration (catagen), and relative quiescence (telogen) [67], over periods of weeks, months, or years, depending on the particular mammalian species [68–70]. HF are formed by an upper, permanent (non-cycling) region where the infundibulum and the bulge reside [71, 72], and a lower, regenerating (cycling) region [73]. Stem cells orchestrate HF's regenerative ability, and HF stem cells (HFSC) with distinct characteristics and potency are located in discrete HF anatomical regions [74]. Among them, bulge HFSC are endowed with the capability to regenerate the epidermis and the pilosebaceous units upon transplantation [75] and in response to skin damage [76]. HFSC give rise to the fast-cycling progenitors localized in the secondary hair germ, a cluster of cells at the bulge's base [77, 78].

Since the discovery of bulge HFSC [71] and their genetic characteristics [72, 75], several findings have provided insight into the regulation of HFSC quiescence

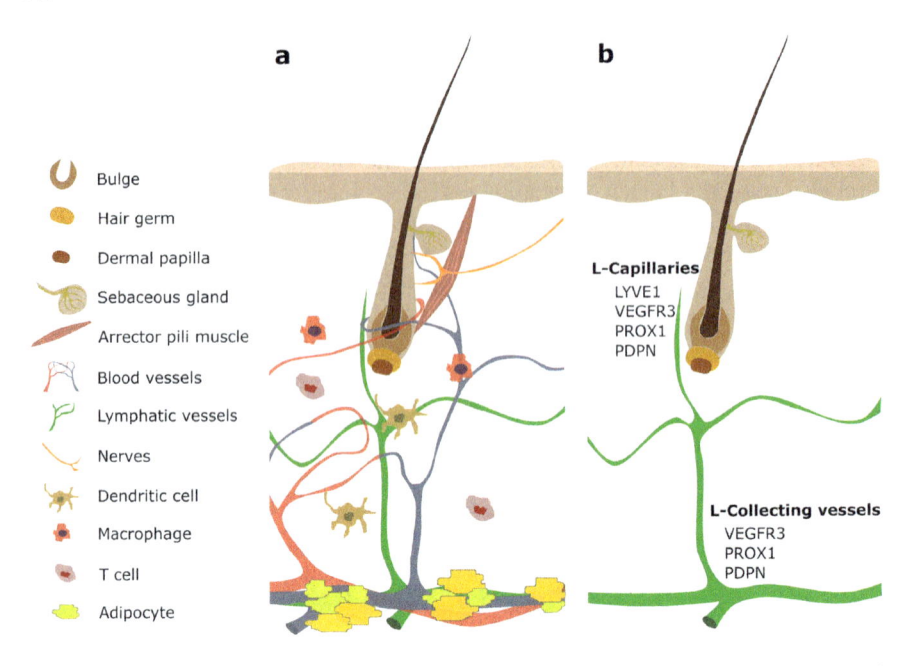

Fig. 7.1 Hair follicle and its niche. (**a**) HFSC interact with diverse cellular elements in their surrounding microenvironment, including the dermal papilla, the dermal sheath, adipocytes, immune cells, sebaceous glands, peripheral nerves, and blood and lymphatic vessels. This crosstalk influences HFSC quiescence, activation, and skin regeneration. (**b**) Telogen HF associated with lymphatic vessels. Lymphatic (L) capillaries and collecting vessels' markers are indicated

and activation by intrinsic factors [79] and the coordinated cellular interactions with their microenvironment [80], where direct, paracrine, and endocrine signaling interactions along with the microenvironment's biophysical characteristics dictate the HFSC niche organization and HFSC activation during tissue regeneration and repair.

At present, the crosstalk between HFSC and diverse cellular elements in their surrounding microenvironment, and the impact on HFSC quiescence, activation, and skin regeneration, is a burgeoning research area. In addition to the HFSC-mesenchymal interactions with the DP [61–64], several other cellular components closely or functionally associate to the HFSC niche, including the dermal sheath, adipocytes, immune cells, sebaceous glands, peripheral nerves as well as blood vessels (Fig. 7.1) [54–56, 81–93].

Dermal lymphatic capillaries have been recently identified as novel components of the HFSC niche in mouse skin [15–17] and human skin [16] (Fig. 7.1). Unlike the parallel distribution of the superficial dermal plexus observed in the ear mouse skin, in mouse back skin, and the human scalp, lymphatic capillaries ascend as blind capillaries from the CD34 + HF bulge area to the upper dermis near the epidermis [15, 16]. These lymphatic capillaries, but not collecting vessels, distribute in a polarized manner, aligned to the anterior side of the permanent region of the HF—opposite to the sebaceous glands [94, 95] and the arrector pili muscle (APM)—where new stem cells form at the initiation of HF growth [15, 16]. Individual HF units interconnect by lymphatic capillaries at the HFSC bulge level and organize into triads (Fig. 7.2). Each row of HF triads connects with other adjacent rows through capillaries stemming from the triads [15]. The lymphatic capillaries connect downwards to collecting LV that drain vertically to larger vessels in the dermis and subcutaneous tissue [3].

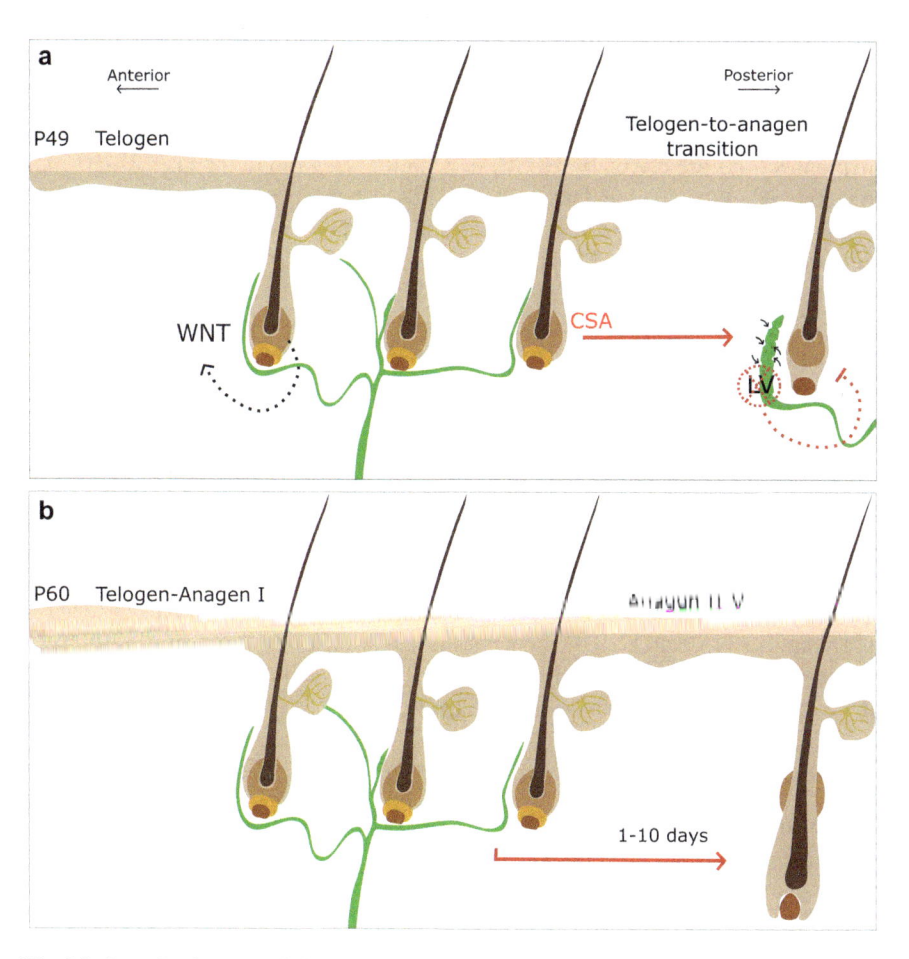

Fig. 7.2 **Lymphatic vessels influence hair follicle regeneration and cycling**. Lymphatic capillaries distribute in a polarized manner, aligned to the HF's anterior side of the permanent region. Individual HF units interconnect by lymphatic capillaries at the level of HFSC bulge and organize into triads. (**a**) HFSC-secreted Wnts are indispensable for the maintenance of the polarized LV association to the HF. At the onset of the second telogen-to-anagen transition, the caliber of lymphatic capillaries associated with the upper region of HF in the proximity of HFSC is temporarily augmented, and the vessels appear fenestrated. The genetic depletion of LV prevents HFSC proliferation and HF growth upon their pharmacological stimulation with cyclosporine A (CSA) in early telogen (P49). (**b**) The genetic ablation of LV during the first and the second telogen (P60) and early anagen leads to precocious HF growth

The LV–HF connections initiate during HF morphogenesis. HF are specified and induced by dermalepithelial interactions, resulting in HF placodes development and the subsequent formation of HF germs and developmental anagen HFs with associated DP [62, 96–98]. The DP is an essential part of the HF and maintains the association with postnatal HFs throughout their lifespan [61–64]. The development and differentiation of HF and the signaling interactions involved in those processes have been extensively addressed in seminal works published elsewhere [61–64, 99–103]. The establishment of LV association to HFs takes place after HF specification, distributing close to HFSCs (marked by Lhx2 + in developing HFs [104] and CD34 + in postnatal HFs [75]) and therefore contributing to the HFSC niche [15, 16]. These interconnections are driven by HFSC, in a similar scenario to the ones observed in previous studies that have underscored the HFSC capability to regulate their microenvironment [105] and organize in the space and time their association to other cellular structures, including the APM apparatus [106, 107] and the vascular venule annulus [57]. It has been shown that HFSC-secreted Wnts are indispensable for the maintenance of the polarized LV association to the HF, at the opposite site of the sebaceous gland and the APM (Fig. 7.2A) [15]. Whether differentially expressed molecules at the anterior/posterior side of the HF govern the specific distribution of LV for the maintenance of the LV association to the HF has still to be defined.

Given the LV localization in the vicinity of the HF, the lymphatic capillaries may constitute a channel for the diffusion and entrance to the HF epithelium of HF-associated immune cells [108], such as memory and regulatory T cells [89, 91], macrophages [87, 88, 109, 110], precursors of dendritic cells [111] and may facilitate the development of Langerhans cells [112]. Intravital imaging demonstrated the presence of a continuous lymphatic flow across parallel HF rows and changes in the LV caliber [15] and LV-HF connections [16] at different stages of the HF cycle, pointing to the relevance of changes in LV drainage capacity in the regulation of HFSC activity and periodicity of the HF cycle.

Lymphatic Vessels Functional Association with Hair Follicle Regeneration and Cycling

With a focus on the HF vasculature, there is extensive evidence of blood vessels' arrangement and reorganization around HF during the HF cycle [54, 55] and their association with the stem cell niches in the skin and some other tissues [57, 113]. Endothelial blood vessel cells establish a signaling crosstalk with HFSC during HF cycle quiescent and regeneration phases. The perifollicular vascularization is stimulated to sustain the anagen phase's physiological development and enhance hair growth and follicle size [54, 55]. More recently, a horizontal vascular plexus has been described to localize beneath the secondary hair germ in late catagen/telogen, delaying anagen progression [56].

The roles of LV in the regulation of the HF cycle have just been recently explored. Two regulation levels have been described: a physical, functional association of HFSC and LV in vivo [15, 16], and a LV–driven coordination of DP activation through paracrine signals, assessed in culture [17]. This chapter focuses on the HFSC–LV level of regulation. The identification of dermal lymphatics as new elements of the HFSC niche and HF's interconnection across the skin [15–17] opened new avenues to investigate their functional connections during skin regeneration [15, 16].

Studies by Gur-Cohen et al. revealed that the genetic targeting of LV or the VEGFR3 signaling blockade of LV during the second telogen promoted the proliferation of quiescent HFSC. The genetic targeting of LV during the first/second telogen and early anagen also led to precocious HF growth (Fig. 7.2B). Overall, these results indicated that LV signals contribute to HFSC quiescence [16]. A different opposite scenario was observed when the pharmacological stimulation of HFSC proliferation and HF growth with cyclosporine A [114, 115] was prevented by the genetic depletion of LV [15] (Fig. 7.2A). Other findings indicated that VEGF-C administration promoted HF growth, and mice conditionally expressing VEGF-C in epidermal cells exhibited a prolonged anagen. In contrast, the expression of sVEGFR3 accelerated the entry into catagen [17], supporting a functional role of LV in HFSC activation and HF growth.

The potential explanation for those seemly opposite roles in HF growth may reside on the use of different systems (mouse strains, HF cycling timing differences, LV depletion protocols, transgene targeting specificity) which might have led to the different effects on HF growth. However, all studies highlight the functional association of lymphatic vessels with HF regeneration and cycling and the relevance of their spatio-temporal connections.

Lymphatic Vessel Remodeling, Drainage Capacity, and Hair Follicles Regeneration

One intriguing aspect of the LV-HFSC functional connection is the temporary reorganization of LV associated with HFSC activation and HF growth. At the onset of the second telogen/anagen transition, the caliber of lymphatic capillaries associated with the upper portion of HF in the proximity of HFSC is augmented in a transitory manner. These caliber changes were accompanied by signs of LV endothelial cell fenestration (Fig. 7.2A) and distinct gene expression (e.g., cytoskeletal and cell adhesion changes), suggesting that an increased LV flow fosters HFSC activation at this particular HF stage [15]. Moreover, the pharmacological increase of LV flow during the refractory second telogen stage stimulated HF growth [15].

Gur-Cohen et al. observed a transitory dissociation of neighboring HF interconnections by lymphatic capillaries at the level of HFSC during anagen II–III. Dilated lymphatic capillaries associated with individual HF units connected to collecting vessels instead of interconnecting neighboring HF (Fig. 7.3). This LV reorganization

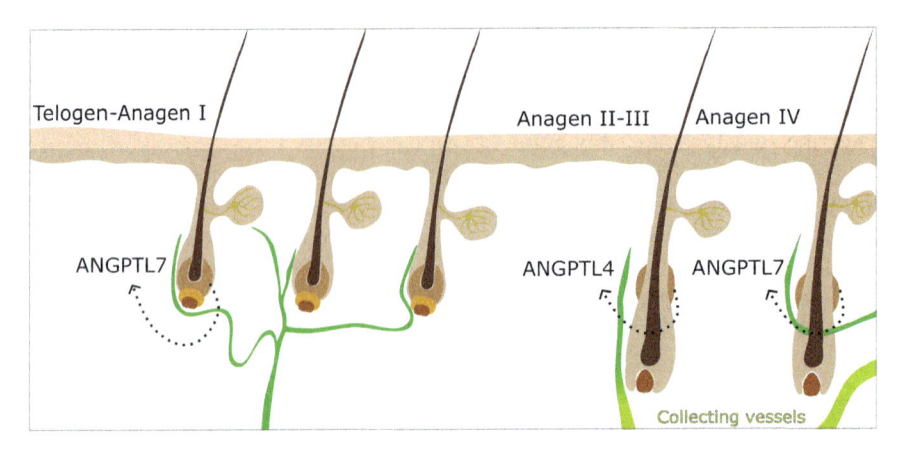

Fig. 7.3 A dynamic HFSC secretome switch regulates lymphatic vessel remodeling and hair follicle growth. Quiescent HFSC express ANGPTL7, whereas proliferating HFSC in anagen II–III show increased ANGPTL4 levels leading to lymphatic remodeling, allowing HF regeneration. During the anagen II–III stages, a transitory dissociation of neighboring HF interconnections mediated by lymphatic capillaries occurs. Dilated lymphatic capillaries reorganize and associate individual HF units to collecting vessels, increasing tissue drainage. At anagen IV, the lymphatic mediated HFSC interconnections are re-established, and HFSC return to quiescence

was associated with signs of reduced vessel permeability and delayed drainage into lymph nodes. These results indicate that a reduction of LV drainage fosters HF down growth at early anagen stages, accompanied by integrin α6 + CD34 + HFSC proliferation. Later, at anagen IV, the HFSC interconnections between HF re-established, HFSC returned to quiescence, and LV drainage resumed (Fig. 7.3).

The transient lymphatic dilations associated with HF regeneration at different stages of the HF cycle highlighted the tissue drainage capacity and vascular exchange's influence in regulating the HF cycle. These processes may modulate the concentration level of molecular inhibitors or activators at specific HF cycle stages [116]. Also, they can potentially regulate the stiffness and mechanotransduction signals at the HFSC niche [117]. Moreover, LV permeability changes also affect the localization of immune cells, such as T cells and macrophages, which regulate the HF cycle. Foxp3 + T regulatory cells, which traffic through LV [118], localize to HF and regulate HFSC proliferation [89], and perifollicular macrophages contribute to the activation of HFSC [87, 88]. Interestingly, LV caliber can be regulated by macrophages [119] through the expression of Wnt ligands [120], and its increment has been observed at the onset of HFSC activation [15] when also macrophage-derived Wnt ligands are expressed [88].

HFSC Secretome Governance of Lymphatic Vessel Remodeling and Hair Follicle Growth

Gene expression analysis of quiescent HFSC (telogen) and proliferating HFSC (anagen II–III) provided insight into the HFSC molecular signatures potentially involved in the HFSC crosstalk with LV [16]. The angiopoietin-like protein 7 (ANGPTL7) was upregulated in quiescent HFSC compared to the proliferating HFSC. Conversely, ANGPTL4 levels remained low during HFSC quiescence and increased in proliferating HFSC (Fig. 7.3). ANGPTL proteins share with the vascular regulators angiopoietins structural similarities and angiogenetic functions. They have recently emerged as regulators of LV remodeling [16], although the downstream mechanisms associated with LV regulation have still to be described.

Besides being involved in angiogenesis, ANGPTL7 is a target gene of the Wnt/βcatenin signaling pathway and plays a role in the hematopoietic stem cell repopulation and glaucoma pathogenesis [121]. ANGPTL4 also has disparate functions ranging from energy homeostasis, redox regulation, lipid, and glucose metabolism to inflammation, wound healing, cell differentiation, oncogenesis, angiogenesis, and vascular permeability [121, 122].

Assessing the involvement of the ANGPTL7/4 switch in HFSC activation and LV dynamic remodeling revealed that while the inducible ANGPTL7 expression in proliferating HFSC prevented LV dissociation and induced the asynchronous entry of HF growth, the inducible expression of ANGPTL4 in resting HFSC led to LV precocious dissociation from the HFSC niche, HFSC proliferation and HF growth [16] (Fig. 7.3). Thus, through the ability of HFSC to coordinate and arrange the microenvironment in which they reside, HFSC affected the LV topology and function. In turn, the dynamic lymphatic capillaries' rearrangement modulates HFSC function. LV express putative ANGPTL7 and ANGPTL4 receptors, and future research will shed light on the downstream signals governing their dynamic remodeling and the LV secreted signals that in turn govern HFSC quiescence and proliferation. It has also been shown that HFSC Wnt ligands favor the polarized association of LV to individual HF at the level of HF bulge [15] (Fig. 7.2A), and the LV transcriptome revealed an upregulation of Wnt inhibitors at HFSC quiescence stages [16]. Thus, these LV Wnt inhibitors might further sustain neighboring HF interconnections while avoiding the individual HF drainage into larger vessels fostering HFSC proliferation and HF growth, as seen in early anagen II–III stages [16].

Potential Implications of LV in Hair Follicle Regeneration and in Diseases

While considerable progress in understanding the role of the lymphatic vasculature in the skin dermis has been made, several questions are now opened in this new exciting field associated with HF regeneration and growth. Further research is needed to understand the molecular basis of the polarized LV-HFSC anchoring, the LV's relevance regulating the molecular concentration of HFSC proliferation inhibitors or activators, including paracrine and endocrine signals. Also, the LV roles in the perifollicular recruitment of immunocytes to the HFSC niche and immune responses are relevant angles to study in this research arena, as mentioned above. The dynamic changes in LV flow regulate all these aspects. Whether quantitative changes in LV flow dynamics exist at defined stages of every phase of the HF cycle may provide a basis to the potential design of regenerative approaches. Damage of the lymphatic system and thus drainage impairment might be responsible for wound-healing defects, hair loss, and inflammation-associated diseases. If so, the reactivation of lymphoangiogenesis and the modulation of lymphatic vasculature flow, by means of lymphoangiogenic factors administration [17] might represent a potential treatment for such conditions.

These future studies also have the potential to shed light on the involvement of LV in other diseases. Defective lymph flow causes hereditary and acquired disorders such as primary and secondary lymphedema [10]. Recent advances have highlighted the relationship between morphological and functional defects in the lymphatic vasculature and several disorders that can have significant implications for the organism, such as obesity, cardiovascular and neurodegenerative disease, atherosclerosis, neurological disorders, glaucoma, and Crohn disease [10]. LV have also been recognized for a long time as the main route of metastatic cell dissemination [123]. Indeed, tumor cells can actively induce LV remodeling and lymphangiogenesis [124] and invade the pre-existing or newly formed LV, reaching lymph nodes. Although regional lymph nodes can constitute a barrier for further cancer cell spread, recent studies indicate that at least some distal metastases are formed by cancer cells that have first transited to them [125].

Moreover, mouse models have been used to demonstrate that tumor cells, including melanoma metastasis, enter blood circulation at the lymph nodes' level to colonize other organs [126, 127]. LV not only constitute a passive route of cancer cell diffusion, but they contribute actively to tumor metastasis and support tumor progression. LV form metastatic cancer cell niches and regulate tumor immunity, influencing the response to anti-tumor immune therapies [7, 124].

The novel interaction between the lymphatic vasculature and the HFSC niche raises whether lymphatic capillaries associate with other stem cell niches in health and disease, including cancer. Indeed, LV contribution to tumor maintenance and in governing cancer stem cell features and behavior has been so far unexplored.

Acknowledgements We thank Laura Mark Jensen for her comments and insight, and all other members of the Perez-Moreno lab, past and present, as well as our many colleagues in the field, for their contributions to this fascinating area of research. We apologize for not discussing many other excellent papers in the field to limit the scope to specific LV connections with HF growth and space limitations. The Perez-Moreno lab appreciates the support of the following foundations: Danish Cancer Foundation (A13956), Novo Nordisk Foundation (28028), NEYE Foundation, Toyota Foundation, Candys Foundation, and Tømmerhandler Vilhelm Bangs Foundation.

References

1. Bolli R (2019) William Harvey and the discovery of the circulation of the blood. Circ Res 124(8):1169–1171
2. Gans H (1962) On the discovery of the lymphatic circulation. Angiology 13:530–536
3. Skobe M, Detmar M (2000) Structure, function, and molecular control of the skin lymphatic system. J Investig Dermatol Symp Proc 5(1):14–19
4. Alitalo K (2011) The lymphatic vasculature in disease. Nat Med 17(11):1371–1380
5. Schineis P, Runge P, Halin C (2019) Cellular traffic through afferent lymphatic vessels. Vascul Pharmacol 112:31–41
6. Steele MM, Lund AW (2021) Afferent lymphatic transport and peripheral tissue immunity. J Immunol 206(2):264–272
7. Petrova TV, Koh GY (2020) Biological functions of lymphatic vessels. Science 369(6500):1–11
8. Jalkanen S, Salmi M (2020) Lymphatic endothelial cells of the lymph node. Nat Rev Immunol 20(9):566–578
9. Bazigou E, Makinen T (2013) Flow control in our vessels: vascular valves make sure there is no way back. Cell Mol Life Sci 70(6):1055–1066
10. Oliver G et al (2020) The lymphatic vasculature in the 21(st) century: novel functional roles in homeostasis and disease. Cell 182(2):270–296
11. Baluk P et al (2007) Functionally specialized junctions between endothelial cells of lymphatic vessels. J Exp Med 204(10):2349–2362
12. Bernier-Latmani J, Petrova TV (2017) Intestinal lymphatic vasculature: structure, mechanisms and functions. Nat Rev Gastroenterol Hepatol 14(9):510–526
13. Louveau A et al (2015) Structural and functional features of central nervous system lymphatic vessels. Nature 523(7560):337–341
14. Aspelund A et al (2015) A dural lymphatic vascular system that drains brain interstitial fluid and macromolecules. J Exp Med 212(7):991–999
15. Peña Jiménez D et al (2019) Lymphatic vessels interact dynamically with the hair follicle stem cell niche during skin regeneration in vivo. EMBO J 38(19):e101688
16. Gur-Cohen S et al (2019) Stem cell-driven lymphatic remodeling coordinates tissue regeneration. Science 366(6470):1218–1225
17. Yoon SY et al (2019) An important role of cutaneous lymphatic vessels in coordinating and promoting anagen hair follicle growth. PLoS One 14(7):e0220341
18. Oliver G and Harvey N (2002) A stepwise model of the development of lymphatic vasculature. Ann N Y Acad Sci 979:159–65; discussion 188–96
19. Johnson NC et al (2008) Lymphatic endothelial cell identity is reversible and its maintenance requires Prox1 activity. Genes Dev 22(23):3282–3291
20. Petrova TV et al (2002) Lymphatic endothelial reprogramming of vascular endothelial cells by the Prox-1 homeobox transcription factor. EMBO J 21(17):4593–4599
21. Wigle JT et al (2002) An essential role for Prox1 in the induction of the lymphatic endothelial cell phenotype. EMBO J 21(7):1505–1513
22. Yang Y et al (2012) Lymphatic endothelial progenitors bud from the cardinal vein and intersomitic vessels in mammalian embryos. Blood 120(11):2340–2348
23. Dyer MA et al (2003) Prox1 function controls progenitor cell proliferation and horizontal cell genesis in the mammalian retina. Nat Genet 34(1):53–58
24. Risebro CA et al (2009) Prox1 maintains muscle structure and growth in the developing heart. Dev 136(3):495–505
25. Sosa-Pineda B, Wigle JT, Oliver G (2000) Hepatocyte migration during liver development requires Prox1. Nat Genet 25(3):254–255
26. Banerji S et al (1999) LYVE-1, a new homologue of the CD44 glycoprotein, is a lymph-specific receptor for hyaluronan. J Cell Biol 144(4):789–801
27. Prevo R et al (2001) Mouse LYVE-1 is an endocytic receptor for hyaluronan in lymphatic endothelium. J Biol Chem 276(22):19420–19430

28. Cho CH et al (2007) Angiogenic role of LYVE-1-positive macrophages in adipose tissue. Circ Res 100(4):e47-57
29. Lim HY et al (2018) Hyaluronan receptor LYVE-1-expressing macrophages maintain arterial tone through hyaluronan-mediated regulation of smooth muscle cell collagen. Immunity 49(6):1191
30. Ensan S et al (2016) Self-renewing resident arterial macrophages arise from embryonic CX3CR1(+) precursors and circulating monocytes immediately after birth. Nat Immunol 17(2):159–168
31. Makinen T et al (2005) PDZ interaction site in ephrinB2 is required for the remodeling of lymphatic vasculature. Genes Dev 19(3):397–410
32. Gale NW et al (2007) Normal lymphatic development and function in mice deficient for the lymphatic hyaluronan receptor LYVE-1. Mol Cell Biol 27(2):595–604
33. Breiteneder-Geleff S et al (1999) Angiosarcomas express mixed endothelial phenotypes of blood and lymphatic capillaries—Podoplanin as a specific marker for lymphatic endothelium. Am J Pathol 154(2):385–394
34. Schacht V et al (2003) T1alpha/podoplanin deficiency disrupts normal lymphatic vasculature formation and causes lymphedema. EMBO J 22(14):3546–3556
35. Fu J et al (2008) Endothelial cell O-glycan deficiency causes blood/lymphatic misconnections and consequent fatty liver disease in mice. J Clin Invest 118(11):3725–3737
36. Jussila L et al (1998) Lymphatic endothelium and Kaposi's sarcoma spindle cells detected by antibodies against the vascular endothelial growth factor receptor-3. Cancer Res 58(8):1599–1604
37. Karaman S, Leppanen VM, and Alitalo K (2018) Vascular endothelial growth factor signaling in development and disease. Development 145(14):dev151019
38. Kaipainen A et al (1995) Expression of the Fms-like tyrosine kinase-4 gene becomes restricted to lymphatic endothelium during development. Proc Natl Acad Sci USA 92(8):3566–3570
39. Kubo H et al (2000) Involvement of vascular endothelial growth factor receptor-3 in maintenance of integrity of endothelial cell lining during tumor angiogenesis. Blood 96(2):546–553
40. Clarijs R et al (2002) Induction of vascular endothelial growth factor receptor-3 expression on tumor microvasculature as a new progression marker in human cutaneous melanoma. Can Res 62(23):7059–7065
41. Herzog Y et al (2001) Differential expression of neuropilin-1 and neuropilin-2 in arteries and veins. Mech Dev 109(1):115–119
42. Xu Y et al (2010) Neuropilin-2 mediates VEGF-C-induced lymphatic sprouting together with VEGFR3. J Cell Biol 188(1):115–130
43. Ulvmar MH, Makinen T (2016) Heterogeneity in the lymphatic vascular system and its origin. Cardiovasc Res 111(4):310–321
44. Ribatti D, Crivellato E (2010) The embryonic origins of lymphatic vessels: an historical review. Br J Haematol 149(5):669–674
45. Sabin FR (1902) On the origin of the lymphatic system from the veins and the development of the lymph hearts and thoracic duct in the pig. Am J Anat 1(3):367–389
46. Huntington GS, McClure CFW (1910) The anatomy and development of the jugular lymph sacs in the domestic cat (Felis domestica). Am J Anat 10(2):177-U93
47. Srinivasan RS et al (2007) Lineage tracing demonstrates the venous origin of the mammalian lymphatic vasculature. Genes Dev 21(19):2422–32
48. Wilting J et al (2006) Dual origin of avian lymphatics. Dev Biol 292(1):165–73
49. Martinez-Corral I et al (2015) Nonvenous origin of dermal lymphatic vasculature. Circ Res 116(10):1649–54
50. Stanczuk L et al (2015) cKit lineage hemogenic endothelium-derived cells contribute to mesenteric lymphatic vessels. Cell Rep 10(10):1708–1721
51. Klotz L et al (2015) Cardiac lymphatics are heterogeneous in origin and respond to injury. Nature 522(7554):62–7

52. Maruyama K et al (2019) Isl1-expressing non-venous cell lineage contributes to cardiac lymphatic vessel development. Dev Biol 452(2):134–143
53. Braverman IM (1989) Ultrastructure and organization of the cutaneous microvasculature in normal and pathologic states. J Invest Dermatol 93(2 Suppl):2S-9S
54. Mecklenburg L et al (2000) Active hair growth (anagen) is associated with angiogenesis. J Invest Dermatol 114(5):909–16
55. Yano K, Brown LF, Detmar M (2001) Control of hair growth and follicle size by VEGF-mediated angiogenesis. J Clin Invest 107(4):409–17
56. Li KN et al (2019) Skin vasculature and hair follicle cross-talking associated with stem cell activation and tissue homeostasis. Elife 8
57. Xiao Y et al (2013) Perivascular hair follicle stem cells associate with a venule annulus. J Invest Dermatol 133(10):2324–2331
58. Xiong Y et al (2019) CD4 T cell sphingosine 1-phosphate receptor (S1PR)1 and S1PR4 and endothelial S1PR2 regulate afferent lymphatic migration. Sci Immunol 4(33):eaav1263
59. Lim HY et al (2013) Lymphatic vessels are essential for the removal of cholesterol from peripheral tissues by SR-BI-mediated transport of HDL. Cell Metab 17(5):671–84
60. Martel C et al (2013) Lymphatic vasculature mediates macrophage reverse cholesterol transport in mice. J Clin Invest 123(4):1571–9
61. Oliver RF, Jahoda CA (1988) Dermal-epidermal interactions. Clin Dermatol 6(4):74–82
62. Driskell RR et al (2011) Hair follicle dermal papilla cells at a glance. J Cell Sci 124(8):1179–1182
63. Sennett R, Rendl M (2012) Mesenchymal-epithelial interactions during hair follicle morphogenesis and cycling. Semin Cell Dev Biol 23(8):917–27
64. Rompolas P et al (2012) Live imaging of stem cell and progeny behaviour in physiological hair-follicle regeneration. Nature 487(7408):496–9
65. Forbes G (1938) Lymphatics of the skin, with a note on lymphatic watershed areas. J Anat 72:399–410
66. Sperling LC (1991) Hair anatomy for the clinician. J Am Acad Dermatol 25(1 Pt 1):1–17
67. Geyfman M et al (2015) Resting no more: re-defining telogen, the maintenance stage of the hair growth cycle. Biol Rev Camb Philos Soc 90(1):1179–96
68. Stenn KS, Paus R (2001) Controls of hair follicle cycling. Physiol Rev 81(1):449–494
69. Muller-Rover S et al (2001) A comprehensive guide for the accurate classification of murine hair follicles in distinct hair cycle stages. J Invest Dermatol 117(1):3–15
70. Oh JW et al (2016) A Guide to Studying Human Hair Follicle Cycling In Vivo. J Invest Dermatol 136(1):34–44
71. Cotsarelis G, Sun TT, Lavker RM (1990) Label-retaining cells reside in the bulge area of pilosebaceous unit: implications for follicular stem cells, hair cycle, and skin carcinogenesis. Cell 61(7):1329–37
72. Tumbar T et al (2004) Defining the epithelial stem cell niche in skin. Science 303(5656):359–63
73. Schneider MR, Schmidt-Ullrich R, Paus R (2009) The hair follicle as a dynamic miniorgan. Curr Biol 19(3):R132-42
74. Woo WM and Oro AE (2011) SnapShot: hair follicle stem cells. Cell 146(2):334–334 e2
75. Blanpain C et al (2004) Self-renewal, multipotency, and the existence of two cell populations within an epithelial stem cell niche. Cell 118(5):635–48
76. Levy V et al (2005) Distinct stem cell populations regenerate the follicle and interfollicular epidermis. Dev Cell 9(6):855–61
77. Greco V et al (2009) A two-step mechanism for stem cell activation during hair regeneration. Cell Stem Cell 4(2):155–69
78. Panteleyev AA (2018) Functional anatomy of the hair follicle: The Secondary Hair Germ. Exp Dermatol 27(7):701–720
79. Lee SA, Li KN, and Tumbar T (2021) Stem cell-intrinsic mechanisms regulating adult hair follicle homeostasis. Exp Dermatol 30(4):430–447

80. Fuchs E, Tumbar T, Guasch G (2004) Socializing with the neighbors: stem cells and their niche. Cell 116(6):769–78
81. Heitman N et al (2020) Dermal sheath contraction powers stem cell niche relocation during hair cycle regression. Science 367(6474):161–166
82. Brownell I et al (2011) Nerve-derived sonic hedgehog defines a niche for hair follicle stem cells capable of becoming epidermal stem cells. Cell Stem Cell 8(5):552–65
83. Martino PA, Heitman N, and Rendl M (2021) The dermal sheath: an emerging component of the hair follicle stem cell niche. Exp Dermatol 30(4):512–521
84. Oshimori N, Fuchs E (2012) Paracrine TGF-beta signaling counterbalances BMP-mediated repression in hair follicle stem cell activation. Cell Stem Cell 10(1):63–75
85. Festa E et al (2011) Adipocyte lineage cells contribute to the skin stem cell niche to drive hair cycling. Cell 146(5):761–71
86. Plikus MV et al (2008) Cyclic dermal BMP signalling regulates stem cell activation during hair regeneration. Nature 451(7176):340–4
87. Wang ECE et al (2019) A Subset of TREM2(+) dermal macrophages secretes oncostatin M to maintain hair follicle stem cell quiescence and inhibit hair growth. Cell Stem Cell 24(4):654–669 e6
88. Castellana D, Paus R, and Perez-Moreno M (2014) Macrophages contribute to the cyclic activation of adult hair follicle stem cells. PLoS Biol 12(12):e1002002
89. Ali N et al (2017) Regulatory T cells in skin facilitate epithelial stem cell differentiation. Cell 169(6):1119–1129 e11
90. Rahmani W, Sinha S, Biernaskie J (2020) Immune modulation of hair follicle regeneration. NPJ Regen Med 5:9
91. Adachi T et al (2015) Hair follicle-derived IL-7 and IL-15 mediate skin-resident memory T cell homeostasis and lymphoma. Nat Med 21(11):1272–1279
92. Wang ECE, Higgins CA (2020) Immune cell regulation of the hair cycle. Exp Dermatol 29(3):322–333
93. Shwartz Y et al (2020) Cell types promoting goosebumps form a niche to regulate hair follicle stem cells. Cell 182(3):578–593.e19
94. Poblet E, Jimenez F, Ortega F (2004) The contribution of the arrector pili muscle and sebaceous glands to the follicular unit structure. J Am Acad Dermatol 51(2):217–22
95. Song WC et al (2007) A study of the secretion mechanism of the sebaceous gland using three-dimensional reconstruction to examine the morphological relationship between the sebaceous gland and the arrector pili muscle in the follicular unit. Br J Dermatol 157(2):325–330
96. Yang Y et al (2017) Derivation of pluripotent stem cells with in vivo embryonic and extraembryonic potency. Cell 169(2):243–257e25
97. Biggs LC, Mikkola ML (2014) Early inductive events in ectodermal appendage morphogenesis. Semin Cell Dev Biol 25–26:11–21
98. Millar SE (2002) Molecular mechanisms regulating hair follicle development. J Invest Dermatol 118(2):216–25
99. Legue E, Nicolas JF (2005) Hair follicle renewal: organization of stem cells in the matrix and the role of stereotyped lineages and behaviors. Development 132(18):4143–54
100. Sequeira I, Nicolas JF (2012) Redefining the structure of the hair follicle by 3D clonal analysis. Development 139(20):3741–51
101. Langbein L et al (2002) A novel epithelial keratin, hK6irs1, is expressed differentially in all layers of the inner root sheath, including specialized huxley cells (Flugelzellen) of the human hair follicle. J Invest Dermatol 118(5):789–99
102. Kaufman CK et al (2003) GATA-3: an unexpected regulator of cell lineage determination in skin. Genes Dev 17(17):2108–22
103. Mesler AL et al (2017) Hair follicle terminal differentiation is orchestrated by distinct early and late matrix progenitors. Cell Rep 19(4):809–821
104. Rhee H, Polak L, Fuchs E (2006) Lhx2 maintains stem cell character in hair follicles. Science 312(5782):1946–9

105. Fuchs E, Blau HM (2020) Tissue stem cells: architects of their niches. Cell Stem Cell 27(4):532–556
106. Fujiwara H et al (2011) The basement membrane of hair follicle stem cells is a muscle cell niche. Cell 144(4):577–89
107. Cheng CC et al (2018) Hair follicle epidermal stem cells define a niche for tactile sensation. Elife 7
108. Christoph T et al (2000) The human hair follicle immune system: cellular composition and immune privilege. Br J Dermatol 142(5):862–73
109. Suzuki S et al (1998) Localization of rat FGF-5 protein in skin macrophage-like cells and FGF-5S protein in hair follicle: possible involvement of two Fgf-5 gene products in hair growth cycle regulation. J Invest Dermatol 111(6):963–72
110. Hardman JA et al (2019) Human perifollicular macrophages undergo apoptosis, express wnt ligands, and switch their polarization during catagen. J Invest Dermatol 139(12):2543–2546e9
111. Nagao K et al (2012) Stress-induced production of chemokines by hair follicles regulates the trafficking of dendritic cells in skin. Nat Immunol 13(8):744–52
112. Wang Y et al (2012) IL-34 is a tissue-restricted ligand of CSF1R required for the development of Langerhans cells and microglia. Nat Immunol 13(8):753–60
113. Rafii S, Butler JM, Ding BS (2016) Angiocrine functions of organ-specific endothelial cells. Nature 529(7586):316–25
114. Paus R, Stenn KS, Link RE (1989) The induction of anagen hair growth in telogen mouse skin by cyclosporine A administration. Lab Invest 60(3):365–9
115. Horsley V et al (2008) NFATc1 balances quiescence and proliferation of skin stem cells. Cell 132(2):299–310
116. Widelitz R, Chuong CM (2016) Quorum sensing and other collective regenerative behavior in organ populations. Curr Opin Genet Dev 40:138–143
117. Lane SW, Williams DA, Watt FM (2014) Modulating the stem cell niche for tissue regeneration. Nat Biotechnol 32(8):795–803
118. Hunter MC, Teijeira A, Halin C (2016) T Cell Trafficking through Lymphatic Vessels. Front Immunol 7:613
119. Gordon EJ et al (2010) Macrophages define dermal lymphatic vessel calibre during development by regulating lymphatic endothelial cell proliferation. Development 137(22):3899–910
120. Muley A et al (2017) Myeloid Wnt ligands are required for normal development of dermal lymphatic vasculature. PLoS One 12(8):e0181549
121. Carbone C et al (2018) Angiopoietin-like proteins in angiogenesis, inflammation and cancer. Int J Mol Sci 19(2)
122. Zhu P et al (2012) Angiopoietin-like 4: a decade of research. Biosci Rep 32(3):211–9
123. Skobe M et al (2001) Induction of tumor lymphangiogenesis by VEGF-C promotes breast cancer metastasis. Nat Med 7(2):192–8
124. Ma Q, Dieterich LC, Detmar M (2018) Multiple roles of lymphatic vessels in tumor progression. Curr Opin Immunol 53:7–12
125. Naxerova K et al (2017) Origins of lymphatic and distant metastases in human colorectal cancer. Science 357(6346):55–60
126. Pereira ER et al (2018) Lymph node metastases can invade local blood vessels, exit the node, and colonize distant organs in mice. Science 359(6382):1403–1407
127. Brown M et al (2018) Lymph node blood vessels provide exit routes for metastatic tumor cell dissemination in mice. Science 359(6382):1408–1411

Part III
Therapeutic Strategies for Hair Follicle Augmentation

Chapter 8
In Vitro and Ex Vivo Hair Follicle Models to Explore Therapeutic Options for Hair Regeneration

Marta Bertolini, Ilaria Piccini, and Kevin J. McElwee

Abstract *Introduction*: Hair follicle regeneration and control of growth cycling is a small but growing field of study. Here we considered some of the more common in vitro and ex vivo hair follicle models that are available for examining follicle growth and cycling, epithelial-mesenchymal signaling, stem cell activity, and follicular neogenesis. *Methods*: Cited literature was selected using the Pubmed database and the associated MESH terms, or conference proceedings. *Results*: A variety of in vitro and ex vivo assay models have been developed over the last 35 years. In vitro research started with simple 2D culture of dermal papilla or dermal sheath cells, but this has now progressed to 3D single cell type aggregate cultures that more accurately reflect the gene and protein expression profiles of dermal papilla and dermal sheath cells in vivo. Combining hair follicle mesenchyme and epithelial cells together in 3D "organoid" cultures enables formation of rudimentary proto-hair follicles. More recently, 3D cultured skin equivalents incorporating hair follicle-like structures have been produced from adult cells as well as induced pluripotent stem cells (iPSCs). Ex vivo models are focused on amputated or full-length hair follicles microdissected from scalp skin, follicular units, or whole scalp skin explant culture. *Conclusions*: Each of the culture systems described holds potential for modelling different aspects of hair follicle growth and regeneration. Potentially, several methods may also be used to provide cells or tissue constructs for treating hair loss. The advantages and limitations of each approach are explored in this review.

Keywords Hair follicle · Regeneration · In vitro · Ex vivo · Models · Telogen · Organ culture · Dermal papilla · Cell culture · Organoids · Stem cell

---

M. Bertolini (✉) · I. Piccini · K. J. McElwee
Monasterium Laboratory - Skin and Hair Research Solutions GmbH, Nano-Bioanalytik Zentrum. Mendelstraße 17, 48149 Münster, Germany
e-mail: M.Bertolini@monasteriumlab.com

K. J. McElwee
Centre for Skin Sciences, University of Bradford, Bradford, UK

Department of Dermatology and Skin Science, University of British Columbia, Vancouver, BC, Canada

Abbreviations

2D	Two dimension
3D	Three dimension
DP	Dermal papilla
DS	Dermal sheath
DSC	Dermal sheath cup
HF	Hair follicle
MSC	Mesenchymal stem cell

KeyPoints

(1) Research with in vitro and ex vivo hair follicle models enables comprehensive control over biological, chemical, and physical parameters while avoiding the expense, time, ethical challenges of rodent and human studies.
(2) Ex vivo models provide the most accurate representation of in vivo hair follicles, while in vitro hair follicle organoid constructs and hair follicle bearing skin equivalents offer significant opportunity for studying hair follicle regeneration.
(3) The limit on time in culture for ex vivo hair follicles and skin explants before tissue breakdown, and the inability so far to produce mature human hair follicles with organoid constructs and skin equivalents, represent obstacles to the investigation of hair follicle cycling and regeneration.

Introduction

The Need for In Vitro and Ex Vivo models of Hair Follicles

The formation of hair follicles (HFs), their growth cycling, and their neogenesis / regeneration, requires a complex and still relatively poorly understood sequence of signals within and between epithelium and mesenchyme cells [38, 185, 252, 279, 280]. How these regulators interact with each other, their relative significance, and how these signals determine the development and size of the complex HF structure, are the focus of considerable research using a variety of clinical and laboratory models. Beyond the study of embryonic HF formation and adult HF growth biology, investigation of HF disorders and the development of therapeutic strategies for them has significantly increased in the new millennium [63, 189].

There are many in vivo rodent models of HF formation and cycling that continue to inform our understanding of HFs and their growth dynamics (e.g. [198, 199]). However, while rodent models are readily available for hair research, the differences in their hair growth patterns, follicle functioning, cycling time periodicity, and differences in some signaling mechanisms, limit the application of animal model data to human HFs and disorders. In addition, the ethical implications of using animals in hair research, particularly for cosmetics studies, have made human HF models much more desirable. While there have been a few attempts to develop in silico models relevant to HF growth and HF disorders [6, 53, 202, 241], so far these models are very restricted in their application and too rudimentary for experimental research use.

As the "gold standard", studies directly on humans can provide significant insights into normal HF development and cycling, as well as our understanding of HF disorders and their treatment [23, 188]. However, ethics boundaries prohibit more invasive and experimental hair research procedures on humans. Both animal model and in vivo human studies tend to be expensive and often require a significant amount of time to complete (e.g. [65]). Consequently, in vitro and ex vivo human HF models have been developed over the last 30 plus years. These models are relatively cheap, quick to set up, and allow comprehensive control over biological, physical and chemical parameters. There are now a wide variety of in vitro and ex vivo HF model systems that can be used for research and determining which model to use can be difficult for those new to the subject.

HF Model Selection Based on Investigation Objectives

Culture of HF cells in 2D has been the main assay system for understanding HFs for many years (Fig. 8.1). It is fast, simple to set up and requires only general cell culture skills (Table 8.1). However, 2D cell culture on plastic substrates is a very artificial environment and inevitably the phenotype and function of cells tend to change over time in response to these exogenous signals. Significantly, the 2D environment may not allow the reproduction of cellular responses that have been demonstrated to occur in vivo [8]. Despite these drawbacks, 2D culture studies using early passage primary human HF cells can provide a significant amount of useful information in a relatively short space of time. Common objectives with 2D culture are often focused on evaluating cell responses to products, particularly drugs, as well as more complex biological extracts. Analyses typically involve quantifying cell proliferation and viability, changes to gene and protein expression, and evaluation of cell metabolism. Readouts frequently involve using standard laboratory techniques to compare cell populations cultured under different conditions in parallel such as; quantitative PCR, immunocytochemistry, ELISA analysis of secreted factors, and sometimes flow cytometry.

While HF bulb matrix cells are highly proliferative in vivo, most mature HF keratinocytes of the growing hair fiber and root sheaths are differentiated and non-proliferative [214, 243]. Somewhat similarly, while dermal sheath (DS) cells multiply and supply the dermal papilla (DP) structure [2, 187, 235], mature DP cells rarely proliferate in vivo [229, 272, 287]. Yet despite these observations, DP and HF outer root sheath keratinocytes proliferate in 2D and in many research studies a common

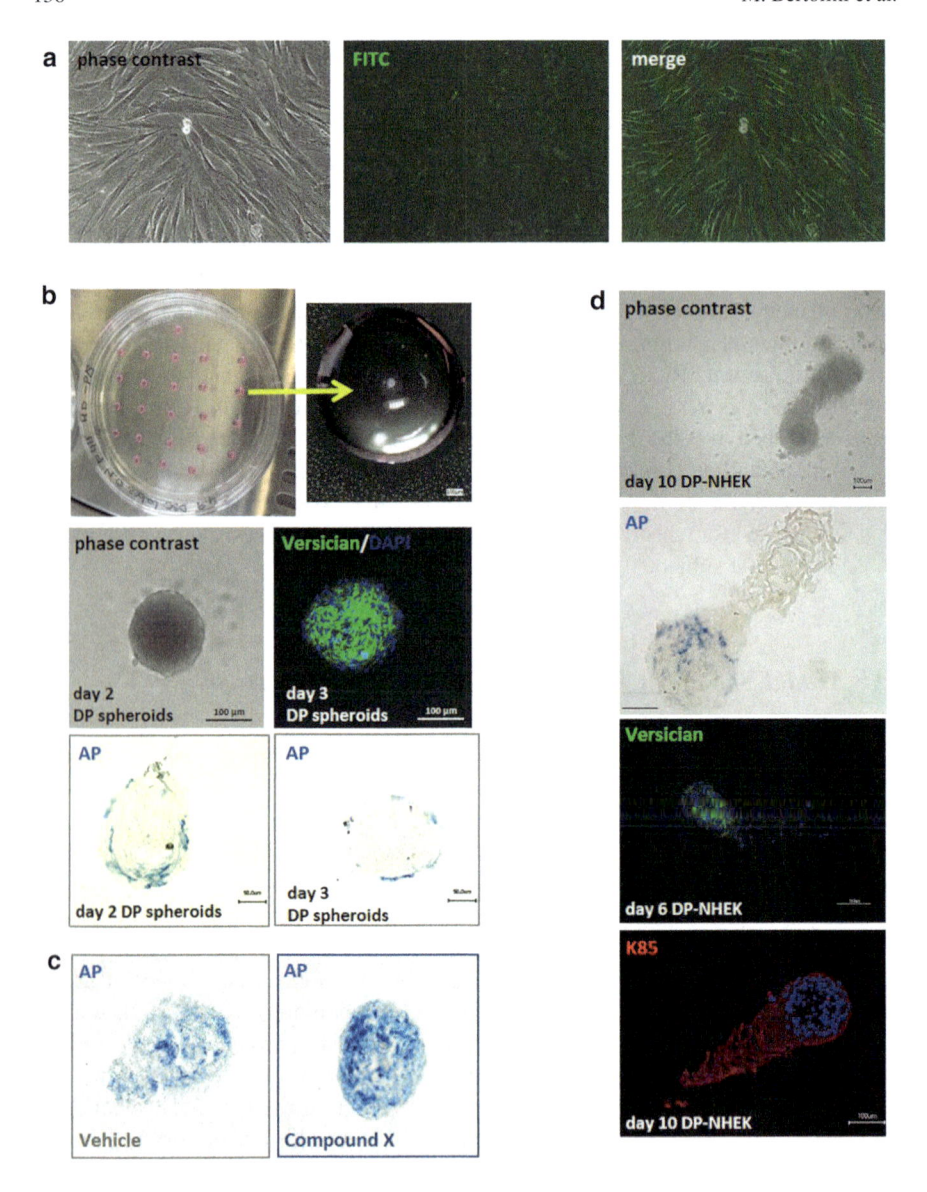

Fig. 8.1 In vitro human hair follicle models. a Representative images showing isolated human DP cells cultured in 2D, fluorescently labelled with a Qtracker 525 cell labeling kit. **b** Representative images showing isolated human DP cells cultured in 3D using the so called "hanging drop" technique, and that the expression of DP inductivity markers (versican and alkaline phosphate activity) is maintained during culture. **c** DP spheroids can be used to screen compounds for their ability to stimulate DP inductivity. **d** Representative images showing HF organoids generated in vitro through the co-culture of one DP spheroid with normal human epidermal keratinocytes. DP inductivity assessed with versican and alkaline phosphatase activity is maintained in DP cells, and hair follicle keratins, such as keratin 85, are produced by the human epidermal keratinocytes, during in vitro culture. AP: Alkaline phosphatase activity, DP: dermal papilla, NHEK: normal human epidermal keratinocytes, K85: keratin 85. Scale bars: 100 μm

8 Human HF In Vitro and Ex Vivo Models

Table 8.1 HF model advantages and disadvantages

Model system	Advantages	Disadvantages
2D cell culture	• Relatively easy to set up using standard culture materials • Technical expertise not required beyond standard cell culture skills • Relatively large numbers of culture parameters can be examined in parallel using cells from a single donor source • DP cells can be obtained from some commercial suppliers, though provenance and quality control are unknown • Readouts can be based on routine laboratory analysis techniques on whole cell population extracts	• Requires tissue dissection skills, flow cytometry, magnetic bead cell separation or similar, to obtain pure populations of cells from donor tissues • Does not accurately reflect the differentiated status and function of cells and their complex interactions in vivo • Rapid loss of HF inductive potential occurs without use of specific culture modifications to promote / maintain HF inductive capacity of DP or lower DS cells beyond early passages (see Table 8.3)
3D single cell type aggregate culture	• Spheroid culture is a superior model compared to cells in 2D culture, if the questions being asked relate to intact DP cell–cell contact behavior [273] • Cells in 3D, and not in contact with substrates, exhibit a more differentiated, non-proliferative state with expression profiles more similar to observations for HF tissue structures in vivo	• The culture system still requires 2D cell culture to produce enough cells prior to 3D aggregate / spheroid formation • The "hanging drop" method for aggregate / spheroid cell culture is arduous, time-consuming, and not easily scalable for large studies • Drug screens require large numbers of cells, which is not always possible to achieve using primary DP cells in aggregate cultures [273]

(continued)

readout for 2D assays with DP or HF keratinocyte cells is to measure cell proliferation. While measurement of cell proliferation in 2D cultures may reflect the role of immature HF cells, it is not an accurate reflection of mature, differentiated HF cell function [81]. In contrast, in 3D assay systems, DP cells are largely non-proliferative and exhibit a gene and protein expression profile more similar to the in vivo situation. HF keratinocytes cultured in 3D also exhibit a phenotype that more accurately reflects their in vivo properties [170]. It has been shown that DP cell type 3D spheroids are more suitable for screening molecules that influence hair growth as compared to 2D

Table 8.1 (continued)

Model system	Advantages	Disadvantages
3D HF organoid culture	• 3D organoids allow for controlled growth conditions with defined culture media and substrates • A single donor cell source can be used to produce numerous organoids for studies involving analysis of several parameters in parallel avoiding inter-individual variability as observed with studies using intact HF culture • Specific modifications of the different cell interactions can be controlled by physical and chemical means as well as by genetic modification of the cells [184]	• Using 3D organoid structures for research removes any modulation influence from the surrounding dermal environment—though this may also be an advantage depending on the study objective • The 3D organoids are relatively simple in construction, with only 2–3 cell types at most, while intact HFs incorporate many cell types each of which may significantly influence HF responses • The epithelial-mesenchymal interactions in vitro do not lead to mature HF formation • Long term maintenance of the HF organoids in a differentiated state can be challenging • The 3D organoids do not exhibit HF cycling in vitro
iPSC HF model culture	• Avoids the use of HF tissue donors as required for other model systems • HF models made from iPSCs avoid ethical issues with the use of embryonic cells • Potentially, the iPSC cell resource is unlimited and could allow for large scale studies for examination of many different parameters in parallel • The organoid structure allows studies on neogenesis and follicle differentiation to be performed	• Due to the prolonged culture period involved in the production process, the model has low throughput • It is challenging to successfully produce the organoid structures using the multistep complex cell culture techniques involved • Questions remain over how closely the iPSC HF constructs reflect in vivo HFs • Concerns have been expressed as to the risk of tumor / teratoma development when using iPSCs [283]

(continued)

Table 8.1 (continued)

Model system	Advantages	Disadvantages
3D cultured skin equivalent	• Allows study of HF formation and development • Potentially an unlimited supply of tissue with the use of iPSCs • Can survive in culture for relatively prolonged periods of time enabling long term study [156] • HF constructs are adjustable in diameter, length, and density using a bioprinting approach [1]	• Skin equivalents produced using iPSCs grow as cyst structures [156] • Does not contain sweat glands or other skin structures [157] • Does not contain all skin cell types such as; endothelial cells, pericytes, and immune cells and consequently does not reflect the complexity of tissues in vivo [156] • Labor intensive involving a complex protocol over a prolonged period of time with considerable variability in results • Human skin equivalents formed from iPSCs produce vellus-like HF structures, while skin equivalents produced from adult human cells exhibit proto-hairs, but terminal HFs are not present
Ex vivo HF culture	• Close equivalence to HFs in vivo as demonstrated in multiple studies (see Table 8.1) • Can examine anagen, catagen and telogen stage HFs using microdissected follicles [76, 79] • Can examine androgen affected intermediate anagen HFs [195] • Can examine HFs with sebaceous glands attached with careful microdissection • Can examine hair growth besides surrogate parameters for HF regeneration • Culture uses serum free medium avoiding the modulatory impact of factors contained in serum and batch to batch variability in serum quality • Can conduct gene silencing studies on the whole HF "organ" using siRNA technology	• Limited donor tissue availability often limits the parameters that can be analyzed in parallel • Inter-individual variability in HF activity can complicate comparative analysis studies • Does not allow for analysis of the influence from surrounding skin tissue and cells on HF regeneration as dissection eliminates neural, vascular and endocrine inputs • Limited study time period of around 7 days for anagen HFs in culture before tissue breakdown occurs and onset of a pseudo-catagen-like state [222]

(continued)

Table 8.1 (continued)

Model system	Advantages	Disadvantages
Ex vivo whole tissue explant culture	• Close equivalence to skin and HFs in vivo • Allows for analysis of the influence from surrounding skin tissue and cells on HF regeneration in a controlled manner • Compounds, drugs, stem cells, or extracellular vesicles can be delivered locally by injection, topically to the skin surface, and systemically by addition to culture media	• Limited donor tissue availability restricts the scope of the studies that can be conducted • Experiments need to be repeated using skin from several donors to investigate inter-individual variations • Limited study time period of around 7 days before tissue breakdown occurs • Not possible to investigate HF morphogenesis

cell cultures [169]. It is possible to use standard analysis techniques on 3D cultures as used with 2D cultures. However, as cells in 3D structures tend to form multiple sub-populations with distinct phenotypes, analysis of 3D assay systems can also involve genomic or proteomic evaluation at the single cell level.

Laboratories with more specialist interests in HF research are now moving towards complex culture systems. The relatively unique regenerative properties exhibited by the epithelial and mesenchymal components of HFs enable their development beyond the basic HF structural model towards tissue "organoids" that exhibit a rudimentary HF phenotype [37, 228]. These organoids are particularly useful for studying HF morphogenesis, but can also be used to evaluate products where responses are expected in both epithelial and mesenchymal components of the HF. For example, drugs with the ability to induce HF development, or stimulate anagen onset and initial hair fiber formation, can be examined with HF organoids. On the downside, it is more complicated to analyze complex structures that incorporate more than one cell type and cell subsets exhibiting various degrees of differentiation. Consequently, readouts from organoid assays tend to be at the cell protein level using immuno-fluorescence labelling that allows analysis of responses in the different cell types, and their observation at different locations, within the organoid structure.

Ex vivo models are widely employed for studying hair growth. This includes the classical Philpott model of microdissected, amputated HFs in serum free culture medium [150, 221–223]. The model has been extensively broadened in the past years to embrace the culture of microdissected full-length HFs, or entire follicular unit extractions (FUE) (e.g. [44]), and human scalp skin organ culture (e.g. [5]). Human HFs and scalp skin explant organ cultures can also be employed to indirectly investigate the potential of novel therapeutic strategies targeting HF regeneration. These models are particularly useful for testing not only drugs and compounds (e.g. [14]), but also cell- and exosome- based modulation of HFs (e.g. [143, 152]). The main advantage of these models is the concomitant investigation of the many different HF cell populations in their own niche and tissue environment; therefore, these models are highly clinically-relevant (Table 8.2). They are also animal product free which is a fundamental prerequisite for cosmetic treatment research as required by many cosmetics companies due to legal restrictions.

However so far, ex vivo models do not easily allow for exploration of HF formation and development. In part, this is due to the ex vivo culturing time limit of a few days before tissue integrity begins to break down. In addition, the investigation of human HF telogen-to-anagen transition has been hampered by the fact that the majority of HFs obtained from human scalp are in anagen VI. Recently, progress has been made in successfully microdissecting human telogen HFs from follicular units harvested with 1 mm micropunches during hair transplant FUE procedures [79] (Fig. 8.2c). Therefore research using cultured telogen HFs has now become a possibility [76]. HFs or scalp skin can also be obtained from balding patients who have naturally higher rates of miniaturized anagen follicles as well as higher rates of telogen HFs (e.g. [47]) providing a particularly clinically relevant model for studies on androgenetic alopecia. Given the low amount of waste tissue available from cosmetic surgeries (i.e. face-lifts or hair transplants), the preferred method of analysis is typically histochemistry or immuno-histochemistry/immuno-fluorescence. Certain specialized laboratories can also perform targeted (qRT-PCR/nanostring), or whole transcriptome, analysis (RNAseq) using the entire tissue, microdissected material, or laser-dissected cell populations from HFs or scalp skin.

Which HF model to select very much depends on the objectives of the research investigation and the level of detailed information required. For example, to examine DP cell ability to induce HF neogenesis, one could begin with simple 2D cell culture and quantitate expression of known markers for HF inductive ability (Table 8.2). However, the data obtained from 2D cultures is somewhat limited in its significance, given the DP cells are in a proliferative phase. Data is more relevant when the DP cells are analysed in a non-proliferative / differentiated state as with 3D spheroids that more accurately reflect the DP cells' in vivo profile. HF organoid models would provide even more direct evidence of HF inductive ability for DP cells prior to the need for an ex vivo or in vivo model. Ex vivo skin organ culture can be used in support of experimenting with the induction potential of HF organoids, namely for the evaluation of HF organoids short-term survival [228]. Ultimately, transferring the DP cells/spheroids/organoids to an in vivo model and determining actual HF induction would be the gold standard.

Studying stem cell activities in HFs is likely best served by using ex vivo intact HFs where bulge epithelial stem cells, melanocyte stem cells, and putative dermal sheath stem cells, are all present in their respective niches. However, here again, HF organoid models could allow for the evaluation of modified stem cells and their ability to interact with other cell types in strictly controlled conditions. To understand cell migration into and from HF structures, or their support by cells external to the HF unit, full skin biopsies, or 3D skin equivalents would be most appropriate to use. Consequently, which in vitro or ex vivo model to use in hair research very much depends on the questions being asked, as well as the availability of tissues and cells, and the technical expertise of laboratory personnel. All these HF models and their applications are discussed in more detail in the following sections.

Table 8.2 Clinical relevance and examples of HF model system applications

Model system	Clinical relevance (scale: 0 to 5 with 5 being most relevant)	Example applications
2D cell culture	1	• Large scale drug response/toxicity screening studies • Evaluation of factors controlling cell differentiation • Evaluation of factors promoting/maintaining DP cell inductive properties
3D single cell type aggregate culture	2	• Drug response/toxicity screening studies on simplified DP-like tissue aggregates • Variability assays comparing cells with distinct genetic/proteomic profiles, aggregated into DP-like structures
3D HF organoid culture	3	• Drug response/toxicity screening studies on simplified HF-like structures • Cell viability studies in complex HF-like structures • Functional in vitro assays examining the effects of cell signals on HF neogenesis • Variability assays evaluating cells in complex HF structures with distinct genetic / proteomic profiles—potentially putting together cells from different epithelial and mesenchymal donors
iPSC HF model culture	3	• In vitro modeling of human HF development and growth cycling • Imaging of cell dynamics over time in complex tissues using labelled cells • In vitro human HF production for subsequent transplantation in vivo

(continued)

Table 8.2 (continued)

Model system	Clinical relevance (scale: 0 to 5 with 5 being most relevant)	Example applications
3D cultured skin equivalent	4	• Studies on follicle formation/neogenesis • Could potentially be used to model skin and HF associated tumors • Patient-specific drug response/toxicity screening where cells from specific patients with known properties are used to produce HFs • In vitro human skin tissue production incorporating HFs for transplantation as a treatment for burns patients and similar
Ex vivo HF culture	5	• Drug response / toxicity screening studies • Cell-based therapy studies by injecting cells in skin organ culture • Response to extracellular vesicles and other complex biological factors • Response to physical factors including light exposure, alterations of oxygen exposure, etc • Knockdown of gene expression at the whole HF "organ" level • Patient-specific modelling of HF associated diseases and disorders

(continued)

2D Cell Culture Models

Microdissected DP and lower DS tissues can induce new HF formation and growth when combined with epithelium [39, 209–211]. With a view to understanding the molecular mechanisms involved in hair growth, and as a potential technique for inducing HFs in vivo, cell cultures from DP and DS tissues would be very useful. Surprisingly, rat DP cells were only first cultured successfully in 1981 [98], and human DP cells in 1984 [191], using standard 2D methodology. HF epithelial cells were also first cultured in 1981 [289]. Subsequently, it was shown that early passage cultured rodent DP cells could induce HF formation after implantation [99, 100].

Table 8.2 (continued)

Model system	Clinical relevance (scale: 0 to 5 with 5 being most relevant)	Example applications
Ex vivo whole tissue explant culture	5	• Drug response/toxicity screening studies • Cell-based therapy studies, including implantation of cell populations • Response to extracellular vesicles and biologic materials such as protein peptides • Patient-specific modelling of HF associated diseases and disorders • Testing of local injection and topical application of factors, as well as systemic administration into the culture media • Drug permeability assays in skin with consideration of HFs as a potential conduit into the skin • Cell viability of DP spheroids or organoids once placed in the skin

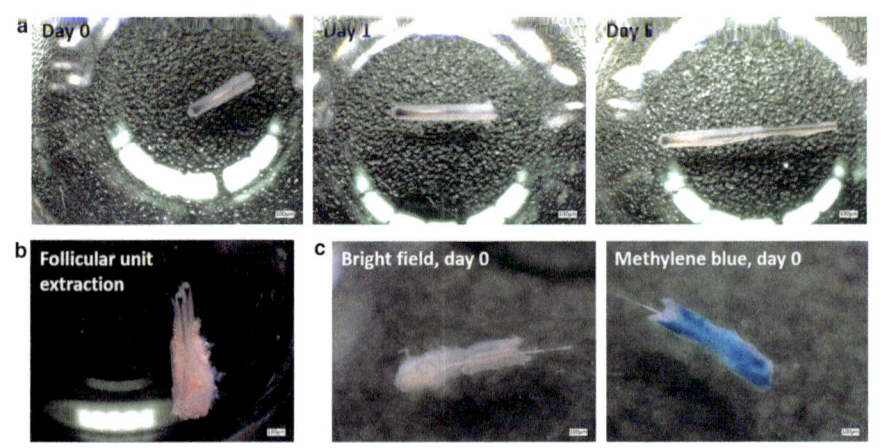

Fig. 8.2 **Human hair follicle organ culture**. a Representative images showing human amputated microdissected HFs immediately after isolation (day 0), and during organ culture at day 1, when usually a treatment is initiated, and at day 5, when newly formed hair shaft and outer root sheath elongation can be appreciated. b Representative images showing a human follicular unit, containing at least two HFs, in culture ex vivo. c Representative images showing a human full-length microdissected telogen HF, immediately after isolation or staining with methylene blue

However DP cells, particularly human DP cells, progressively lose their ability to induce HFs with longer time duration in 2D culture and higher passage number [80, 85, 168].

Studies confirm gradual change in DP cell gene and protein profile with increased passage number [80, 305]. Consequently, many modifications of standard 2D culture systems (Fig. 8.1a) have been suggested to improve DP cell HF inductive properties and/or to prolong their inductive capacity through longer culture duration. Co-culture with epithelial cells or other "feeder" cells is possible, or alternatively using conditioned media from supporting cells (Table 8.3). A few studies have used substrates or alterations to environmental parameters to improve DP cell cultures. Others indicate adding chemokines or drugs that have been shown to support HF growth and maintenance of DP HF inductive ability (Table 8.3). Despite the issues around long term 2D culture of DP and DS cells, this standard culture technique remains a popular platform for research studies on these cells.

A wide range of agents have been evaluated for their impact on DP cell signaling using 2D culture including plant extracts, cell and tissue extracts, chemokines, hormones, drugs, peptides, inorganic salts and metals, and electromagnetic fields [182] (Table 8.3). Cell proliferation, apoptosis and production of factors related to HF biology are generally used as the primary end-point markers in analyzing 2D cell cultures in many studies. As the 2D environment favors cell proliferation over differentiation, some studies progress to investigation of 2D cultured DP HF inductive ability using in vivo models. Notably, studies suggest DP and lower DS "cup" (DSC) cells can be produced by 2D culture that retain HF inductive capacity that could be used to promote hair growth in humans [233, 275].

The trichogenicity of DP and DS derived cells is the main focus of most 2D HF cell culture publications. Relatively few studies examine the properties of HF epithelial cells in 2D culture and their status in HF regeneration. A variety of studies demonstrate human outer root sheath (ORS), bulge, or matrix cells can be cultured successfully [41, 172, 196, 208, 288]. However, it has been shown that hair-inductive activity of cultured keratinocytes is inversely correlated with the time duration of cultivation. The ability of cultured human ORS cells to form HFs, in combination with freshly isolated mouse dermal cells after implantation to nude mice, is progressively reduced with increasing time in 2D culture, due in part to changes in expression of FOXA2 [13], p. 2) Trichogenicity of high-passage rat or human ORS cells can be restored by coculture with rat or human DP cells respectively [12, 26, 171]. Combining cultured ORS cells with DP cells allows for formation of rudimentary HF organoids in vitro [101]. Notably, human interfollicular epidermal keratinocytes with stem-like features also promote HF inductive features in co-culture with DP cells and enable HF formation with co-grafting to nude mice [3]. The data suggest that specific subsets of keratinocyte cells grown in culture are more effective for HF regeneration. However, further research is needed to fully characterize the most relevant cell populations.

Table 8.3 Factors added to 2D cultures to maintain or improve HF DP cell properties

Category	Factors evaluated	References
Growth factors, proteins, etc	Angiogenin	Zhou et al. [310]
	BMP2	Ohyama et al. [207], Rendl et al. [242]
	BMP4	Rendl et al. [242]
	BMP6	Rendl et al. [242]
	BMP7	Rendl et al. [242]
	Cyclophilin A	Kim and Choe [130]
	Erythropoietin	Kang et al. [116]
	FGF2	Goodman and Ledbetter [62], Kiso et al. [138], Ohyama et al. [207], Osada et al. [212]
	Humanin	Kim et al. [126]
	IL-1alpha	Boivin et al. [21]
	Lactoferrin	Huang et al. [88]
	PDGF	González et al. [61], Goodman and Ledbetter [62], Kamp et al. [113], p. 2003, Kiso et al. [138], Yue et al. [303]
	VEGF	Li et al. [167]
	WNT10b	Zhou et al. [311]
Drugs and similar	1-Ascorbic acid 2-phosphate	Kwack et al. [144]
	3 Deoxysappanchalcone compound	Kim et al. [133]
	6-bromoindirubin-39-oxime (BIO)	Ohyama et al. [207], Yamauchi and Kurosaka [293]
	Adenosine	Hwang et al. [93]
	Arachidonic acid	Munkhbayar et al. [197]
	Cilostazol	Choi et al. [35]
	Ciprofloxacin	Kiratipaiboon et al. [137]
	Costunolide	Kim et al. [127]
	Finasteride	Rattanachitthawat et al. [240]
	JAK Inhibitor effect on DP cells and immune privilege restoration	Kim et al. [128]
	Lithium chloride	Kang et al. [120], Leirós et al. [164], Sun et al. [266]
	Minoxidil	Han et al. [66], Kwack et al. [142], Lachgar et al. [148], Li et al. [166]

(continued)

Table 8.3 (continued)

Category	Factors evaluated	References
	Mycophenolic acid	Jeong et al. [103], Ryu et al. [247]
	Norgalanthamine	Yoon et al. [297]
	Retinol	Yoo et al. [296]
	Shikimic acid	Choi et al. [36]
	Sildenafil	Choi et al. [34]
	Sinapic acid	Woo et al. [290]
	Valproic acid	Jo et al. [105], Lee et al. [160]
	Vanillic acid	Kang et al. [117]
Biological extracts	Aconiti Ciliare Tuber	Park et al. [217]
	Bidens pilosa	Hughes et al. [90]
	Chaga mushrooms	Sagayama et al. [248]
	Cinnamomum osmophloeum Kanehira Leaf	Wen et al. [286]
	Ecklonia cava	Bak et al. [11], Shin et al. [257]
	Epigallocatechin-3-gallate (green tea)	Kwon et al. [146], Shin et al. [258]
	Erica multiflora	Kawano et al. [123]
	Fermented fish oil	Kang et al. [122]
	Ginseng	Lee et al. [157], Park et al. [215, 218], Shin et al. [256], Zhang et al. [304]
	Hottuynia cordata	Kim et al. [134]
	Icariin flavanoid	Su et al. [264]
	Ishige sinicola	Kang et al. [118]
	Malva verticillata	Lee et al. [153], Ryu et al. [246]
	Miscanthus sinensis var. purpurascens	Jeong et al. [102]
	Quercitrin	Kim et al. [131]
	Rice bran mineral extract	Kim et al. [135]
	Rumex japonicus Houtt	Lee et al. [154]
	Salvia plebeia	Jin et al. [104]
	Sargassum muticum	Kang et al. [121]
	Trapa japonica Fruit	Nam et al. [200, 201]
	Undariopsis peterseniana	Kang et al. [119]
Incubation parameters	Low-oxygen cell culture conditions	Kanayama et al. [114], Ye et al. [295], Zheng et al. [307]

(continued)

Table 8.3 (continued)

Category	Factors evaluated	References
	Low-frequency electromagnetic fields	Ki et al. [125]
	1,763 MHz radiofrequency-irradiation	Yoon et al. [298]
	Extracellular matrix substrates	Young et al. [301]
	Micro-current electrical stimulation	Hwang et al. [91]
Feeder cells and conditioned medium	Conditioned medium from human umbilical cord blood-derived MSCs	Oh et al. [204]
	Amniotic fluid-derived mesenchymal stem cell-conditioned medium	Park et al. [216]
	Keratinocyte feeder cells	Inamatsu et al. [94]
	Keratinocyte cell conditioned medium	Inamatsu et al. [94], Qiao et al. [233]
	Ovine DP cells co-cultured with human DP cells	Rufaut et al. [245], Sari et al. [251]
Cell and plasma extracts	Platelet-rich plasma (PRP)	Shen et al. [254], Wang et al. [284], Xiao et al. [291]
	MSC-derived extracellular vesicles	Rajendran et al. [236]
	Macrophage-derived extracellular vesicles	Rajendran et al. [238]

3D Single Cell Type Aggregate Culture Models

In vitro models can potentially be used to study molecular mechanisms leading to HF induction, growth and maintenance. A key challenge however, is the failure of DP cells to maintain their HF inductive properties over time in 2D culture [80, 81]. Several approaches have been suggested to overcome this issue including addition of specific factors to culture media to maintain DP cell expression profiles (as above), and promoting cell–cell contact as a way to maintain DP properties. Perhaps the most well-known of these methodologies is described as the culture of 3D "spheroids" of DP cells (Fig. 8.1b). The idea of culturing DP cells in 3D has been around for some time, but has gained popularity with several research laboratories more recently [212].

As cell–cell contact in 3D prevents cell proliferation [82], DP cells must first be cultured in standard 2D cultures to increase cell numbers. When the number of cells is achieved (usually within 3–5 passages), a 3D "hanging drop" culture technique is initiated to enable re-differentiation of the cells [273]. Studies show that by plating cultured DP cells in hanging drops of cell culture media, the cells are forced into contact with each other, partly due to the effects of gravity and the dynamic forces of the cell culture media meniscus surface tension [173]. The absence of contact with culture flasks avoids DP cell differentiation into fibroblasts that may be caused by culture on standard plastic substrates [42, 270]. The result is the formation of dermal spheroids morphologically similar to intact papillae [82]. Analysis of the dermal spheroids shows that forced aggregation of cultured DP cells elicits cell gene and protein profiles more similar to that observed in intact DP [80]. Most notably, transplantation of the 3D spheroids into rodent models induces hair growth [80, 115, 173].

There may be alternatives to using hanging drops to promote 3D aggregates including culturing DP cells on poorly adhesive hydrophilic polyvinyl alcohol (PVA) coated surfaces and similar [86, 89, 300], or by incorporating cells into microcapsules using matrigel [193], alginate [174], or other scaffolds that allow for 3D cell culture and DP cell aggregation [16, 179, 306]. Some variations using hanging drop methodology include using composite spheroids with DP cells and adipose-derived stem cells that may further promote HF inductive ability [87]. Intriguingly, sphere formation of lung mesenchymal cells also promotes some HF inducing capacity [255].

The 3D hanging drop culture system has been used to evaluate compounds for their ability to increase HF induction by the spheroids (Fig. 8.1c) including Flavonoid Silibinin [30], Wnt activator CHIR99021 [299], a stabilised form of β-catenin [308], JAK-STAT signalling pathway inhibitors [68], minoxidil [64], the monoterpenoid compound Loliolide [161], and over-expression of alkaline phosphatase [141]. Typically, compounds are evaluated for their ability to promote factors in the DP spheroids that have previously been shown to be correlated to hair growth, such as versican (Fig. 8.1b), VEGF, IGF1, KGF, and Wnt factors. Alkaline phosphatase expression or activity is often evaluated as an indicator of HF inductive ability of DP and DS cells (Fig. 8.1c) (eg. [187, 230]). Though the functional significance of alkaline phosphatase in not completely understood, its overexpression promotes Wnt signalling in DP spheres [143]. The gold standard of implanting the treated spheroids into an in vivo rodent model to evaluate hair growth induction is also sometimes evaluated [299].

3D HF Organoid Models

Following on from the development of DP cell spheroids as a method of maintaining the HF inductive gene and protein expression signature of DP cells, combining keratinocytes with DP cells could further enable essential epithelial-mesenchymal interactions typical of human HFs (Fig. 8.1d). Development of these composite 3D "organoids" has long been a goal for both academic and industrial laboratories. While one could use ex vivo HF culture, or whole skin explant culture, the limited supply of human tissue and the expense of obtaining it, severely restricts HF research (Table

8.1). Consequently, the formation of proto-HF organoids in culture should significantly facilitate research into HF development. In principle, HF organoid models allow for the evaluation of drugs and other external factors on HF neogenesis and growth, may function as a discovery tool for identifying new target genes and protein products for candidate hair growth-modulatory agents, and could also have potential for generating structures in vitro that could then be transplanted to patients for the treatment of hair loss [75, 184, 232].

Several variations in the method of HF organoid production have been reported. Early techniques involved culturing dermal fibroblasts in a pseudodermis of matrigel or similar basement membrane matrix and then adding DP cells and outer root sheath derived HF keratinocytes into, or on top of, the pseudodermis [74]. By bringing the HF mesenchyme and epithelial cells physically in contact together on a suitable substrate, the keratinocytes form spheroid epithelial cell aggregates. These aggregates exhibit some characteristics typically observed in HF formation, such as keratinocyte expression of cytokeratin 6, while DP cells expressed versican [74]. Similarly, cultured DP cells transferred to ultra-low attachment plates formed DP-like aggregates. Co-cultivation of the DP cell aggregates with keratinocytes obtained from the HF outer root sheath created organoids that underwent stages of HF development and produced hair fiber in some instances [175]. Similar results are obtained when DP spheroids are co-cultured with normal human epidermal keratinocytes (NHEK) (Fig. 8.1d) or hair matrix keratinocytes [228].

More recent protocols have focused on producing 3D combinations of epithelial and mesenchymal cells. This requires lower cell numbers and makes preparation easier and more reproducible [75, 175]. The 3D hanging drop technique has also been used to produce HF epithelial-mesenchymal cell spheroids [112, 232]. These HF germ-like structures exhibit cell signalling patterns typical of HF neogenesis and early growth, including Wnt pathway activation and expression of follicular markers [111]. Further variations on the principle of combining epithelial and mesenchymal cells to produce proto-hair organoids include a microwell "mould" array made of oxygen-permeable silicone and hydrogels to hold the cell composites together [110], cell-encapsulated collagen drops [109], and microwells fabricated from polyethylene glycol diacrylate hydrogels [269]. Most recently, epithelial cells, DP cells, and vascular endothelial cells have been combined into HF germs, with the vascular endothelial cells localizing to the proto-DP, which enabled improved HF formation after transplantation to immunodeficient mice [107]. In a novel variation, DP cell spheroids have been cultured with plucked anagen hair shafts to produce rudimentary organoids for analysis of HF induction activity [230].

These bioengineered HF organoids have been used to evaluate the effects of platelet rich plasma (PRP) [108], hyaluronic acid [111], and Wnt factors [43, 263]. The phenotypic identity of HF-resident epithelial stem cells has also been investigated using recombinations of different cell subsets into organoid structures [268]. One disadvantage of using 3D epithelial-mesenchymal structures, however, is that it removes any modulation influence from the surrounding dermal environment (Table 8.1). Given non-follicular dermal fibroblasts, immune cells, and adipose cells are likely to affect hair growth and cycling, 3D composite structures are still relatively simplistic models in comparison to intact HFs or whole skin explants cultured ex vivo.

Ideally, one would want to form fully mature HFs in culture for research and potentially for transplantation as a regenerated tissue treatment for hair loss. Thus far, composite keratinocyte-mesenchymal HF organoids can form "proto-hairs" in vitro, but fail to achieve full follicle formation unless are implanted into an in vivo model [110, 232, 263]. There are likely two primary challenges to be addressed before composite cell organoids can be successfully used to make mature HFs grow hair in vitro; a much more detailed understanding of the molecular signalling that regulates cell fate and HF growth dynamics is needed, and a better understanding of the tissue architecture and arrangement, along with the ratio and orientation of cells, is required [158].

iPSC HF Models

The apparent loss of HF inductive ability as DP cells are cultured to progressively higher passages is an issue that is challenging to overcome. This limitation typically requires cultured DP cells to be utilised at passage 3–5 according to published reports [80, 85, 168]. While there are in vitro methods to maintain DP cells through higher passages, alternative approaches have been suggested using stem cells. Given stem cells can self-renew indefinitely and generate cells representing any tissue, they could potentially provide an unlimited source of cells for research or therapy [136]. Human embryonic stem cells differentiate into hair-inducing DP-like cells in culture, and express markers similar to those found in adult human DP cells including nestin, versican, and alkaline phosphatase. When transplanted to immunodeficient mice the cells are able to induce HF formation [60]. Somewhat similarly, Weber and colleagues have developed a 3D model with co-culture of human neonatal foreskin keratinocytes and foetal scalp dermal cells that permits their self-organization into hair peg-like structures in droplet cultures [285]. However, the ethical implications of using embryonic stem cells and human fetal cells likely limit the practical use of such an approach.

One alternative may be to use induced pluripotent stem cells (iPSCs). IPSCs have been developed for both keratinocyte and DP components of the HF. In early studies, human iPSCs derived by promoting the expression of 3–4 stem cell factors (Oct4/Pou5f1, Sox2, Klf4, ± Myc) in precursor cells exhibited an epithelial cell profile. When co-cultured with DP cells a follicular keratinocyte profile became apparent and co-transplantation to immunodeficient mice enabled HF formation [281]. IPSCs can be differentiated into epithelial stem cells with expression of the relevant CD200 + /ITGA6 + markers by careful control of epidermal growth factor (EGF), retinoic acid (RA) and BMP signalling using a multi-step cell culture protocol. These cells can be further differentiated towards K15 + trichogenic HF bulge stem cells and ultimately into mature keratinocytes [294].

DP-like cells can also be produced from iPSCs, though the methodology needs further development to produce fully functioning DP structures. In recent studies, iPSCs could be differentiated into induced mesenchymal cells with a bone marrow stromal cell phenotype. An CD271 + /CD90 + cell subset could then be programmed to DP-like cells using retinoic acid and DP cell activating culture medium [213, 282]. Also of note, human iPSC-derived melanocytes could successfully incorporate into HFs and produce melanin and pigmented hair fiber [59, 177, 178]. In principle, all of the key cells required for follicle neogenesis and pigmented hair fiber production can be obtained from iPSCs. As such, it should be possible to bioengineer functional HFs in 3D culture entirely using iPSCs. Studies with mouse iPSCs demonstrate well developed HF structures can be produced [155], though equivalent HF formation has yet to be achieved with human cells [206]. This concept is discussed further in Chap. 6.

3D Cultured Skin Equivalents Incorporating HFs

While 3D skin equivalents have been engineered from constituent cell types and iPSCs they generally fail to incorporate HFs [22, 59, 97]. Only relatively recently have laboratories attempted to reconstruct skin in 3D which includes HF structures. The production of full skin equivalents provides a key advantage over study with isolated intact HFs or HF organoids in that the tissue environment around HFs can also be examined for its role in supporting HF growth.

Mouse iPSC derived embryoid bodies can be developed in culture and transplanted to immunodeficient mice where they successfully form hair bearing skin [267, 274]. Following on from initial studies with mouse pluripotent stem cells [155], Lee and colleagues developed a 3D skin equivalent containing HFs using human iPSCs [156, 157]. These skin equivalents contain more developed vellus-like HF structures that progressively form in vitro from around 70–120 days and can be maintained for around 150 days. However, the skin equivalents develop in vitro as cysts rather than as healthy adult skin [156, 157]. Intriguingly, when transplanted to immunodeficient mice, the cultured iPSC cysts form stratified skin with emergent hair, though hair growth tends to be variable and follicles are disorientated [159]. The observations suggest the in vitro cyst-structure can be converted to more normal skin in response to appropriate signals. Potentially with a better understanding of the signalling events and structural factors that lead to cyst development, it may be possible to control the iPSC skin equivalents and move them towards a more standard representation of skin in vitro [234].

Using adult human cells, rather than embryonic cells or iPSCs, to produce HF bearing skin equivalents presents a greater challenge. Microfabricated plastic molds with HF-shaped projections have been used to print tubular wells into collagen matrix containing dermal fibroblasts. DP cells were then seeded into the collagen microwells with keratinocytes seeded above to produce the basic epithelial-mesenchymal HF structure and inter-follicular tissue [1]. The DP cells exhibited expression profiles consistent with their HF inductive ability while keratinocytes developed differentiated morphology resembling HF keratinocytes. Culture of these constructs for 3 weeks enabled elongated HFs to form along with improved organization of inner and outer root sheaths. Transient transfection of *Lef-1* or *Fli-1* in cultured DP cells further improved HF formation [1]. However, terminal hair production was not observed until the constructs were transplanted to immunodeficient mice.

Vahav and colleagues cultured human DP cell spheroids which were then incorporated into human skin reconstructed from cultured epidermal cells grown on a dermal fibroblast-populated hydrogel, cultured at the air–liquid interface [276]. Though mature follicles did not form in vitro, the spheroids did elicit hair peg-like structures with some degree of inner and outer root sheath differentiation. As such, models produced from constituent cell types suffer from the same problem as HF organoids in that the follicles formed are typically proto-follicles and not fully differentiated [25]. Some factors are clearly missing which would promote full follicle formation during 3D culture of human cells. Whether the factors are locally produced cell signalling cytokines and chemokines, hormones from other locations in the body, extracellular substrates, immune cells or other cells not currently incorporated into engineered skin equivalents, or something else entirely, is still unknown. However, the in vitro culture of these constructs could provide an ideal approach to investigate drugs and other factors that may play a role in human HF development.

Ex Vivo HF Organ Culture

This well established method to investigate HF physiology is used to examine the effects of treatments on HFs in a clinically-relevant ex vivo setting. Detailed methodology for using the assay system is available from other publications [44, 150, 222]. It is based on the isolation of growing anagen VI HFs from scalp skin obtained as waste material from hair transplantation or face-lift surgeries [221–223] (Fig. 8.2a). The HFs are then placed in serum free medium and can be cultured for up to 7 days while maintaining an anagen hair growth state. The HFs continue to produce a hair shaft until 7–10 days when the HFs spontaneously undergo a pseudo-catagen process (Fig. 8.2a). Therefore, this assay can be used to identify strategies for premature catagen promotion or inhibition, and modulation of hair shaft growth rates (Table 8.4). Anagen maintenance ex vivo is evaluated by defining the hair cycle stage of each HF, based on widely accepted macroscopic and microscopic criteria [44, 140, 150], counting proliferative and apoptotic hair matrix keratinocytes, and investigating the expression of hair growth-associated growth factors, such as Wnt proteins [14, 203]. The system classically uses HFs that are amputated (Fig. 8.2a) and hence do not retain the bulge region epithelial stem cells nor the sebaceous gland. However with careful dissection, full length HFs, or HFs with sebaceous glands attached, as well as entire follicular unit extractions (Fig. 8.2b), can be utilised where needed [44, 150, 222].

Table 8.4 Factors evaluated for effects on hair growth using ex vivo HF culture

Category	Factors evaluated	References
Growth factors, proteins, etc	1,25-dihydroxyvitamin D3 effect on HF growth	Harmon and Nevins [70]
	Adenosine promotion of HF growth	Lisztes et al. [176]
	Alpha-melanocyte-stimulating hormone (α-MSH) protective effect against chemotherapy induced catagen	Böhm et al. [20]
	Brain-derived neurotrophic factor (BDNF) promotion of catagen in HFs	Peters et al. [220]
	Beta 1 integrin promotion of HF growth	Kloepper et al. [139]
	Cardiotrophin inhibition of hair growth	Yu et al. [302]
	Prolongation of anagen by circadian clock factor CRY1	Buscone et al. [24]
	Epidermal growth factor (EGF) induction of 'catagen-like' effect	Hoffmann et al. [83], Philpott et al. [223], Philpott and Kealey [224]
	Erythropoietin (EPO) impact on hair growth and pigmentation	Bodó et al. [18, 19], Kang et al. [116]
	Hepatocyte growth factor (HGF) promotion of hair growth	Lee et al. [162], Nicu et al. [203]
	IL-1α, IL-1β, inhibition of HF growth	Harmon and Nevins [71], Hoffmann et al. [83], Philpott et al. [225], Xiong and Harmon [292]
	Insulin-like growth factors IGF-I / IGF-II stimulation of HF growth	Alam et al. [4], Philpott et al. [223, 226]
	Interferon-gamma (IFNγ) promotion of catagen in HFs	Ito et al. [95]
	Macrophage stimulating protein (MSP) promotion of hair growth	McElwee et al. [186]
	Oncostatin M (OSM) inhibition of hair growth	Yu et al. [302]
	Osteopontin-derived peptide inhibition of HF growth	Alam et al. [5]
	P-cadherin silencing induction of HF catagen regression	Samuelov et al. [250]
	P-cadherin silencing in relation to reduced HF melanogenesis	Samuelov et al. [249]
	Substance P promotion of catagen in HFs	Peters et al. [219]
	Transforming growth factors TGFβ1 / TGFβ2 promotion of catagen in HFs	Hoffmann et al. [83], Philpott et al. [223], Soma et al. [261, 262]

(continued)

Table 8.4 (continued)

Category	Factors evaluated	References
	Tumor necrosis factor-alpha (TNFα) inhibition of HF growth	Hoffmann et al. [83], Philpott et al. [225], Soma et al. [261]
Drugs and similar	12-O-tetradecanoyl-phorbol-13-acetate (TPA), protein kinase C (PKC) activator, inhibition of HF growth	Harmon et al. [72]
	All-trans retinoic acid (ATRA, tretinoin) promotion of catagen in HFs	Foitzik et al. [52]
	Androgen promotion of HF growth in women's intermediate facial follicles	Miranda et al. [194]
	β-catenin inhibitor (21H7) inhibition of HF growth	Chen et al. [27]
	β-catenin activator (IM12) promotion of HF growth	Chen et al. [27]
	Capsaicin promotion of catagen in HFs via TRPV1	Bodó et al. [17]
	Corticotropin-releasing hormone (CRH) promotion of catagen in HFs	Fischer et al. [47], Ito et al. [96]
	Cyclosporine promotion of HF growth	Hawkshaw et al. [78], Taylor et al. [271]
	Dibutyryl-cAMP (db-cAMP), protein kinase A (PKA) activator, inhibition of HF growth	Harmon and Nevins [69]
	Dopamine promotion of catagen in HFs	Langan et al. [149]
	Estrogen 17-beta-estradiol (E2) modulation of hair growth	Conrad et al. [40]
	Galanin promotion of catagen in HFs	Holub et al. [84]
	High mobility group box protein 1 (HMGB1) stimulation of HF growth	Hwang et al. [92]
	Insulin stimulation of HF growth	Philpott et al. [226]
	L-carnitine, L-tartrate promotion of hair growth	Foitzik et al. [50]
	Minoxidil modulation of HF growth	Kwon et al. [147], Magerl et al. [183], Philpott et al. [223], Shorter et al. [260]
	PPARγ modulator, N-Acetyl-GED-0507-34-Levo, protection against lichen planopilaris-associated HF damage	Chéret et al. [32]
	Prolactin promotion of catagen in HFs	Foitzik et al. [51], Langan et al. [151]

(continued)

Table 8.4 (continued)

Category	Factors evaluated	References
	Prostaglandins and modulation of HF growth	Garza et al. [55], Joo et al. [106], Purba et al. [231]
	SFRP1 antagonist inhibition of catagen in HFs	Bertolini et al. [14]
	Testosterone and dihydrotestosterone (DHT) inhibition of HF growth	Chen et al. (2020)), Fischer et al. [48, 49]
	Thyroid hormone T4 prolongation of anagen in HFs	van Beek et al. [277]
	Thyrotropin-releasing hormone (TRH) prolongation of anagen in HFs	Gáspár et al. [56]
	Thymulin prolongation of anagen in HFs	Meier et al. [190]
	Tolbutamide, K(ATP) channel blocker, inhibition of HF growth	Shorter et al. [260]
	WAY-316606, SFRP1 antagonist, enhancement of hair shaft production	Hawkshaw et al. [77]
Biological extracts	Caffeine prolongation of anagen in HFs	Fischer et al. [47–49]
	Chrysanthemum zawadskii extract prolongation of anagen in HFs	Kim et al. [132]
	Korean red ginseng protective effect against chemotherapy induced catagen in HFs	Keum et al. [124]
	Oriental melon leaf extract prolongation of anagen in HFs	Pi et al. [227]
	Synthetic sandalwood odorant promotion of HF growth via OR2AT4	Chéret et al. [31]
	Spermidine prolongation of anagen in HFs	Ramot et al. [239]
	JAK-STAT inhibitors, ruxolitinib and tofacitinib, promotion of HF growth	Harel et al. [68]
Incubation parameters	655-nm red light promotion of hair shaft elongation	Han et al. [67]
	Intense pulsed light (IPL) promotion of catagen in HFs	Roosen et al. [244]
	UV light induced catagen promotion in HFs	Lu et al. [180]
Feeder cells and conditioned medium	Dermal papilla-derived exosome improvement of HF growth	Kwack et al. [143]
	Activated dermal fibroblast derived extracellular vesicles stimulation of HF growth via Norrin	le Riche et al. [152]
	Human fibroblast-derived extracellular vesicles promotion of HF growth	Rajendran et al. [237, 238]
	Dermal papilla-derived extracellular vesicles promotion of HF growth	Chen et al. [28, 29]

(continued)

8 Human HF In Vitro and Ex Vivo Models

Table 8.4 (continued)

Category	Factors evaluated	References
	Regulation of HF immune privilege by human hematopoietic mesenchymal stem cells	Kim et al. [129]

In this model, factors for evaluation can be only delivered in the medium, effectively to mimic systemic administration (Table 8.4). Recently this model has been used to assess the effect of extracellular vesicles (exosomes) on human hair growth, produced from dermal fibroblasts [152], macrophages [237, 238], and DP cells [28]. Potentially, physical parameters can also be evaluated with this system. For example, exposure of ex vivo HFs to 655-nm red light has been shown to promote hair shaft elongation [67], and UVB irradiation stimulates premature catagen induction [180].

Although this model does not allow induction of HF formation per se, given that this process is associated with DP and DS cell inductivity, the anagen VI HF organ culture can serve as support for such investigations. In fact, one of the advantages of the ex vivo HF model is that targeted cells remain in their respective environment niches and can cross-talk with neighbouring cell populations. It is particularly critical to maintain epithelial-mesenchymal interaction for HF regeneration and growth. During HF organ culture, DP and DS cell inductivity is preserved for as long as the HF remains in anagen VI, and can be evaluated by assessing the expression or activities of known inductivity markers such as alkaline phosphatase, versican, noggin, HGF, and FGF7 (Fig. 8.3) [5]. Key signalling pathways involved in HF regeneration are also maintained and modulated during organ-culture. Recently it was shown that cyclosporin, an immunosuppressant hair growth promoter, stimulated hair growth ex vivo by activating Wnt signalling through the inhibition of SFRP1 [77]. This novel finding led to the discovery of additional drugs targeting SFRP1 [77], as well as novel cosmetic strategies [14]. Importantly, the size and the diameter of the DP, which may correlate with inductive potential [89], and HF size [45], can also be analysed. Usually, this is determined by measuring the diameter at the

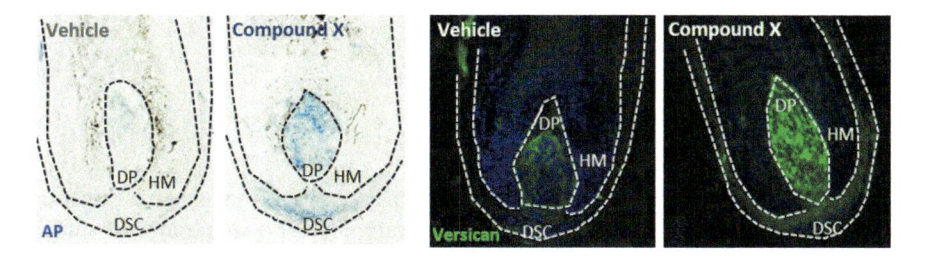

Fig. 8.3 Assessment of DP inductivity in human microdissected hair follicle or scalp skin organ culture. Representative images showing alkaline phosphatase activity (AP, light blue) and Versican expression (green) in vehicle and compound X treated HFs. DP: dermal papilla, HM: hair matrix, DSC: dermal sheath cup

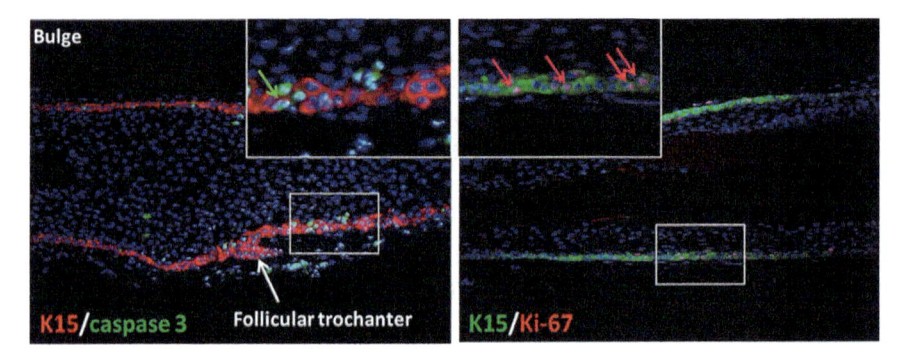

Fig. 8.4 Assessment of bulge stem cell activities in human microdissected hair follicle or scalp skin organ culture. Representative images showing an apoptotic, caspase 3+, bulge keratin 15+ stem cell (green arrow) and proliferating, Ki-67+, bulge keratin 15+ stem cells (pink arrows) in situ. K15: Keratin 15

level of Auber´s line [9] during organ culture, or after sectioning. However, optimal coherence tomography may be used to visualize the DP in 3D, to provide a more accurate assessment [163].

Similarly, it is not possible to investigate telogen-to-anagen transition ex vivo. However, by microdissecting full-length rather than amputated HFs [44], it is possible to investigate the activation of bulge stem cells within their niche (Fig. 8.4), the key initial step for new growth cycle initiation. For example, it was shown that stimulation of cannabinoid receptor-1-mediated signaling promotes the proliferation of K15+ cells in the bulge ex vivo [265]. Further, PDG2 has been shown to selectively inhibit bulge stem cell proliferation in HF organ culture [231]. Advances in the identification of telogen HFs after microdissection [79] have finally allowed the characterization of signaling molecules expressed during telogen in human HFs [76]. Thus, it is conceivable that signaling involved in telogen-to-anagen conversion may be investigated upon treatment with compounds or drugs using ex vivo telogen HF organ culture (Fig. 8.2c). As mentioned above, improved access to human catagen and telogen HFs may be provided by the use of samples derived from male or female pattern hair loss patients undergoing hair transplantation [76, 79].

Ex Vivo Skin Explant Organ Culture

Biopsies can be obtained from human skin samples derived from elective surgeries. Using these skin biopsies in explant organ culture enables a wide range of studies on subjects including skin aging, pigmentation, wound healing and scarring, hyperproliferative diseases, and skin cancer, as well as normal skin physiology [15, 309]. Biopsies can be obtained from the scalp containing fully intact HFs, with careful consideration of the angle of HF growth to avoid HF transection. Just as for microdissected HFs, these skin samples can be cultured ex vivo at the air / liquid interface in

Fig. 8.5 Human skin organ culture. Representative images showing biopsies from human heathy scalp and abdominal skin, obtained as waste from cosmetic surgeries, during organ culture

the absence of serum for up to 7 days [44, 181]. Scalp skin explant organ culture, which in healthy individuals contains mainly anagen VI HFs, is typically used for hair growth studies. Notably, it has been suggested that skin explants containing vellus hair follicles may survive in culture for longer, though the small size of the follicles limits their practical use [165]. Hair cycle staging is performed based on morphological and immunohistological criteria previously assessed clinically in vivo [205]. All parameters that can be analysed in HF organ culture can also be investigated in this model, including anagen / catagen conversion, cell activation and apoptosis [57, 58, 73], DP inductivity (Fig. 8.3), and bulge stem cell activation (Fig. 8.4).

Similar to HF organ culture, addition of factors to the skin explant culture media can be considered as a "systemic" application, potentially having an effect on all cells in the explant. However, some care has to be taken with skin biopsy size given permeability into the explant may vary from the periphery to the center of the tissue, depending on the molecular weight of the factor under investigation (Fig. 8.3).

The great advantage of the scalp skin explant model is that drugs, chemicals, and even physical parameters can be administered to the skin surface. Studies have investigated the effects of topically applied nicotinamide and its effects on hair growth using the skin explant model [73]. Further, the effects of UV light exposure and protection from tissue damage using topical caffeine has been studied [58]. Therefore, this model allows drug permeability studies and evaluation of external exposure to environmental factors. In addition, high molecular weight drugs, biological extracts, or protein peptides, can be injected into the skin explants to investigate their effects. As one example, intra-dermal injection into cultured skin explants of an osteopontin derived peptide elicited a significantly higher percentage of catagen HFs as compared to administration of the peptide in the culture media [5]. Importantly, this model is useful also to validate strategies for HF regeneration targeting perifollicular cells, such as immune cells, which remain viable in the skin biopsies during organ culture [57].

The use of non-hairy skin explant organ culture also enables short-term investigation of cell viability for cell-based therapy aiming at restoring hair density, using mesenchymal stem cells, DS cells, or DP cells, as these can be easily injected intra-dermally into skin biopsies (Fig. 8.6) [228]. Recently, in our laboratory, we succeeded in placing 3D HF organoids into human skin explants to assess viability (Fig. 8.7) as well as to examine DP cell inductivity and keratin production ex vivo [228]. As

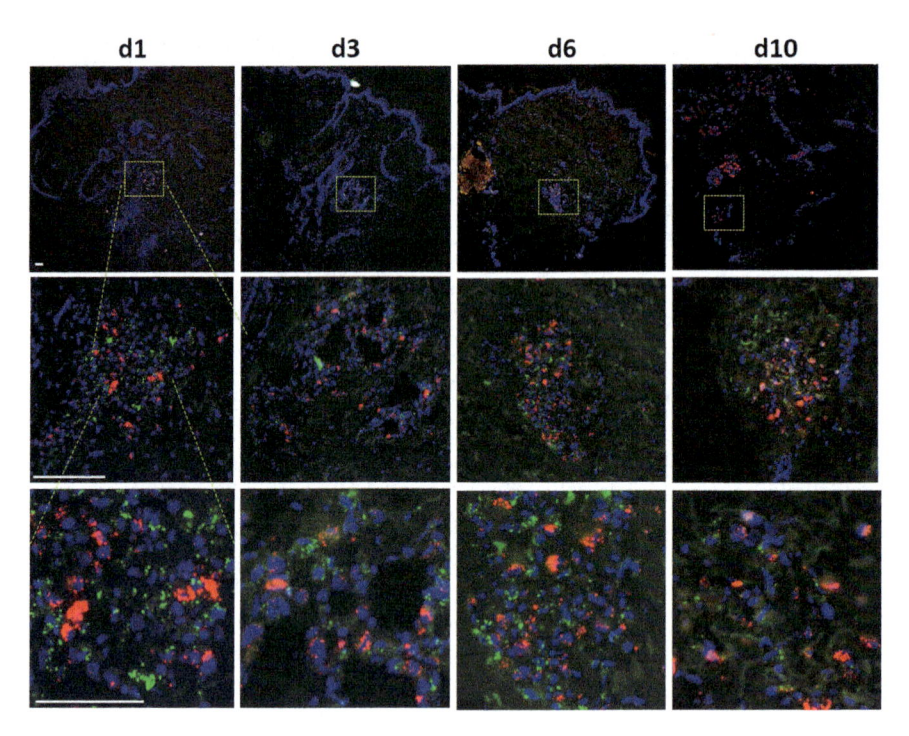

Fig. 8.6 Assessment of DP cell and hair matrix keratinocyte viability after injection into human skin ex vivo. DP cells (green cells) and hair matrix keratinocytes (HMx-red cells) were isolated from human scalp HFs and fluorescently labelled respectively with Qtracker 525 or Qtracker 605. A mixture of 85,000 green labelled DP cells and 65,000 red labelled HMx cells in a volume of 25 μl PBS was injected in 6 mm punch biopsies deriving from and upper-arm skin sample. Skin biopsies were cultured ex vivo up to 10 days and survival of DP and HMx cells were assessed by sectioning the samples and qualitatively observing the fluorescent signal in unfixed sections. Scale bars: 100 μm

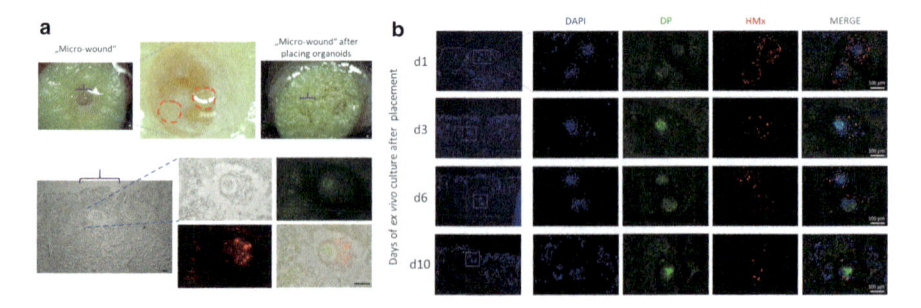

Fig. 8.7 Assessment of HF organoid viability in human skin ex vivo. a HF organoids were generated in vitro by seeding isolated labelled hair matrix keratinocytes (red) with spheroids of labelled DP cells (green). After 48 h of in vitro cultures, the HF organoids were placed in 1 mm micro-wounds in human skin punch biopsies. Skin biopsies were cultured ex vivo for 10 days and survival of the HF organoids was assessed by sectioning the samples and qualitatively observing the fluorescent signal in unfixed sections. Scale bars: 100 μm

such, skin explant organ culture is a very flexible model system for examining HF growth modulation, as well as neogenesis and regeneration.

Challenges and the Future of Hair Research Using HF Models

Research on all aspects of HF development, growth, cycling, their disorders and treatment, is becoming more common with a progressive increase in the annual rate of HF related studies published in the medical literature [189]. Consequently, demand for in vitro and ex vivo HF models continues to grow. Research on HFs without resorting to the use of animals or clinical studies affords many advantages as described above, however, there are still some significant limitations and problems with their use.

Most laboratories use primary tissues and cells for their in vitro and ex vivo HF research studies. Tissue supply is very limited, to the extent that in some countries laboratories only receive tissue from a clinic in rotation with other research groups. Several commercial companies claim to provide primary DP cells for a fee, but the provenance and phenotype of these cells is open to question and other cell types, such as DS and HF keratinocytes, do not seem to be readily commercially available. Similarly, some companies can provide whole scalp skin biopsies, however, it is expensive and the time to complete the logistics involved tends to reduce the quality of the tissue samples on arrival at the laboratory.

A few reports have developed and examined immortalized DP cells for use in HF research [54, 145, 259] including immortalized DP cells from balding scalp [33], but the cell lines are poorly characterized and not freely accessible by other laboratories. While there are many immortalized keratinocyte cell lines available, none are derived from the HF unit. Potentially though, if well characterized HF derived cell lines can be developed and made widely available, it may be possible to progress HF organoid culture towards a more standardized, routine process that can be used in large scale research and screening studies. Alternatively, iPSC use in producing HF organoids could be the answer. However, the current process of culturing, differentiating, and then combining iPSC cell types together into HF organoids is long and complicated and would need to be greatly simplified for more routine HF research use.

Because of the multi-year anagen phase of human scalp HFs, and the relatively short telogen and very brief catagen phases, it is difficult to study human HFs in culture across all of their growth cycle stages [222]. The current ex vivo parameters for mature HFs and whole tissue explants are suboptimal in that they do not allow for hair cycle research in any great detail. Human HFs can be cultured for up to 7 days before a pseudo-catagen disintegration of the HF occurs [222]. Similarly, the structural integrity of whole tissue explants containing HFs tends to break down after 7 days. Extending the in vitro and ex vivo lifespan of HFs, organoids, and whole tissue explants would allow for much more complex studies to be performed over longer time periods better reflecting the in vivo situation.

Despite these issues, use of the current in vitro and ex vivo HF models described above is likely to increase in the foreseeable future. For now, organ culture of intact human HFs is still the gold standard for much research and development where close equivalence to in vivo HFs is needed. In the future as demand increases, so HF models will be further developed and improved. Given the promising potential of iPSC models, a full skin and HF equivalent can be anticipated, though when such an assay system will be routinely available is unknown. Human 3D skin equivalents are already being used to examine pigment-modulating components of cosmetic products [10], the effects of organic environmental toxins on hyperpigmentation [192], and to characterize the impact of microbiota on skin [46], for example. As 3D skin equivalents become more sophisticated with the incorporation of HFs and other skin structures, so we can expect very detailed in vitro models of human HFs and their disorders to become widely used in research [8]. The ultimate goal is a true organ-on-a-chip bioreactor that will enable systemic inputs to be applied to HF functions over prolonged observation periods [7, 253, 278]. The development of such an assay system will finally allow for research on HF regeneration and interventional approaches without the need for in vivo studies on human subjects.

Acknowledgements The authors thank Sabrina Höfling for the support with image editing, and Christine Collin-Djangoné and Khalid Bakkar from L´Oreal for the great collaboration during the project aiming at assessing the induction and morphogenic potential of human hair matrix cells and dermal papilla fibroblasts ex vivo.

Funding sources that supported the work: Writing of this review was supported by Monasterium Laboratory, Münster, as an educational service to the skin research community. All images shown in this chapter were collected from Monasterium Laboratory. Images in Figure 6, and 7 derive from a project commissioned by L´Oreal to Monasterium Laboratory.

Conflict of Interest Disclosures: MB is CSO & Deputy General Manager, IP is Team Leader, while KM is a Senior Consultant for Monasterium Laboratory, Skin & Hair Research Solutions GmbH. KM is also Chief Scientific Officer for Replicel Life Sciences Inc (Canada) and director of McElwee Consulting (UK). None of these commercial interests have influenced the authors' views communicated here.

References

1. Abaci HE, Coffman A, Doucet Y, Chen J, Jacków J, Wang E, Guo Z, Shin JU, Jahoda CA, Christiano AM (2018) Tissue engineering of human hair follicles using a biomimetic developmental approach. Nat Commun 9:5301. https://doi.org/10.1038/s41467-018-07579-y
2. Abbasi S, Biernaskie J (2019) Injury modifies the fate of hair follicle dermal stem cell progeny in a hair cycle-dependent manner. Exp Dermatol 28:419–424. https://doi.org/10.1111/exd.13924

3. Abreu CM, Pirraco RP, Reis RL, Cerqueira MT, Marques AP (2021) Interfollicular epidermal stem-like cells for the recreation of the hair follicle epithelial compartment. Stem Cell Res Ther 12:62. https://doi.org/10.1186/s13287-020-02104-9
4. Ahn S-Y, Pi L-Q, Hwang ST, Lee W-S (2012) Effect of IGF-I on hair growth is related to the anti-apoptotic effect of IGF-I and up-regulation of PDGF-A and PDGF-B. Ann Dermatol 24:26–31. https://doi.org/10.5021/ad.2012.24.1.26
5. Alam M, Bertolini M, Gherardini J, Keren A, Ponce L, Chéret J, Alenfall J, Dunér P, Nilsson AH, Gilhar A, Paus R (2020) An osteopontin-derived peptide inhibits human hair growth at least in part by decreasing fibroblast growth factor-7 production in outer root sheath keratinocytes. Br J Dermatol 182:1404–1414. https://doi.org/10.1111/bjd.18479
6. Al-Nuaimi Y, Goodfellow M, Paus R, Baier G (2012) A prototypic mathematical model of the human hair cycle. J Theor Biol 310:143–159. https://doi.org/10.1016/j.jtbi.2012.05.027
7. Ataç B, Wagner I, Horland R, Lauster R, Marx U, Tonevitsky AG, Azar RP, Lindner G (2013) Skin and hair on-a-chip: in vitro skin models versus ex vivo tissue maintenance with dynamic perfusion. Lab Chip 13:3555–3561. https://doi.org/10.1039/c3lc50227a
8. Atwood SX, Plikus MV (2021) Fostering a healthy culture: Biological relevance of in vitro and ex vivo skin models. Exp Dermatol. https://doi.org/10.1111/exd.14296
9. Auber L (1952) VII.—the anatomy of follicles producing wool-fibres, with special reference to keratinization. Earth Environ Sci Trans R Soc Edinb 62:191–254.https://doi.org/10.1017/S0080456800009285
10. Bae I-H, Lee ES, Yoo JW, Lee SH, Ko JY, Kim YJ, Lee TR, Kim D-Y, Lee CS (2019) Mannosylerythritol lipids inhibit melanogenesis via suppressing ERK-CREB-MiTF-tyrosinase signalling in normal human melanocytes and a three-dimensional human skin equivalent. Exp Dermatol 28:738–741. https://doi.org/10.1111/exd.13836
11. Bak SS, Ahn BN, Kim JA, Shin SH, Kim JC, Kim MK, Sung YK, Kim SK (2013) Ecklonia cava promotes hair growth. Clin Exp Dermatol 38:904–910. https://doi.org/10.1111/ced.12120
12. Bak SS, Kwack MH, Shin HS, Kim JC, Kim MK, Sung YK (2018) Restoration of hair-inductive activity of cultured human follicular keratinocytes by co-culturing with dermal papilla cells. Biochem Biophys Res Commun 505:360–364. https://doi.org/10.1016/j.bbrc.2018.09.125
13. Bak S-S, Park JM, Oh JW, Kim JC, Kim MK, Sung YK (2020) Knockdown of FOXA2 impairs hair-inductive activity of cultured human follicular keratinocytes. Front Cell Dev Biol 8:575382.https://doi.org/10.3389/fcell.2020.575382
14. Bertolini M, Chéret J, Pinto D, Hawkshaw N, Ponce L, Erdmann H, Jimenez F, Funk W, Paus R (2021) A novel nondrug SFRP1 antagonist inhibits catagen development in human hair follicles ex vivo. Br J Dermatol 184:371–373. https://doi.org/10.1111/bjd.19552
15. Bertolini M, Ramot Y, Gherardini J, Heinen G, Chéret J, Welss T, Giesen M, Funk W, Paus R (2020) Theophylline exerts complex anti-ageing and anti-cytotoxicity effects in human skin ex vivo. Int J Cosmet Sci 42:79–88. https://doi.org/10.1111/ics.12589
16. Betriu N, Jarrosson-Moral C, Semino CE (2020) Culture and Differentiation of human hair follicle dermal papilla cells in a soft 3D self-assembling peptide scaffold. Biomolecules 10.https://doi.org/10.3390/biom10050684
17. Bodó E, Bíró T, Telek A, Czifra G, Griger Z, Tóth BI, Mescalchin A, Ito T, Bettermann A, Kovács L, Paus R (2005) A hot new twist to hair biology: involvement of vanilloid receptor-1 (VR1/TRPV1) signaling in human hair growth control. Am J Pathol 166:985–998. https://doi.org/10.1016/S0002-9440(10)62320-6
18. Bodó E, Kromminga A, Funk W, Laugsch M, Duske U, Jelkmann W, Paus R (2007) Human hair follicles are an extrarenal source and a nonhematopoietic target of erythropoietin. FASEB J Off Publ Fed Am Soc Exp Biol 21:3346–3354. https://doi.org/10.1096/fj.07-8628com
19. Bodó E, Wiersma F, Funk W, Kromminga A, Jelkmann W, Paus R (2010) Does erythropoietin modulate human hair follicle melanocyte activities in situ? Exp Dermatol 19:65–67. https://doi.org/10.1111/j.1600-0625.2009.00938.x

20. Böhm M, Bodó E, Funk W, Paus R (2014) α-Melanocyte-stimulating hormone: a protective peptide against chemotherapy-induced hair follicle damage? Br J Dermatol 170:956–960
21. Boivin WA, Jiang H, Utting OB, Hunt DWC (2006) Influence of interleukin-1alpha on androgen receptor expression and cytokine secretion by cultured human dermal papilla cells. Exp Dermatol 15:784–793. https://doi.org/10.1111/j.1600-0625.2006.00462.x
22. Boyce ST, Kagan RJ, Yakuboff KP, Meyer NA, Rieman MT, Greenhalgh DG, Warden GD (2002) Cultured skin substitutes reduce donor skin harvesting for closure of excised, full-thickness burns. Ann Surg 235:269–279. https://doi.org/10.1097/00000658-200202000-00016
23. Breitkopf T, Leung G, Yu M, Wang E, McElwee KJ (2013) The basic science of hair biology: what are the causal mechanisms for the disordered hair follicle? Dermatol Clin 31:1–19. https://doi.org/10.1016/j.det.2012.08.006
24. Buscone S, Mardaryev AN, Westgate GE, Uzunbajakava NE, Botchkareva NV (2021) Cryptochrome 1 is modulated by blue light in human keratinocytes and exerts positive impact on human hair growth. Exp Dermatol 30:271–277. https://doi.org/10.1111/exd.14231
25. Castro AR, Logarinho E (2020) Tissue engineering strategies for human hair follicle regeneration: how far from a hairy goal? Stem Cells Transl Med 9:342–350. https://doi.org/10.1002/sctm.19-0301
26. Chan C-C, Fan SM-Y, Wang W-H, Mu Y-F, Lin S-J (2015) A two-stepped culture method for efficient production of trichogenic keratinocytes. Tissue Eng Part C Methods 21:1070–1079. https://doi.org/10.1089/ten.TEC.2015.0033
27. Chen X, Liu B, Li Y, Han L, Tang X, Deng W, Lai W, Wan M (2020) Dihydrotestosterone regulates hair growth through the Wnt/β-catenin pathway in C57BL/6 mice and in vitro organ culture. Front Pharmacol 10:1528.https://doi.org/10.3389/fphar.2019.01528
28. Chen Y, Huang J, Chen R, Yang L, Wang J, Liu B, Du L, Yi Y, Jia J, Xu Y, Chen Q, Ngondi DG, Miao Y, Hu Z (2020) Sustained release of dermal papilla-derived extracellular vesicles from injectable microgel promotes hair growth. Theranostics 10:1454–1478. https://doi.org/10.7150/thno.39566
29. Chen Y, Huang J, Liu Z, Chen R, Fu D, Yang L, Wang J, Du L, Wen L, Miao Y, Hu Z (2020) miR-140-5p in small extracellular vesicles from human papilla cells stimulates hair growth by promoting proliferation of outer root sheath and hair matrix cells. Front Cell Dev Biol 8:593638.https://doi.org/10.3389/fcell.2020.593638
30. Cheon HI, Bae S, Ahn KJ (2019) Flavonoid silibinin increases hair-inductive property via Akt and Wnt/β-catenin signaling activation in 3-dimensional-spheroid cultured human dermal papilla cells. J Microbiol Biotechnol 29:321–329. https://doi.org/10.4014/jmb.1810.10050
31. Chéret J, Bertolini M, Ponce L, Lehmann J, Tsai T, Alam M, Hatt H, Paus R (2018) Olfactory receptor OR2AT4 regulates human hair growth. Nat Commun 9:3624.https://doi.org/10.1038/s41467-018-05973-0
32. Chéret J, Piccini I, Hardman-Smart J, Ghatak S, Alam M, Lehmann J, Jimenez F, Erdmann H, Poblet E, Botchkareva N, Paus R, Bertolini M (2020) Preclinical evidence that the PPARγ modulator, N-Acetyl-GED-0507-34-Levo, may protect human hair follicle epithelial stem cells against lichen planopilaris-associated damage. J Eur Acad Dermatol Venereol JEADV 34:e195–e197. https://doi.org/10.1111/jdv.16114
33. Chew EGY, Tan JHJ, Bahta AW, Ho BS-Y, Liu X, Lim TC, Sia YY, Bigliardi PL, Heilmann S, Wan ACA, Nöthen MM, Philpott MP, Hillmer AM (2016) Differential expression between human dermal papilla cells from balding and non-balding scalps reveals new candidate genes for androgenetic alopecia. J Invest Dermatol 136:1559–1567. https://doi.org/10.1016/j.jid.2016.03.032
34. Choi H-I, Kang B-M, Jang J, Hwang ST, Kwon O (2018) Novel effect of sildenafil on hair growth. Biochem Biophys Res Commun 505:685–691. https://doi.org/10.1016/j.bbrc.2018.09.164
35. Choi H-I, Kim DY, Choi S-J, Shin C-Y, Hwang ST, Kim KH, Kwon O (2018) The effect of cilostazol, a phosphodiesterase 3 (PDE3) inhibitor, on human hair growth with the dual promoting mechanisms. J Dermatol Sci 91:60–68. https://doi.org/10.1016/j.jdermsci.2018.04.005

36. Choi M, Choi S-J, Jang S, Choi H-I, Kang B-M, Hwang ST, Kwon O (2019) Shikimic acid, a mannose bioisostere, promotes hair growth with the induction of anagen hair cycle. Sci Rep 9:17008.https://doi.org/10.1038/s41598-019-53612-5
37. Chueh S-C, Lin S-J, Chen C-C, Lei M, Wang LM, Widelitz R, Hughes MW, Jiang T-X, Chuong CM (2013) Therapeutic strategy for hair regeneration: hair cycle activation, niche environment modulation, wound-induced follicle neogenesis, and stem cell engineering. Expert Opin Biol Ther 13:377–391. https://doi.org/10.1517/14712598.2013.739601
38. Chuong CM, Widelitz RB, Ting-Berreth S, Jiang TX (1996) Early events during avian skin appendage regeneration: dependence on epithelial-mesenchymal interaction and order of molecular reappearance. J Invest Dermatol 107:639–646
39. Cohen J (1961) The transplantation of individual rat and guineapig whisker papillae. J Embryol Exp Morphol 9:117–127
40. Conrad F, Ohnemus U, Bodo E, Biro T, Tychsen B, Gerstmayer B, Bosio A, Schmidt-Rose T, Altgilbers S, Bettermann A, Saathoff M, Meyer W, Paus R (2005) Substantial sex-dependent differences in the response of human scalp hair follicles to estrogen stimulation in vitro advocate gender-tailored management of female versus male pattern balding. J Investig Dermatol Symp Proc 10:243–246. https://doi.org/10.1111/j.1087-0024.2005.10115.x
41. Detmar M, Schaart FM, Blume U, Orfanos CE (1993) Culture of hair matrix and follicular keratinocytes. J Invest Dermatol 101:130S-134S. https://doi.org/10.1111/1523-1747.ep12363168
42. Discher DE, Janmey P, Wang Y-L (2005) Tissue cells feel and respond to the stiffness of their substrate. Science 310:1139–1143. https://doi.org/10.1126/science.1116995
43. Dong L, Hao H, Liu J, Tong C, Ti D, Chen D, Chen L, Li M, Liu H, Fu X, Han W (2017) Wnt1a maintains characteristics of dermal papilla cells that induce mouse hair regeneration in a 3D preculture system. J Tissue Eng Regen Med 11:1479–1489. https://doi.org/10.1002/term.2046
44. Edelkamp J, Gherardini J, Bertolini M (2020) Methods to study human hair follicle growth ex vivo: human microdissected hair follicle and human full thickness skin organ culture. Methods Mol Biol Clifton NJ 2154:105–119. https://doi.org/10.1007/978-1-0716-0648-3_9
45. Elliott K, Stephenson TJ, Messenger AG (1999) Differences in hair follicle dermal papilla volume are due to extracellular matrix volume and cell number: implications for the control of hair follicle size and androgen responses. J Invest Dermatol 113:873–877. https://doi.org/10.1046/j.1523-1747.1999.00797.x
46. Emmert H, Rademacher F, Gläser R, Harder J (2020) Skin microbiota analysis in human 3D skin models-"Free your mice". Exp Dermatol 29:1133–1139. https://doi.org/10.1111/exd.14164
47. Fischer TW, Bergmann A, Kruse N, Kleszczynski K, Skobowiat C, Slominski AT, Paus R (2021) New effects of caffeine on corticotropin-releasing hormone (CRH)-induced stress along the intrafollicular classical hypothalamic-pituitary-adrenal (HPA) axis (CRH-R1/2, IP3-R, ACTH, MC-R2) and the neurogenic non-HPA axis (substance P, p75NTR and TrkA) in ex vivo human male androgenetic scalp hair follicles. Br J Dermatol 184:96–110. https://doi.org/10.1111/bjd.19115
48. Fischer TW, Herczeg-Lisztes E, Funk W, Zillikens D, Bíró T, Paus R (2014) Differential effects of caffeine on hair shaft elongation, matrix and outer root sheath keratinocyte proliferation, and transforming growth factor-β2/insulin-like growth factor-1-mediated regulation of the hair cycle in male and female human hair follicles in vitro. Br J Dermatol 171:1031–1043. https://doi.org/10.1111/bjd.13114
49. Fischer TW, Hipler UC, Elsner P (2007) Effect of caffeine and testosterone on the proliferation of human hair follicles in vitro. Int J Dermatol 46:27–35. https://doi.org/10.1111/j.1365-4632.2007.03119.x
50. Foitzik K, Hoting E, Förster T, Pertile P, Paus R (2007) L-carnitine-L-tartrate promotes human hair growth in vitro. Exp Dermatol 16:936–945. https://doi.org/10.1111/j.1600-0625.2007.00611.x

51. Foitzik K, Krause K, Conrad F, Nakamura M, Funk W, Paus R (2006) Human scalp hair follicles are both a target and a source of prolactin, which serves as an autocrine and/or paracrine promoter of apoptosis-driven hair follicle regression. Am J Pathol 168:748–756. https://doi.org/10.2353/ajpath.2006.050468
52. Foitzik K, Spexard T, Nakamura M, Halsner U, Paus R (2005) Towards dissecting the pathogenesis of retinoid-induced hair loss: all-trans retinoic acid induces premature hair follicle regression (catagen) by upregulation of transforming growth factor-beta2 in the dermal papilla. J Invest Dermatol 124:1119–1126. https://doi.org/10.1111/j.0022-202X.2005.23686.x
53. Fong P, Tong HHY, Ng KH, Lao CK, Chong CI, Chao CM (2015) In silico prediction of prostaglandin D2 synthase inhibitors from herbal constituents for the treatment of hair loss. J Ethnopharmacol 175:470–480. https://doi.org/10.1016/j.jep.2015.10.005
54. Fukuda T, Takahashi K, Takase S, Orimoto A, Eitsuka T, Nakagawa K, Kiyono T (2020) Human derived immortalized dermal papilla cells with a constant expression of testosterone receptor. Front Cell Dev Biol 8:157.https://doi.org/10.3389/fcell.2020.00157
55. Garza LA, Liu Y, Yang Z, Alagesan B, Lawson JA, Norberg SM, Loy DE, Zhao T, Blatt HB, Stanton DC, Carrasco L, Ahluwalia G, Fischer SM, FitzGerald GA, Cotsarelis G (2012) Prostaglandin D2 inhibits hair growth and is elevated in bald scalp of men with androgenetic alopecia. Sci Transl Med 4:126ra34. https://doi.org/10.1126/scitranslmed.3003122
56. Gáspár E, Hardenbicker C, Bodó E, Wenzel B, Ramot Y, Funk W, Kromminga A, Paus R (2010) Thyrotropin releasing hormone (TRH): a new player in human hair-growth control. FASEB J Off Publ Fed Am Soc Exp Biol 24:393–403. https://doi.org/10.1096/fj.08-126417
57. Gherardini J, Uchida Y, Hardman JA, Chéret J, Mace K, Bertolini M, Paus R (2020) Tissue-resident macrophages can be generated de novo in adult human skin from resident progenitor cells during substance P-mediated neurogenic inflammation ex vivo. PloS One 15:e0227817.https://doi.org/10.1371/journal.pone.0227817
58. Gherardini J, Wegner J, Chéret J, Ghatak S, Lehmann J, Alam M, Jimenez F, Funk W, Böhm M, Botchkareva NV, Ward C, Paus R, Bertolini M (2019) Transepidermal UV radiation of scalp skin ex vivo induces hair follicle damage that is alleviated by the topical treatment with caffeine. Int J Cosmet Sci 41:164–182. https://doi.org/10.1111/ics.12521
59. Gledhill K, Guo Z, Umegaki-Arao N, Higgins CA, Itoh M, Christiano AM (2015) Melanin transfer in human 3D skin equivalents generated exclusively from induced pluripotent stem cells. PloS One 10:e0136713.https://doi.org/10.1371/journal.pone.0136713
60. Gnedeva K, Vorotelyak E, Cimadamore F, Cattarossi G, Giusto E, Terskikh VV, Terskikh AV (2015) Derivation of hair-inducing cell from human pluripotent stem cells. PloS One 10:e0116892.https://doi.org/10.1371/journal.pone.0116892
61. González R, Moffatt G, Hagner A, Sinha S, Shin W, Rahmani W, Chojnacki A, Biernaskie J (2017) Platelet-derived growth factor signaling modulates adult hair follicle dermal stem cell maintenance and self-renewal. NPJ Regen Med 2:11.https://doi.org/10.1038/s41536-017-0013-4
62. Goodman LV, Ledbetter SR (1992) Secretion of stromelysin by cultured dermal papilla cells: differential regulation by growth factors and functional role in mitogen-induced cell proliferation. J Cell Physiol 151:41–49. https://doi.org/10.1002/jcp.1041510108
63. Guo H, Gao WV, Endo H, McElwee KJ (2017) Experimental and early investigational drugs for androgenetic alopecia. Expert Opin Investig Drugs 26:917–932. https://doi.org/10.1080/13543784.2017.1353598
64. Gupta AC, Chawla S, Hegde A, Singh D, Bandyopadhyay B, Lakshmanan CC, Kalsi G, Ghosh S (2018) Establishment of an in vitro organoid model of dermal papilla of human hair follicle. J Cell Physiol 233:9015–9030. https://doi.org/10.1002/jcp.26853
65. Halloy J, Bernard BA, Loussouarn G, Goldbeter A (2000) Modeling the dynamics of human hair cycles by a follicular automaton. Proc Natl Acad Sci U S A 97:8328–8333. https://doi.org/10.1073/pnas.97.15.8328
66. Han JH, Kwon OS, Chung JH, Cho KH, Eun HC, Kim KH (2004) Effect of minoxidil on proliferation and apoptosis in dermal papilla cells of human hair follicle. J Dermatol Sci 34:91–98. https://doi.org/10.1016/j.jdermsci.2004.01.002

67. Han L, Liu B, Chen X, Chen H, Deng W, Yang C, Ji B, Wan M (2018) Activation of Wnt/β-catenin signaling is involved in hair growth-promoting effect of 655-nm red light and LED in in vitro culture model. Lasers Med Sci 33:637–645. https://doi.org/10.1007/s10103-018-2455-3
68. Harel S, Higgins CA, Cerise JE, Dai Z, Chen JC, Clynes R Christiano AM (2015) Pharmacologic inhibition of JAK-STAT signaling promotes hair growth. Sci Adv 1:e1500973.https://doi.org/10.1126/sciadv.1500973
69. Harmon CS, Nevins TD (1997) Evidence that activation of protein kinase A inhibits human hair follicle growth and hair fibre production in organ culture and DNA synthesis in human and mouse hair follicle organ culture. Br J Dermatol 136:853–858
70. Harmon CS, Nevins TD (1994) Biphasic effect of 1,25-dihydroxyvitamin D3 on human hair follicle growth and hair fiber production in whole-organ cultures. J Invest Dermatol 103:318–322. https://doi.org/10.1111/1523-1747.ep12394788
71. Harmon CS, Nevins TD (1993) IL-1 alpha inhibits human hair follicle growth and hair fiber production in whole-organ cultures. Lymphokine Cytokine Res 12:197–203
72. Harmon CS, Nevins TD, Bollag WB (1995) Protein kinase C inhibits human hair follicle growth and hair fibre production in organ culture. Br J Dermatol 133:686–693. https://doi.org/10.1111/j.1365-2133.1995.tb02739.x
73. Haslam IS, Hardman JA, Paus R (2018) Topically applied nicotinamide inhibits human hair follicle growth ex vivo. J Invest Dermatol 138:1420–1422. https://doi.org/10.1016/j.jid.2017.12.019
74. Havlickova B, Bíró T, Mescalchin A, Arenberger P, Paus R (2004) Towards optimization of an organotypic assay system that imitates human hair follicle-like epithelial-mesenchymal interactions. Br J Dermatol 151:753–765. https://doi.org/10.1111/j.1365-2133.2004.06184.x
75. Havlickova B, Bíró T, Mescalchin A, Tschirschmann M, Mollenkopf H, Bettermann A, Pertile P, Lauster R, Bodó E, Paus R (2009) A human folliculoid microsphere assay for exploring epithelial- mesenchymal interactions in the human hair follicle. J Invest Dermatol 129:972–983. https://doi.org/10.1038/jid.2008.315
76. Hawkshaw NJ, Hardman JA, Alam M, Jimenez F, Paus R (2020) Deciphering the molecular morphology of the human hair cycle; Wnt signalling during the telogen-anagen transformation. Br J Dermatol 182:1184–1193. https://doi.org/10.1111/bjd.18356
77. Hawkshaw NJ, Hardman JA, Haslam IS, Shahmalak A, Gilhar A, Lim X, Paus R (2018) Identifying novel strategies for treating human hair loss disorders: cyclosporine A suppresses the Wnt inhibitor, SFRP1, in the dermal papilla of human scalp hair follicles. PLoS Biol 16:e2003705.https://doi.org/10.1371/journal.pbio.2003705
78. Hawkshaw NJ, Haslam IS, Ansell DM, Shamalak A, Paus R (2015) Re-evaluating cyclosporine A as a hair growth-promoting agent in human scalp hair follicles. J Invest Dermatol 135:2129–2132. https://doi.org/10.1038/jid.2015.121
79. Hernandez I, Alam M, Platt C, Hardman J, Smart E, Poblet E, Bertolini M, Paus R, Jimenez F (2018) A technique for more precise distinction between catagen and telogen human hair follicles ex vivo. J Am Acad Dermatol 79:558–559. https://doi.org/10.1016/j.jaad.2018.02.009
80. Higgins CA, Chen JC, Cerise JE, Jahoda CAB, Christiano AM (2013) Microenvironmental reprogramming by three-dimensional culture enables dermal papilla cells to induce de novo human hair-follicle growth. Proc Natl Acad Sci U S A 110:19679–19688. https://doi.org/10.1073/pnas.1309970110
81. Higgins CA, Christiano AM (2014) Regenerative medicine and hair loss: how hair follicle culture has advanced our understanding of treatment options for androgenetic alopecia. Regen Med 9:101–111. https://doi.org/10.2217/rme.13.87
82. Higgins CA, Richardson GD, Ferdinando D, Westgate GE, Jahoda CAB (2010) Modelling the hair follicle dermal papilla using spheroid cell cultures. Exp Dermatol 19:546–548. https://doi.org/10.1111/j.1600-0625.2009.01007.x
83. Hoffmann R, Eicheler W, Huth A, Wenzel E, Happle R (1996) Cytokines and growth factors influence hair growth in vitro. Possible implications for the pathogenesis and treatment of alopecia areata. Arch Dermatol Res 288:153–156. https://doi.org/10.1007/BF02505825

84. Holub BS, Kloepper JE, Tóth BI, Bíro T, Kofler B, Paus R (2012) The neuropeptide galanin is a novel inhibitor of human hair growth. Br J Dermatol 167:10–16. https://doi.org/10.1111/j.1365-2133.2012.10890.x
85. Horne KA, Jahoda CA, Oliver RF (1986) Whisker growth induced by implantation of cultured vibrissa dermal papilla cells in the adult rat. J Embryol Exp Morphol 97:111–124
86. Hsieh C-H, Wang J-L, Huang Y-Y (2011) Large-scale cultivation of transplantable dermal papilla cellular aggregates using microfabricated PDMS arrays. Acta Biomater 7:315–324. https://doi.org/10.1016/j.actbio.2010.08.012
87. Huang C-F, Chang Y-J, Hsueh Y-Y, Huang C-W, Wang D-H, Huang T-C, Wu Y-T, Su F-C, Hughes M, Chuong C-M, Wu C-C. (2016) Assembling composite dermal papilla spheres with adipose-derived stem cells to enhance hair follicle induction. Sci Rep 6:26436.https://doi.org/10.1038/srep26436
88. Huang H-C, Lin H, Huang M-C (2019) Lactoferrin promotes hair growth in mice and increases dermal papilla cell proliferation through Erk/Akt and Wnt signaling pathways. Arch Dermatol Res 311:411–420. https://doi.org/10.1007/s00403-019-01920-1
89. Huang Y-C, Chan C-C, Lin W-T, Chiu H-Y, Tsai R-Y, Tsai T-H, Chan J-Y, Lin S-J (2013) Scalable production of controllable dermal papilla spheroids on PVA surfaces and the effects of spheroid size on hair follicle regeneration. Biomaterials 34:442–451. https://doi.org/10.1016/j.biomaterials.2012.09.083
90. Hughes K, Ho R, Greff S, Filaire E, Ranouille E, Chazaud C, Herbette G, Butaud J-F, Berthon J-Y, Raharivelomanana P (2020) Hair growth activity of three plants of the Polynesian Cosmetopoeia and their regulatory effect on dermal papilla cells. Mol Basel Switz 25.https://doi.org/10.3390/molecules25194360
91. Hwang D, Lee H, Lee J, Lee M, Cho S, Kim T, Kim H (2021) Micro-current stimulation has potential effects of hair growth-promotion on human hair follicle-derived papilla cells and animal model. Int J Mol Sci 22.https://doi.org/10.3390/ijms22094361
92. Hwang J-H, Chu H, Ahn Y, Kim J, Kim D-Y (2019) HMGB1 promotes hair growth via the modulation of prostaglandin metabolism. Sci Rep 9:6660.https://doi.org/10.1038/s41598-019-43242-2
93. Hwang K-A, Hwang Y L, Lee M H, Kim N-R, Roh S-S, Lee Y, Kim CD, Lee J-H, Choi K-C (2012) Adenosine stimulates growth of dermal papilla and lengthens the anagen phase by increasing the cysteine level via fibroblast growth factors 2 and 7 in an organ culture of mouse vibrissae hair follicles. Int J Mol Med 29:195–201. https://doi.org/10.3892/ijmm.2011.817
94. Inamatsu M, Matsuzaki T, Iwanari H, Yoshizato K (1998) Establishment of rat dermal papilla cell lines that sustain the potency to induce hair follicles from afollicular skin. J Invest Dermatol 111:767–775. https://doi.org/10.1046/j.1523-1747.1998.00382.x
95. Ito N, Ito T, Kromminga A, Bettermann A, Takigawa M, Kees F, Straub RH, Paus R (2005) Human hair follicles display a functional equivalent of the hypothalamic-pituitary-adrenal axis and synthesize cortisol. FASEB J 19:1332–1334
96. Ito T, Ito N, Saathoff M, Bettermann A, Takigawa M, Paus R (2005) Interferon-γ is a potent inducer of catagen-like changes in cultured human anagen hair follicles. Br J Dermatol 152:623–631
97. Itoh M, Umegaki-Arao N, Guo Z, Liu L, Higgins CA, Christiano AM (2013) Generation of 3D skin equivalents fully reconstituted from human induced pluripotent stem cells (iPSCs). PloS One 8:e77673. https://doi.org/10.1371/journal.pone.0077673
98. Jahoda C, Oliver RF (1981) The growth of vibrissa dermal papilla cells in vitro. Br J Dermatol 105:623–627. https://doi.org/10.1111/j.1365-2133.1981.tb00971.x
99. Jahoda CA, Horne KA, Oliver RF (1984) Induction of hair growth by implantation of cultured dermal papilla cells. Nature 311:560–562. https://doi.org/10.1038/311560a0
100. Jahoda CA, Reynolds AJ, Oliver RF (1993) Induction of hair growth in ear wounds by cultured dermal papilla cells. J Invest Dermatol 101:584–590. https://doi.org/10.1111/1523-1747.ep12366039
101. Jang S, Ohn J, Kang BM, Park M, Kim KH, Kwon O (2020) "Two-Cell Assemblage" assay: a simple in vitro method for screening hair growth-promoting compounds. Front Cell Dev Biol 8:581528.https://doi.org/10.3389/fcell.2020.581528

102. Jeong GH, Boisvert WA, Xi M-Z, Zhang Y-L, Choi Y-B, Cho S, Lee S, Choi C, Lee B-H (2018) Effect of miscanthus sinensis var. purpurascens flower extract on proliferation and molecular regulation in human dermal papilla cells and stressed C57BL/6 mice. Chin J Integr Med 24:591–599. https://doi.org/10.1007/s11655-017-2755-7
103. Jeong KH, Joo HJ, Kim JE, Park YM, Kang H (2015) Effect of mycophenolic acid on proliferation of dermal papilla cells and induction of anagen hair follicles. Clin Exp Dermatol 40:894–902. https://doi.org/10.1111/ced.12650
104. Jin G-R, Zhang Y-L, Yap J, Boisvert WA, Lee B-H (2020) Hair growth potential of Salvia plebeia extract and its associated mechanisms. Pharm Biol 58:400–409. https://doi.org/10.1080/13880209.2020.1759654
105. Jo SJ, Choi S-J, Yoon S-Y, Lee JY, Park W-S, Park P-J, Kim KH, Eun HC, Kwon O (2013) Valproic acid promotes human hair growth in in vitro culture model. J Dermatol Sci 72:16–24. https://doi.org/10.1016/j.jdermsci.2013.05.007
106. Joo HW, Kang YR, Kwack MH, Sung YK (2016) 15-deoxy prostaglandin J2, the nonenzymatic metabolite of prostaglandin D2, induces apoptosis in keratinocytes of human hair follicles: a possible explanation for prostaglandin D2-mediated inhibition of hair growth. Naunyn Schmiedebergs Arch Pharmacol 389:809–813. https://doi.org/10.1007/s00210-016-1257-z
107. Kageyama T, Chun Y-S, Fukuda J (2021) Hair follicle germs containing vascular endothelial cells for hair regenerative medicine. Sci Rep 11:624.https://doi.org/10.1038/s41598-020-79722-z
108. Kageyama T, Nanmo A, Yan L, Nittami T, Fukuda J (2020) Effects of platelet-rich plasma on in vitro hair follicle germ preparation for hair regenerative medicine. J Biosci Bioenghttps://doi.org/10.1016/j.jbiosc.2020.08.005
109. Kageyama T, Yan L, Shimizu A, Maruo S, Fukuda J (2019) Preparation of hair beads and hair follicle germs for regenerative medicine. Biomaterials 212:55–63. https://doi.org/10.1016/j.biomaterials.2019.05.003
110. Kageyama T, Yoshimura C, Myasnikova D, Kataoka K, Nittami T, Maruo S, Fukuda J (2018) Spontaneous hair follicle germ (HFG) formation in vitro, enabling the large-scale production of HFGs for regenerative medicine. Biomaterials 154:291–300. https://doi.org/10.1016/j.biomaterials.2017.10.056
111. Kalabusheva E, Terskikh V, Vorotelyak E (2017) Hair germ model in vitro via human postnatal keratinocyte-dermal papilla interactions: impact of hyaluronic acid. Stem Cells Int 2017:9271869. https://doi.org/10.1155/2017/9271869
112. Kalabusheva EP, Vorotelyak EA (2020) Generation of hair follicle germs in vitro using human postnatal skin cells. Methods Mol Biol Clifton NJ 2154:153–163. https://doi.org/10.1007/978-1-0716-0648-3_13
113. Kamp H, Geilen CC, Sommer C, Blume-Peytavi U (2003) Regulation of PDGF and PDGF receptor in cultured dermal papilla cells and follicular keratinocytes of the human hair follicle. Exp Dermatol 12:662–672. https://doi.org/10.1034/j.1600-0625.2003.00089.x
114. Kanayama K, Takada H, Saito N, Kato H, Kinoshita K, Shirado T, Mashiko T, Asahi R, Mori M, Tashiro K, Sunaga A, Kurisaki A, Yoshizato K, Yoshimura K (2020) Hair regeneration potential of human dermal sheath cells cultured under physiological oxygen. Tissue Eng Part A.https://doi.org/10.1089/ten.TEA.2019.0329
115. Kang BM, Kwack MH, Kim MK, Kim JC, Sung YK (2012) Sphere formation increases the ability of cultured human dermal papilla cells to induce hair follicles from mouse epidermal cells in a reconstitution assay. J Invest Dermatol 132:237–239. https://doi.org/10.1038/jid.2011.250
116. Kang BM, Shin SH, Kwack MH, Shin H, Oh JW, Kim J, Moon C, Moon C, Kim JC, Kim MK, Sung YK (2010) Erythropoietin promotes hair shaft growth in cultured human hair follicles and modulates hair growth in mice. J Dermatol Sci 59:86–90. https://doi.org/10.1016/j.jdermsci.2010.04.015
117. Kang J-I, Choi YK, Koh Y-S, Hyun J-W, Kang J-H, Lee KS, Lee CM, Yoo E-S, Kang H-K (2020) Vanillic acid stimulates anagen signaling via the PI3K/Akt/ β-catenin pathway in dermal papilla cells. Biomol Ther 28:354–360. https://doi.org/10.4062/biomolther.2019.206

118. Kang J-I, Kim E-J, Kim M-K, Jeon Y-J, Kang S-M, Koh Y-S, Yoo E-S, Kang H-K (2013) The promoting effect of Ishige sinicola on hair growth. Mar Drugs 11:1783–1799. https://doi.org/10.3390/md11061783
119. Kang J-I, Kim M-K, Lee J-H, Jeon Y-J, Hwang E-K, Koh Y-S, Hyun J-W, Kwon S-Y, Yoo E-S, Kang H-K (2017) Undariopsis peterseniana promotes hair growth by the activation of Wnt/β-catenin and ERK pathways. Mar Drugs 15.https://doi.org/10.3390/md15050130
120. Kang J-I, Kim S-C, Kim M-K, Boo H-J, Kim E-J, Im G-J, Kim YH, Hyun J-W, Kang J-H, Koh Y-S, Park D-B, Yoo E-S, Kang H-K (2015) Effects of dihydrotestosterone on rat dermal papilla cells in vitro. Eur J Pharmacol 757:74–83. https://doi.org/10.1016/j.ejphar.2015.03.055
121. Kang J-I, Yoo E-S, Hyun J-W, Koh Y-S, Lee NH, Ko M-H, Ko C-S, Kang H-K (2016) Promotion effect of apo-9'-fucoxanthinone from Sargassum muticum on hair growth via the activation of Wnt/β-catenin and VEGF-R2. Biol Pharm Bull 39:1273–1283. https://doi.org/10.1248/bpb.b16-00024
122. Kang J-I, Yoon H-S, Kim SM, Park JE, Hyun YJ, Ko A, Ahn Y-S, Koh YS, Hyun JW, Yoo E-S, Kang H-K (2018) Mackerel-derived fermented fish oil promotes hair growth by anagen-stimulating pathways. Int J Mol Sci 19. https://doi.org/10.3390/ijms19092770
123. Kawano M, Han J, Kchouk ME, Isoda H (2009) Hair growth regulation by the extract of aromatic plant Erica multiflora. J Nat Med 63:335–339. https://doi.org/10.1007/s11418-009-0324-x
124. Keum DI, Pi L-Q, Hwang ST, Lee W-S (2016) Protective effect of Korean Red Ginseng against chemotherapeutic drug-induced premature catagen development assessed with human hair follicle organ culture model. J Ginseng Res 40:169–175. https://doi.org/10.1016/j.jgr.2015.07.004
125. Ki G-E, Kim Y-M, Lim H-M, Lee E-C, Choi Y-K, Seo Y-K (2020) Extremely low-frequency electromagnetic fields increase the expression of anagen-related molecules in human dermal papilla cells via GSK-3β/ERK/Akt signaling pathway. Int J Mol Sci 21.https://doi.org/10.3390/ijms21030784
126. Kim J, Kim SR, Choi Y-H, Shin JY, Kim CD, Kang N-G, Park BC, Lee S (2020) Quercitrin stimulates hair growth with enhanced expression of growth factors via activation of MAPK/CREB signaling pathway. Mol Basel Switz 25, https://doi.org/10.3390/molecules25174004
127. Kim J, Shin JY, Choi Y-H, Jang M, Nam YJ, Lee SY, Jeon J, Jin MH, Lee S (2019) Hair growth promoting effect of hottuynia cordata extract in cultured human hair follicle dermal papilla cells. Biol Pharm Bull 42:1665–1673. https://doi.org/10.1248/bpb.b19-00254
128. Kim JE, Lee YJ, Park HR, Lee DG, Jeong KH, Kang H (2020) The effect of JAK inhibitor on the survival, anagen re-entry, and hair follicle immune privilege restoration in human dermal papilla cells. Int J Mol Sci 21.https://doi.org/10.3390/ijms21145137
129. Kim JE, Oh JH, Woo YJ, Jung JH, Jeong KH, Kang H (2020) Effects of mesenchymal stem cell therapy on alopecia areata in cellular and hair follicle organ culture models. Exp Dermatol 29:265–272. https://doi.org/10.1111/exd.13812
130. Kim K, Choe W (2019) Recombinant human cyclophilin A stimulates hair follicle cells via Wnt/β-catenin signaling pathway. Biotechnol Lett 41:1451–1458. https://doi.org/10.1007/s10529-019-02751-w
131. Kim SM, Kang J-I, Yoon H-S, Choi YK, Go JS, Oh SK, Ahn M, Kim J, Koh YS, Hyun JW, Yoo E-S, Kang H-K (2020) HNG, a humanin analogue, promotes hair growth by inhibiting anagen-to-catagen transition. Int J Mol Sci 21.https://doi.org/10.3390/ijms21124553
132. Kim Y-D, Pi L-Q, Lee W-S (2020) Effect of chrysanthemum zawadskii extract on dermal papilla cell proliferation and hair growth. Ann Dermatol 32:395–401. https://doi.org/10.5021/ad.2020.32.5.395
133. Kim YE, Choi HC, Lee I-C, Yuk DY, Lee H, Choi BY (2016) 3-Deoxysappanchalcone promotes proliferation of human hair follicle dermal papilla cells and hair growth in C57BL/6 mice by modulating WNT/β-catenin and STAT signaling. Biomol Ther 24:572–580. https://doi.org/10.4062/biomolther.2016.183

134. Kim YE, Choi HC, Nam G, Choi BY (2019) Costunolide promotes the proliferation of human hair follicle dermal papilla cells and induces hair growth in C57BL/6 mice. J Cosmet Dermatol 18:414–421. https://doi.org/10.1111/jocd.12674
135. Kim Y-M, Kwon S-J, Jang H-J, Seo Y-K (2017) Rice bran mineral extract increases the expression of anagen-related molecules in human dermal papilla through Wnt/catenin pathway. Food Nutr Res 61:1412792. https://doi.org/10.1080/16546628.2017.1412792
136. Kimbrel EA, Lanza R (2020) Next-generation stem cells—ushering in a new era of cell-based therapies. Nat Rev Drug Discov 19:463–479. https://doi.org/10.1038/s41573-020-0064-x
137. Kiratipaiboon C, Tengamnuay P, Chanvorachote P (2016) Ciprofloxacin improves the stemness of human dermal papilla cells. Stem Cells Int 2016:5831276. https://doi.org/10.1155/2016/5831276
138. Kiso M, Hamazaki TS, Itoh M, Kikuchi S, Nakagawa H, Okochi H (2015) Synergistic effect of PDGF and FGF2 for cell proliferation and hair inductive activity in murine vibrissal dermal papilla in vitro. J Dermatol Sci 79:110–118. https://doi.org/10.1016/j.jdermsci.2015.04.007
139. Kloepper JE, Hendrix S, Bodó E, Tiede S, Humphries MJ, Philpott MP, Fässler R, Paus R (2008) Functional role of beta 1 integrin-mediated signalling in the human hair follicle. Exp Cell Res 314:498–508. https://doi.org/10.1016/j.yexcr.2007.10.030
140. Kloepper JE, Sugawara K, Al-Nuaimi Y, Gáspár E, van Beek N, Paus R (2010) Methods in hair research: how to objectively distinguish between anagen and catagen in human hair follicle organ culture. Exp Dermatol 19:305–312. https://doi.org/10.1111/j.1600-0625.2009.00939.x
141. Kwack MH, Jang YJ, Won GH, Kim MK, Kim JC, Sung YK (2019) Overexpression of alkaline phosphatase improves the hair-inductive capacity of cultured human dermal papilla spheres. J Dermatol Sci 95:126–129. https://doi.org/10.1016/j.jdermsci.2019.07.008
142. Kwack MH, Kang BM, Kim MK, Kim JC, Sung YK (2011) Minoxidil activates β-catenin pathway in human dermal papilla cells: a possible explanation for its anagen prolongation effect. J Dermatol Sci 62:154–159. https://doi.org/10.1016/j.jdermsci.2011.01.013
143. Kwack MH, Seo CH, Gangadaran P, Ahn B-C, Kim MK, Kim JC, Sung YK (2019) Exosomes derived from human dermal papilla cells promote hair growth in cultured human hair follicles and augment the hair-inductive capacity of cultured dermal papilla spheres. Exp Dermatol 28:854–857. https://doi.org/10.1111/exd.13927
144. Kwack MH, Shin SH, Kim SR, Im SU, Han IS, Kim MK, Kim JC, Sung YK (2009) l-Ascorbic acid 2-phosphate promotes elongation of hair shafts via the secretion of insulin-like growth factor-1 from dermal papilla cells through phosphatidylinositol 3-kinase. Br J Dermatol 160:1157–1162. https://doi.org/10.1111/j.1365-2133.2009.09108.x
145. Kwack MH, Yang JM, Won GH, Kim MK, Kim JC, Sung YK (2018) Establishment and characterization of five immortalized human scalp dermal papilla cell lines. Biochem Biophys Res Commun 496:346–351. https://doi.org/10.1016/j.bbrc.2018.01.058
146. Kwon OS, Han JH, Yoo HG, Chung JH, Cho KH, Eun HC, Kim KH (2007) Human hair growth enhancement in vitro by green tea epigallocatechin-3-gallate (EGCG). Phytomedicine Int J Phytother Phytopharm 14:551–555. https://doi.org/10.1016/j.phymed.2006.09.009
147. Kwon OS, Oh JK, Kim MH, Park SH, Pyo HK, Kim KH, Cho KH, Eun HC (2006) Human hair growth ex vivo is correlated with in vivo hair growth: selective categorization of hair follicles for more reliable hair follicle organ culture. Arch Dermatol Res 297:367–371. https://doi.org/10.1007/s00403-005-0619-z
148. Lachgar S, Charveron M, Gall Y, Bonafe JL (1998) Minoxidil upregulates the expression of vascular endothelial growth factor in human hair dermal papilla cells. Br J Dermatol 138:407–411. https://doi.org/10.1046/j.1365-2133.1998.02115.x
149. Langan EA, Lisztes E, Bíró T, Funk W, Kloepper JE, Griffiths CEM, Paus R (2013) Dopamine is a novel, direct inducer of catagen in human scalp hair follicles in vitro. Br J Dermatol 168:520–525. https://doi.org/10.1111/bjd.12113
150. Langan EA, Philpott MP, Kloepper JE, Paus R (2015) Human hair follicle organ culture: theory, application and perspectives. Exp Dermatol 24:903–911. https://doi.org/10.1111/exd.12836

151. Langan EA, Ramot Y, Goffin V, Griffiths CEM, Foitzik K, Paus R (2010) Mind the (gender) gap: does prolactin exert gender and/or site-specific effects on the human hair follicle? J Invest Dermatol 130:886–891. https://doi.org/10.1038/jid.2009.340
152. le Riche A, Aberdam E, Marchand L, Frank E, Jahoda C, Petit I, Bordes S, Closs B, Aberdam D (2019) Extracellular vesicles from activated dermal fibroblasts stimulate hair follicle growth through dermal papilla-secreted norrin. Stem Cells Dayt Ohio 37:1166–1175. https://doi.org/10.1002/stem.3043
153. Lee EY, Choi E-J, Kim JA, Hwang YL, Kim C-D, Lee MH, Roh SS, Kim YH, Han I, Kang S (2016) Malva verticillata seed extracts upregulate the Wnt pathway in human dermal papilla cells. Int J Cosmet Sci 38:148–154. https://doi.org/10.1111/ics.12268
154. Lee H, Kim N-H, Yang H, Bae SK, Heo Y, Choudhary I, Kwon YC, Byun JK, Yim HJ, Noh BS, Heo J-D, Kim E, Kang C (2016) The hair growth-promoting effect of rumex japonicus Houtt. Extract Evid-Based Complement Altern Med ECAM 2016:1873746. https://doi.org/10.1155/2016/1873746
155. Lee J, Böscke R, Tang P-C, Hartman BH, Heller S, Koehler KR (2018) Hair follicle development in mouse pluripotent stem cell-derived skin organoids. Cell Rep 22:242–254. https://doi.org/10.1016/j.celrep.2017.12.007
156. Lee J, Koehler KR (2021) Skin organoids: a new human model for developmental and translational research. Exp Dermatolhttps://doi.org/10.1111/exd.14292
157. Lee J, Rabbani CC, Gao H, Steinhart MR, Woodruff BM, Pflum ZE, Kim A, Heller S, Liu Y, Shipchandler TZ, Koehler KR (2020) Hair-bearing human skin generated entirely from pluripotent stem cells. Nature 582:399–404. https://doi.org/10.1038/s41586-020-2352-3
158. Lee LF, Chuong C-M (2009) Building complex tissues: high-throughput screening for molecules required in hair engineering. J Invest Dermatol 129:815–817. https://doi.org/10.1038/jid.2008.434
159. Lee N-E, Park S-D, Hwang H, Choi S-H, Lee RM, Nam SM, Choi JH, Rhim H, Cho I-H, Kim H-C, Hwang S-H, Nah S-Y (2020) Effects of a gintonin-enriched fraction on hair growth: an in vitro and in vivo study. J Ginseng Res 44:168–177. https://doi.org/10.1016/j.jgr.2019.05.013
160. Lee S-H, Yoon J, Shin SH, Zahoor M, Kim HJ, Park PJ, Park W S, Min DS, Kim H-Y, Choi K-Y (2012) Valproic acid induces hair regeneration in murine model and activates alkaline phosphatase activity in human dermal papilla cells. PloS One 7:e34152.https://doi.org/10.1371/journal.pone.0034152
161. Lee YR, Bae S, Kim JY, Lee J, Cho D-H, Kim H-S, An I-S, An S (2019) Monoterpenoid loliolide regulates hair follicle inductivity of human dermal papilla cells by activating the Akt/β-catenin signaling pathway. J Microbiol Biotechnol 29:1830–1840. https://doi.org/10.4014/jmb.1908.08018
162. Lee YR, Yamazaki M, Mitsui S, Tsuboi R, Ogawa H (2001) Hepatocyte growth factor (HGF) activator expressed in hair follicles is involved in in vitro HGF-dependent hair follicle elongation. J Dermatol Sci 25:156–163. https://doi.org/10.1016/s0923-1811(00)00124-9
163. Lehmann J, Kemper B, Barroso Peña A, Schnekenburger J, Paus R, Bertolini M, Chéret J (2019) Image gallery: optical coherence tomography for intravital human hair follicle analyses ex vivo. Br J Dermatol 180:e141.https://doi.org/10.1111/bjd.17518
164. Leirós GJ, Attorresi AI, Balañá ME (2012) Hair follicle stem cell differentiation is inhibited through cross-talk between Wnt/β-catenin and androgen signalling in dermal papilla cells from patients with androgenetic alopecia. Br J Dermatol 166:1035–1042. https://doi.org/10.1111/j.1365-2133.2012.10856.x
165. Li L, Margolis LB, Paus R, Hoffman RM (1992) Hair shaft elongation, follicle growth, and spontaneous regression in long-term, gelatin sponge-supported histoculture of human scalp skin. Proc Natl Acad Sci U S A 89:8764–8768. https://doi.org/10.1073/pnas.89.18.8764
166. Li M, Marubayashi A, Nakaya Y, Fukui K, Arase S (2001) Minoxidil-induced hair growth is mediated by adenosine in cultured dermal papilla cells: possible involvement of sulfonylurea receptor 2B as a target of minoxidil. J Invest Dermatol 117:1594–1600. https://doi.org/10.1046/j.0022-202x.2001.01570.x

167. Li W, Man X-Y, Li C-M, Chen J-Q, Zhou J, Cai S-Q, Lu Z-F, Zheng M (2012) VEGF induces proliferation of human hair follicle dermal papilla cells through VEGFR-2-mediated activation of ERK. Exp Cell Res 318:1633–1640. https://doi.org/10.1016/j.yexcr.2012.05.003
168. Lichti U, Weinberg WC, Goodman L, Ledbetter S, Dooley T, Morgan D, Yuspa SH (1993) In vivo regulation of murine hair growth: insights from grafting defined cell populations onto nude mice. J Invest Dermatol 101:124S-129S. https://doi.org/10.1111/1523-1747.ep12363165
169. Lim TC, Leong MF, Lu H, Du C, Wan ACA (2021) Can an in vitro hair drug model be developed using dermal papilla cells alone? Exp Dermatol https://doi.org/10.1111/exd.14297
170. Limat A, Breitkreutz D, Hunziker T, Klein CE, Noser F, Fusenig NE, Braathen LR (1994) Outer root sheath (ORS) cells organize into epidermoid cyst-like spheroids when cultured inside Matrigel: a light-microscopic and immunohistological comparison between human ORS cells and interfollicular keratinocytes. Cell Tissue Res 275:169–176. https://doi.org/10.1007/BF00305384
171. Limat A, Hunziker T, Waelti ER, Inaebnit SP, Wiesmann U, Braathen LR (1993) Soluble factors from human hair papilla cells and dermal fibroblasts dramatically increase the clonal growth of outer root sheath cells. Arch Dermatol Res 285:205–210. https://doi.org/10.1007/BF00372010
172. Limat A, Noser FK (1986) Serial cultivation of single keratinocytes from the outer root sheath of human scalp hair follicles. J Invest Dermatol 87:485–488. https://doi.org/10.1111/1523-1747.ep12455548
173. Lin B, Miao Y, Wang J, Fan Z, Du L, Su Y, Liu B, Hu Z, Xing M (2016) Surface tension guided hanging-drop: producing controllable 3d spheroid of high-passaged human dermal papilla cells and forming inductive microtissues for hair-follicle regeneration. ACS Appl Mater Interfaces 8:5906–5916. https://doi.org/10.1021/acsami.6b00202
174. Lin C-M, Li Y, Ji Y-C, Keng H, Cai X-N, Zhang J-K (2008) Microencapsulated human hair dermal papilla cells: a substitute for dermal papilla? Arch Dermatol Res 300:531–535. https://doi.org/10.1007/s00403-008-0852-3
175. Lindner G, Horland R, Wagner I, Ataç B, Lauster R (2011) De novo formation and ultrastructural characterization of a fiber-producing human hair follicle equivalent in vitro. J Biotechnol 152:108–112. https://doi.org/10.1016/j.jbiotec.2011.01.019
176. Lisztes E, Tóth BI, Bertolini M, Szabó IL, Zákány N, Oláh A, Szöllősi AG, Paus R, Bíró T (2020) Adenosine promotes human hair growth and inhibits catagen transition in vitro: role of the outer root sheath keratinocytes. J Invest Dermatol 140:1085-1088.e6. https://doi.org/10.1016/j.jid.2019.08.456
177. Liu L-P, Guo N-N, Li Y-M, Zheng Y-W (2020) Generation of human iMelanocytes from induced pluripotent stem cells through a suspension culture system. STAR Protoc 1:100004.https://doi.org/10.1016/j.xpro.2019.100004
178. Liu L-P, Li Y-M, Guo N-N, Li S, Ma X, Zhang Y-X, Gao Y, Huang J-L, Zheng D-X, Wang L-Y, Xu H, Hui L, Zheng Y-W (2019) Therapeutic potential of patient iPSC-Derived iMelanocytes in autologous transplantation. Cell Rep 27:455-466.e5. https://doi.org/10.1016/j.celrep.2019.03.046
179. Liu Y, Lin C, Zeng Y, Li H, Cai B, Huang K, Yuan Y, Li Y (2016) Comparison of calcium and barium microcapsules as scaffolds in the development of artificial dermal papillae. BioMed Res Int 2016:9128535.https://doi.org/10.1155/2016/9128535
180. Lu Z, Fischer TW, Hasse S, Sugawara K, Kamenisch Y, Krengel S, Funk W, Berneburg M, Paus R (2009) Profiling the response of human hair follicles to ultraviolet radiation. J Invest Dermatol 129:1790–1804
181. Lu Z, Hasse S, Bodo E, Rose C, Funk W, Paus R (2007) Towards the development of a simplified long-term organ culture method for human scalp skin and its appendages under serum-free conditions. Exp Dermatol 16:37–44
182. Madaan A, Verma R, Singh AT, Jaggi M (2018) Review of hair follicle dermal papilla cells as in vitro screening model for hair growth. Int J Cosmet Sci 40:429–450. https://doi.org/10.1111/ics.12489

183. Magerl M, Paus R, Farjo N, Müller-Röver S, Peters EMJ, Foitzik K, Tobin DJ (2004) Limitations of human occipital scalp hair follicle organ culture for studying the effects of minoxidil as a hair growth enhancer. Exp Dermatol 13:635–642. https://doi.org/10.1111/j.0906-6705.2004.00207.x
184. Mahjour SB, Ghaffarpasand F, Wang H (2012) Hair follicle regeneration in skin grafts: current concepts and future perspectives. Tissue Eng Part B Rev 18:15–23. https://doi.org/10.1089/ten.teb.2011.0064
185. McElwee K, Hoffmann R (2000) Growth factors in early hair follicle morphogenesis. Eur J Dermatol 10:341–350
186. McElwee KJ, Huth A, Kissling S, Hoffmann R (2004) Macrophage-stimulating protein promotes hair growth ex vivo and induces anagen from telogen stage hair follicles in vivo. J Invest Dermatol 123:34–40. https://doi.org/10.1111/j.0022-202X.2004.22712.x
187. McElwee KJ, Kissling S, Wenzel E, Huth A, Hoffmann R (2003) Cultured peribulbar dermal sheath cells can induce hair follicle development and contribute to the dermal sheath and dermal papilla. J Invest Dermatol 121:1267–1275. https://doi.org/10.1111/j.1523-1747.2003.12568.x
188. McElwee KJ, Sinclair R (2008) Hair physiology and its disorders. Drug Discov Today Dis Mech 5:e163–e171
189. McElwee KJ, Tosti A (2020) New developments in hair research. Exp Dermatol 29:204–207. https://doi.org/10.1111/exd.14078
190. Meier N, Langan D, Hilbig H, Bodó E, Farjo NP, Farjo B, Armbruster FP, Paus R (2012) Thymic peptides differentially modulate human hair follicle growth. J Invest Dermatol 132:1516–1519. https://doi.org/10.1038/jid.2012.2
191. Messenger AG (1984) The culture of dermal papilla cells from human hair follicles. Br J Dermatol 110:685–689. https://doi.org/10.1111/j.1365-2133.1984.tb04705.x
192. Mi T, Dong Y, Santhanam U, Huang N (2019) Niacinamide and 12-hydroxystearic acid prevented benzo(a)pyrene and squalene peroxides induced hyperpigmentation in skin equivalent. Exp Dermatol 28:742–746. https://doi.org/10.1111/exd.13811
193. Miao Y, Sun YB, Liu BC, Jiang JD, Hu ZQ (2014) Controllable production of transplantable adult human high-passage dermal papilla spheroids using 3D matrigel culture. Tissue Eng Part A 20:2329–2338. https://doi.org/10.1089/ten.TEA.2013.0547
194. Miranda BH, Charlesworth MR, Tobin DJ, Sharpe DT, Randall VA (2018) Androgens trigger different growth responses in genetically identical human hair follicles in organ culture that reflect their epigenetic diversity in life. FASEB J Off Publ Fed Am Soc Exp Biol 32:795–806. https://doi.org/10.1096/fj.201700260RR
195. Miranda BH, Tobin DJ, Sharpe DT, Randall VA (2010) Intermediate hair follicles: a new more clinically relevant model for hair growth investigations. Br J Dermatol 163:287–295. https://doi.org/10.1111/j.1365-2133.2010.09867.x
196. Moll I (1995) Proliferative potential of different keratinocytes of plucked human hair follicles. J Invest Dermatol 105:14–21. https://doi.org/10.1111/1523-1747.ep12312406
197. Munkhbayar S, Jang S, Cho A-R, Choi S-J, Shin CY, Eun HC, Kim KH, Kwon O (2016) Role of arachidonic acid in promoting hair growth. Ann Dermatol 28:55–64. https://doi.org/10.5021/ad.2016.28.1.55
198. Nakamura M, Schneider MR, Schmidt-Ullrich R, Paus R (2013) Mutant laboratory mice with abnormalities in hair follicle morphogenesis, cycling, and/or structure: an update. J Dermatol Sci 69:6–29. https://doi.org/10.1016/j.jdermsci.2012.10.001
199. Nakamura M, Sundberg JP, Paus R (2001) Mutant laboratory mice with abnormalities in hair follicle morphogenesis, cycling, and/or structure: annotated tables. Exp Dermatol 10:369–390. https://doi.org/10.1034/j.1600-0625.2001.100601.x
200. Nam G-H, Jo K-J, Park Y-S, Kawk HW, Yoo J-G, Jang JD, Kang SM, Kim S-Y, Kim Y-M (2019) Bacillus/Trapa japonica Fruit Extract Ferment Filtrate enhances human hair follicle dermal papilla cell proliferation via the Akt/ERK/GSK-3β signaling pathway. BMC Complement Altern Med 19:104.https://doi.org/10.1186/s12906-019-2514-8

201. Nam GH, Jo K-J, Park Y-S, Kawk HW, Yoo J-G, Jang JD, Kang SM, Kim S-Y, Kim Y-M (2019) The peptide AC 2 isolated from Bacillus-treated Trapa japonica fruit extract rescues DHT (dihydrotestosterone)-treated human dermal papilla cells and mediates mTORC1 signaling for autophagy and apoptosis suppression. Sci Rep 9:16903.https://doi.org/10.1038/s41598-019-53347-3
202. Nelms MD, Ates G, Madden JC, Vinken M, Cronin MTD, Rogiers V, Enoch SJ (2015) Proposal of an in silico profiler for categorisation of repeat dose toxicity data of hair dyes. Arch Toxicol 89:733–741. https://doi.org/10.1007/s00204-014-1277-8
203. Nicu C, O'Sullivan JDB, Ramos R, Timperi L, Lai T, Farjo N, Farjo B, Pople J, Bhogal R, Hardman JA, Plikus MV, Ansell DM, Paus R (2021) Dermal adipose tissue secretes HGF to promote human hair growth and pigmentation. J Invest Dermatol.https://doi.org/10.1016/j.jid.2020.12.019
204. Oh HA, Kwak J, Kim BJ, Jin HJ, Park WS, Choi SJ, Oh W, Um S (2020) Migration inhibitory factor in conditioned medium from human umbilical cord blood-derived mesenchymal stromal cells stimulates hair growth. Cells 9.https://doi.org/10.3390/cells9061344
205. Oh JW, Kloepper J, Langan EA, Kim Y, Yeo J, Kim MJ, Hsi TC, Rose C, Yoon GS, Lee SJ, Seykora J, Kim JC, Sung YK, Kim M, Paus R, Plikus MV (2016) A guide to studying human hair follicle cycling in vivo. J Invest Dermatol 136:34–44. https://doi.org/10.1038/jid.2015.354
206. Ohyama M (2019) Use of human intra-tissue stem/progenitor cells and induced pluripotent stem cells for hair follicle regeneration. Inflamm Regen 39:4.https://doi.org/10.1186/s41232-019-0093-1
207. Ohyama M, Kobayashi T, Sasaki T, Shimizu A, Amagai M (2012) Restoration of the intrinsic properties of human dermal papilla in vitro. J Cell Sci 125:4114–4125. https://doi.org/10.1242/jcs.105700
208. Ohyama M, Terunuma A, Tock CL, Radonovich MF, Pise-Masison CA, Hopping SB, Brady JN, Udey MC, Vogel JC (2006) Characterization and isolation of stem cell–enriched human hair follicle bulge cells. J Clin Invest 116:249–260
209. Oliver RF (1970) The induction of hair follicle formation in the adult hooded rat by vibrissa dermal papillae. J Embryol Exp Morphol 23:219–236
210. Oliver RF (1967) Ectopic regeneration of whiskers in the hooded rat from implanted lengths of vibrissa follicle wall. J Embryol Exp Morphol 17:27–34
211. Oliver RF (1967) The experimental induction of whisker growth in the hooded rat by implantation of dermal papillae. J Embryol Exp Morphol 18:43–51
212. Osada A, Iwabuchi T, Kishimoto J, Hamazaki TS, Okochi H (2007) Long-term culture of mouse vibrissal dermal papilla cells and de novo hair follicle induction. Tissue Eng 13:975–982. https://doi.org/10.1089/ten.2006.0304
213. Ouchi T, Morikawa S, Shibata S, Fukuda K, Okuno H, Fujimura T, Kuroda T, Ohyama M, Akamatsu W, Nakagawa T, Okano H (2016) LNGFR+THY-1+ human pluripotent stem cell-derived neural crest-like cells have the potential to develop into mesenchymal stem cells. Differ Res Biol Divers 92:270–280. https://doi.org/10.1016/j.diff.2016.04.003
214. Panteleyev AA, Jahoda CA, Christiano AM (2001) Hair follicle predetermination. J Cell Sci 114:3419–3431
215. Park G-H, Park K, Cho H, Lee S-M, Han JS, Won CH, Chang SE, Lee MW, Choi JH, Moon KC, Shin H, Kang YJ, Lee DH (2015) Red ginseng extract promotes the hair growth in cultured human hair follicles. J Med Food 18:354–362. https://doi.org/10.1089/jmf.2013.3031
216. Park J, Jun EK, Son D, Hong W, Jang J, Yun W, Yoon BS, Song G, Kim IY, You S (2019) Over-expression of Nanog in amniotic fluid-derived mesenchymal stem cells accelerates dermal papilla cell activity and promotes hair follicle regeneration. Exp Mol Med 51:72.https://doi.org/10.1038/s12276-019-0266-7
217. Park P-J, Moon B-S, Lee S-H, Kim S-N, Kim A-R, Kim H-J, Park W-S, Choi K-Y, Cho E-G, Lee TR (2012) Hair growth-promoting effect of Aconiti Ciliare Tuber extract mediated by the activation of Wnt/β-catenin signaling. Life Sci 91:935–943. https://doi.org/10.1016/j.lfs.2012.09.008

218. Park S, Shin W-S, Ho J (2011) Fructus panax ginseng extract promotes hair regeneration in C57BL/6 mice. J Ethnopharmacol 138:340–344. https://doi.org/10.1016/j.jep.2011.08.013
219. Peters EM, Liotiri S, Bodó E, Hagen E, Bíró T, Arck PC, Paus R (2007) Probing the effects of stress mediators on the human hair follicle: substance P holds central position. Am J Pathol 171:1872–1886
220. Peters EMJ, Hansen MG, Overall RW, Nakamura M, Pertile P, Klapp BF, Arck PC, Paus R (2005) Control of human hair growth by neurotrophins: brain-derived neurotrophic factor inhibits hair shaft elongation, induces catagen, and stimulates follicular transforming growth factor beta2 expression. J Invest Dermatol 124:675–685. https://doi.org/10.1111/j.0022-202X.2005.23648.x
221. Philpott M, Green MR, Kealey T (1989) Studies on the biochemistry and morphology of freshly isolated and maintained rat hair follicles. J Cell Sci 93(Pt 3):409–418
222. Philpott MP (2018) Culture of the human pilosebaceous unit, hair follicle and sebaceous gland. Exp Dermatol 27:571–577. https://doi.org/10.1111/exd.13669
223. Philpott MP, Green MR, Kealey T (1990) Human hair growth in vitro. J Cell Sci 97(Pt 3):463–471
224. Philpott MP, Kealey T (1994) Effects of EGF on the morphology and patterns of DNA synthesis in isolated human hair follicles. J Invest Dermatol 102:186–191. https://doi.org/10.1111/1523-1747.ep12371760
225. Philpott MP, Sanders DA, Bowen J, Kealey T (1996) Effects of interleukins, colony-stimulating factor and tumour necrosis factor on human hair follicle growth in vitro: a possible role for interleukin-1 and tumour necrosis factor-alpha in alopecia areata. Br J Dermatol 135:942–948. https://doi.org/10.1046/j.1365-2133.1996.d01-1099.x
226. Philpott MP, Sanders DA, Kealey T (1994) Effects of insulin and insulin-like growth factors on cultured human hair follicles: IGF-I at physiologic concentrations is an important regulator of hair follicle growth in vitro. J Invest Dermatol 102:857–861. https://doi.org/10.1111/1523-1747.ep12382494
227. Pi L-Q, Lee W-S, Min SH (2016) Hot water extract of oriental melon leaf promotes hair growth and prolongs anagen hair cycle: In vivo and in vitro evaluation. Food Sci Biotechnol 25:375–380. https://doi.org/10.1007/s10068-016-0080-0
228. Piccini I, Bakkar K, Collin-Djangoné C, Langan EA, Gherardini J, Paus R, Bertolini M (2019) 613 assessment of the induction and morphogenic potential of human hair matrix cells and dermal papilla fibroblasts ex vivo. J Invest Dermatol 139:S320.https://doi.org/10.1016/j.jid.2019.07.617
229. Pierard GE, de la Brassinne M (1975) Modulation of dermal cell activity during hair growth in the rat. J Cutan Pathol 2:35–41. https://doi.org/10.1111/j.1600-0560.1975.tb00829.x
230. Platt CI, Chéret J, Paus R (2021) Towards developing an organotypic model for the preclinical study and manipulation of human hair matrix-dermal papilla interactions. Arch Dermatol Reshttps://doi.org/10.1007/s00403-020-02178-8
231. Purba TS, Peake M, Farjo B, Farjo N, Bhogal RK, Jenkins G, Paus R (2017) Divergent proliferation patterns of distinct human hair follicle epithelial progenitor niches in situ and their differential responsiveness to prostaglandin D2. Sci Rep 7:15197.https://doi.org/10.1038/s41598-017-15038-9
232. Qiao J, Turetsky A, Kemp P, Teumer J (2008) Hair morphogenesis in vitro: formation of hair structures suitable for implantation. Regen Med 3:683–692. https://doi.org/10.2217/17460751.3.5.683
233. Qiao J, Zawadzka A, Philips E, Turetsky A, Batchelor S, Peacock J, Durrant S, Garlick D, Kemp P, Teumer J (2009) Hair follicle neogenesis induced by cultured human scalp dermal papilla cells. Regen Med 4:667–676. https://doi.org/10.2217/rme.09.50
234. Qiu W, Gu P-R, Chuong C-M, Lei M (2020) Skin cyst: a pathological dead-end with a new twist of morphogenetic potentials in organoid cultures. Front Cell Dev Biol 8:628114.https://doi.org/10.3389/fcell.2020.628114
235. Rahmani W, Abbasi S, Hagner A, Raharjo E, Kumar R, Hotta A, Magness S, Metzger D, Biernaskie J (2014) Hair follicle dermal stem cells regenerate the dermal sheath, repopulate

the dermal papilla, and modulate hair type. Dev Cell 31:543–558. https://doi.org/10.1016/j. devcel.2014.10.022

236. Rajendran RL, Gangadaran P, Bak SS, Oh JM, Kalimuthu S, Lee HW, Baek SH, Zhu L, Sung YK, Jeong SY, Lee S-W, Lee J, Ahn B-C (2017) Extracellular vesicles derived from MSCs activates dermal papilla cell in vitro and promotes hair follicle conversion from telogen to anagen in mice. Sci Rep 7:15560.https://doi.org/10.1038/s41598-017-15505-3
237. Rajendran RL, Gangadaran P, Kwack MH, Oh JM, Hong CM, Sung YK, Lee J, Ahn B-C (2021) Human fibroblast-derived extracellular vesicles promote hair growth in cultured human hair follicles. FEBS Lett.https://doi.org/10.1002/1873-3468.14050
238. Rajendran RL, Gangadaran P, Seo CH, Kwack MH, Oh JM, Lee HW, Gopal A, Sung YK, Jeong SY, Lee S-W, Lee J, Ahn B-C (2020) Macrophage-derived extracellular vesicle promotes hair growth. Cells 9.https://doi.org/10.3390/cells9040856
239. Ramot Y, Tiede S,. Bíró T, Abu Bakar MH, Sugawara K, Philpott MP, Harrison W, Pietilä M, Paus R (2011) Spermidine promotes human hair growth and is a novel modulator of human epithelial stem cell functions. PloS One 6:e22564.https://doi.org/10.1371/journal.pone.0022564
240. Rattanachitthawat N, Pinkhien T, Opanasopit P, Ngawhirunpat T, Chanvorachote P (2019) Finasteride enhances stem cell signals of human dermal papilla cells. Vivo Athens Greece 33:1209–1220. https://doi.org/10.21873/invivo.11592
241. Raza SI, Muhammad D, Jan A, Ali RH, Hassan M, Ahmad W, Rashid S (2014) In silico analysis of missense mutations in LPAR6 reveals abnormal phospholipid signaling pathway leading to hypotrichosis. PloS One 9:e104756.https://doi.org/10.1371/journal.pone.0104756
242. Rendl M, Polak L, Fuchs E (2008) BMP signaling in dermal papilla cells is required for their hair follicle-inductive properties. Genes Dev 22:543–557. https://doi.org/10.1101/gad.1614408
243. Reynolds AJ, Jahoda CAB (1993) Hair fibre progenitor cells: developmental status and interactive potential. Semin Dev Biol 4:241–250. https://doi.org/10.1006/sedb.1993.1027
244. Roosen GF, Westgate GE, Philpott M, Berretty PJM, Nuijs TAM, Bjerring P (2008) Temporary hair removal by low fluence photoepilation: histological study on biopsies and cultured human hair follicles. Lasers Surg Med 40:520–528. https://doi.org/10.1002/lsm.20668
245. Rufaut NW, Nixon AJ, Goldthorpe NT, Wallace OAM, Pearson AJ, Sinclair RD (2013) An in vitro model for the morphogenesis of hair follicle dermal papillae. J Invest Dermatol 133:2085–2088. https://doi.org/10.1038/jid.2013.132
246. Ryu HS, Jeong J, Lee CM, Lee KS, Lee J-N, Park S-M, Lee Y-M (2021) Activation of hair cell growth factors by linoleic acid in malva verticillata seed. Mol Basel Switz 26.https://doi.org/10.3390/molecules26082117
247. Ryu S, Lee Y, Hyun MY, Choi SY, Jeong KH, Park YM, Kang H, Park KY, Armstrong CA, Johnson A, Song PI, Kim BJ (2014) Mycophenolate antagonizes IFN-γ-induced catagen-like changes via β-catenin activation in human dermal papilla cells and hair follicles. Int J Mol Sci 15:16800–16815. https://doi.org/10.3390/ijms150916800
248. Sagayama K, Tanaka N, Fukumoto T, Kashiwada Y (2019) Lanostane-type triterpenes from the sclerotium of Inonotus obliquus (Chaga mushrooms) as proproliferative agents on human follicle dermal papilla cells. J Nat Med 73:597–601. https://doi.org/10.1007/s11418-019-01280-0
249. Samuelov L, Sprecher E, Sugawara K, Singh SK, Tobin DJ, Tsuruta D, Bíró T, Kloepper JE, Paus R (2013) Topobiology of human pigmentation: P-cadherin selectively stimulates hair follicle melanogenesis. J Invest Dermatol 133:1591–1600. https://doi.org/10.1038/jid.2013.18
250. Samuelov L, Sprecher E, Tsuruta D, Bíró T, Kloepper JE, Paus R (2012) P-cadherin regulates human hair growth and cycling via canonical Wnt signaling and transforming growth factor-β2. J Invest Dermatol 132:2332–2341. https://doi.org/10.1038/jid.2012.171
251. Sari ARP, Rufaut NW, Jones LN, Sinclair RD (2016) The effect of ovine secreted soluble factors on human dermal papilla cell aggregation. Int J Trichol 8:103–110. https://doi.org/10.4103/0974-7753.188963

252. Saxena N, Mok K-W, Rendl M (2019) An updated classification of hair follicle morphogenesis. Exp Dermatol 28:332–344. https://doi.org/10.1111/exd.13913
253. Sharma A, Sances S, Workman MJ, Svendsen CN (2020) Multi-lineage human iPSC-derived platforms for disease modeling and drug discovery. Cell Stem Cell 26:309–329. https://doi.org/10.1016/j.stem.2020.02.011
254. Shen H, Cheng H, Chen H, Zhang J (2017) Identification of key genes induced by platelet-rich plasma in human dermal papilla cells using bioinformatics methods. Mol Med Rep 15:81–88. https://doi.org/10.3892/mmr.2016.5988
255. Shimizu R, Okabe K, Kubota Y, Nakamura-Ishizu A, Nakajima H, Kishi K (2011) Sphere formation restores and confers hair-inducing capacity in cultured mesenchymal cells. Exp Dermatol 20:679–681. https://doi.org/10.1111/j.1600-0625.2011.01281.x
256. Shin DH, Cha YJ, Yang KE, Jang I-S, Son C-G, Kim BH, Kim JM (2014) Ginsenoside Rg3 up-regulates the expression of vascular endothelial growth factor in human dermal papilla cells and mouse hair follicles. Phytother Res PTR 28:1088–1095. https://doi.org/10.1002/ptr.5101
257. Shin H, Cho A-R, Kim DY, Munkhbayer S, Choi S-J, Jang S, Kim SH, Shin H-C, Kwon O (2016) Enhancement of human hair growth using ecklonia cava polyphenols. Ann Dermatol 28:15–21. https://doi.org/10.5021/ad.2016.28.1.15
258. Shin S, Kim K, Lee MJ, Lee J, Choi S, Kim K-S, Ko J-M, Han H, Kim SY, Youn HJ, Ahn KJ, An I-S, An S, Cha HJ (2016) Epigallocatechin gallate-mediated alteration of the MicroRNA expression profile in 5α-dihydrotestosterone-treated human dermal papilla cells. Ann Dermatol 28:327–334. https://doi.org/10.5021/ad.2016.28.3.327
259. Shin SH, Park SY, Kim MK, Kim JC, Sung YK (2011) Establishment and characterization of an immortalized human dermal papilla cell line. BMB Rep 44:512–516. https://doi.org/10.5483/bmbrep.2011.44.8.512
260. Shorter K, Farjo NP, Picksley SM, Randall VA (2008) Human hair follicles contain two forms of ATP-sensitive potassium channels, only one of which is sensitive to minoxidil. FASEB J Off Publ Fed Am Soc Exp Biol 22:1725–1736. https://doi.org/10.1096/fj.07-099424
261. Soma T, Ogo M, Suzuki J, Takahashi T, Hibino T (1998) Analysis of apoptotic cell death in human hair follicles in vivo and in vitro. J Invest Dermatol 111:948–954. https://doi.org/10.1046/j.1523-1747.1998.00408.x
262. Soma T, Tsuji Y, Hibino T (2002) Involvement of transforming growth factor-beta2 in catagen induction during the human hair cycle. J Invest Dermatol 118:993–997. https://doi.org/10.1046/j.1523-1747.2002.01746.x
263. Su Y, Wen J, Zhu J, Xie Z, Liu C, Ma C, Zhang Q, Xu X, Wu X (2019) Pre-aggregation of scalp progenitor dermal and epidermal stem cells activates the WNT pathway and promotes hair follicle formation in in vitro and in vivo systems. Stem Cell Res Ther 10:403.https://doi.org/10.1186/s13287-019-1504-6
264. Su Y-S, Fan Z-X, Xiao S-E, Lin B-J, Miao Y, Hu Z-Q, Liu H (2017) Icariin promotes mouse hair follicle growth by increasing insulin-like growth factor 1 expression in dermal papillary cells. Clin Exp Dermatol 42:287–294. https://doi.org/10.1111/ced.13043
265. Sugawara K, Zákány N, Tiede S, Purba T, Harries M, Tsuruta D, Bíró T, Paus R (2021) Human epithelial stem cell survival within their niche requires "tonic" cannabinoid receptor 1-signalling-lessons from the hair follicle. Exp Dermatol 30:479–493. https://doi.org/10.1111/exd.14294
266. Sun H, Sui Z, Wang D, Ba H, Zhao H, Zhang L, Li C (2020) Identification of interactive molecules between antler stem cells and dermal papilla cells using an in vitro co-culture system. J Mol Histol 51:15–31. https://doi.org/10.1007/s10735-019-09853-9
267. Takagi R, Ishimaru J, Sugawara A, Toyoshima K-E, Ishida K, Ogawa M, Sakakibara K, Asakawa K, Kashiwakura A, Oshima M, Minamide R, Sato A, Yoshitake T, Takeda A, Egusa H, Tsuji T (2016) Bioengineering a 3D integumentary organ system from iPS cells using an in vivo transplantation model. Sci Adv 2:e1500887.https://doi.org/10.1126/sciadv.1500887
268. Takeo M, Asakawa K, Toyoshima K, Ogawa M, Tong J, Irié T, Yanagisawa M, Sato A, Tsuji T (2021) Expansion and characterization of epithelial stem cells with potential for cyclical hair regeneration. Sci Rep 11.https://doi.org/10.1038/s41598-020-80624-3

269. Tan JJY, Common JE, Wu C, Ho PCL, Kang L (2019) Keratinocytes maintain compartmentalization between dermal papilla and fibroblasts in 3D heterotypic tri-cultures. Cell Prolif 52:e12668.https://doi.org/10.1111/cpr.12668
270. Tan JJY, Tee JK, Chou KO, Yong SYA, Pan J, Ho HK, Ho PCL, Kang L (2018) Impact of substrate stiffness on dermal papilla aggregates in microgels. Biomater Sci 6:1347–1357. https://doi.org/10.1039/c8bm00248g
271. Taylor M, Ashcroft AT, Messenger AG (1993) Cyclosporin A prolongs human hair growth in vitro. J Invest Dermatol 100:237–239. https://doi.org/10.1111/1523-1747.ep12468979
272. Tobin DJ, Gunin A, Magerl M, Handijski B, Paus R (2003) Plasticity and cytokinetic dynamics of the hair follicle mesenchyme: implications for hair growth control. J Invest Dermatol 120:895–904. https://doi.org/10.1046/j.1523-1747.2003.12237.x
273. Topouzi H, Logan NJ, Williams G, Higgins CA (2017) Methods for the isolation and 3D culture of dermal papilla cells from human hair follicles. Exp Dermatol 26:491–496. https://doi.org/10.1111/exd.13368
274. Toyoshima K-E, Ogawa M, Tsuji T (2019) Regeneration of a bioengineered 3D integumentary organ system from iPS cells. Nat Protoc 14:1323–1338. https://doi.org/10.1038/s41596-019-0124-z
275. Tsuboi R, Niiyama S, Irisawa R, Harada K, Nakazawa Y, Kishimoto J (2020) Autologous cell-based therapy for male and female pattern hair loss using dermal sheath cup cells: a randomized placebo-controlled double-blinded dose-finding clinical study. J Am Acad Dermatol 83:109–116. https://doi.org/10.1016/j.jaad.2020.02.033
276. Vahav I, van den Broek LJ, Thon M, Monsuur HN, Spiekstra SW, Atac B, Scheper RJ, Lauster R, Lindner G, Marx U, Gibbs S (2020) Reconstructed human skin shows epidermal invagination towards integrated neopapillae indicating early hair follicle formation in vitro. J Tissue Eng Regen Med 14:761–773. https://doi.org/10.1002/term.3039
277. van Beek N, Bodó E, Kromminga A, Gáspár E, Meyer K, Zmijewski MA, Slominski A, Wenzel BE, Paus R (2008) Thyroid hormones directly alter human hair follicle functions: anagen prolongation and stimulation of both hair matrix keratinocyte proliferation and hair pigmentation. J Clin Endocrinol Metab 93:4381–4388. https://doi.org/10.1210/jc.2008-0283
278. van den Broek LJ, Bergers LIJC, Reijnders CMA, Gibbs S (2017) Progress and future prospectives in skin-on-chip development with emphasis on the use of different cell types and technical challenges. Stem Cell Rev Rep 13:418–429. https://doi.org/10.1007/s12015-017-9737-1
279. Van Scott EJ, Ekel TM (1958) Geometric relationships between the matrix of the hair bulb and its dermal papilla in normal and alopecic scalp. J Invest Dermatol 31:281–287
280. Van Scott EJ, Ekel TM, Auerbach R (1963) Determinants of rate and kinetics of cell division in scalp hair. J Invest Dermatol 41:269–273
281. Veraitch O, Kobayashi T, Imaizumi Y, Akamatsu W, Sasaki T, Yamanaka S, Amagai M, Okano H, Ohyama M (2013) Human induced pluripotent stem cell-derived ectodermal precursor cells contribute to hair follicle morphogenesis in vivo. J Invest Dermatol 133:1479–1488. https://doi.org/10.1038/jid.2013.7
282. Veraitch O, Mabuchi Y, Matsuzaki Y, Sasaki T, Okuno H, Tsukashima A, Amagai M, Okano H, Ohyama M (2017) Induction of hair follicle dermal papilla cell properties in human induced pluripotent stem cell-derived multipotent LNGFR(+)THY-1(+) mesenchymal cells. Sci Rep 7:42777.https://doi.org/10.1038/srep42777
283. Wang Y, Liu J, Tan X, Li G, Gao Y, Liu X, Zhang L, Li Y (2013) Induced pluripotent stem cells from human hair follicle mesenchymal stem cells. Stem Cell Rev Rep 9:451–460. https://doi.org/10.1007/s12015-012-9420-5
284. Wang Y, Tang L, Zhu F, Jia M (2017) Platelet-rich plasma promotes cell viability of human hair dermal papilla cells (HHDPCs) in vitro. Int J Clin Exp Pathol 10:11703–11709
285. Weber EL, Woolley TE, Yeh C-Y, Ou K-L, Maini PK, Chuong C-M (2019) Self-organizing hair peg-like structures from dissociated skin progenitor cells: new insights for human hair follicle organoid engineering and Turing patterning in an asymmetric morphogenetic field. Exp Dermatol 28:355–366. https://doi.org/10.1111/exd.13891

286. Wen T-C, Li Y-S, Rajamani K, Harn H-J, Lin S-Z, Chiou T-W (2018) Effect of cinnamomum osmophloeum kanehira leaf aqueous extract on dermal papilla cell proliferation and hair growth. Cell Transplant 27:256–263. https://doi.org/10.1177/0963689717741139
287. Wessells NK, Roessner KD (1965) Nonproliferation in dermal condensations of mouse vibrissae and pelage hairs. Dev Biol 12:419–433. https://doi.org/10.1016/0012-1606(65)90007-2
288. Weterings PJ, Roelofs HM, Vermorken AJ, Bloemendal H (1983) Serial cultivation of human scalp hair follicle keratinocytes. Acta Derm Venereol 63:315–320
289. Weterings PJ, Vermorken AJ, Bloemendal H (1981) A method for culturing human hair follicle cells. Br J Dermatol 104:1–5. https://doi.org/10.1111/j.1365-2133.1981.tb01704.x
290. Woo H, Lee S, Kim S, Park D, Jung E (2017) Effect of sinapic acid on hair growth promoting in human hair follicle dermal papilla cells via Akt activation. Arch Dermatol Res 309:381–388. https://doi.org/10.1007/s00403-017-1732-5
291. Xiao S-E, Miao Y, Wang J, Jiang W, Fan Z-X, Liu X-M, Hu Z-Q (2017) As a carrier-transporter for hair follicle reconstitution, platelet-rich plasma promotes proliferation and induction of mouse dermal papilla cells. Sci Rep 7:1125.https://doi.org/10.1038/s41598-017-01105-8
292. Xiong Y, Harmon CS (1997) Interleukin-1beta is differentially expressed by human dermal papilla cells in response to PKC activation and is a potent inhibitor of human hair follicle growth in organ culture. J Interferon Cytokine Res Off J Int Soc Interferon Cytokine Res 17:151–157. https://doi.org/10.1089/jir.1997.17.151
293. Yamauchi K, Kurosaka A (2009) Inhibition of glycogen synthase kinase-3 enhances the expression of alkaline phosphatase and insulin-like growth factor-1 in human primary dermal papilla cell culture and maintains mouse hair bulbs in organ culture. Arch Dermatol Res 301:357–365. https://doi.org/10.1007/s00403-009-0929-7
294. Yang R, Zheng Y, Burrows M, Liu S, Wei Z, Nace A, Guo W, Kumar S, Cotsarelis G, Xu X (2014) Generation of folliculogenic human epithelial stem cells from induced pluripotent stem cells. Nat Commun 5:3071.https://doi.org/10.1038/ncomms4071
295. Ye J, Tang X, Long Y, Chu Z, Zhou Q, Lin B (2021) The effect of hypoxia on the proliferation capacity of dermal papilla cell by regulating lactate dehydrogenase. J Cosmet Dermatol 20:684–690. https://doi.org/10.1111/jocd.13578
296. Yoo HG, Chang I-Y, Pyo HK, Kang YJ, Lee SH, Kwon OS, Cho KH, Eun HC, Kim KH (2007) The additive effects of minoxidil and retinol on human hair growth in vitro. Biol Pharm Bull 30:21–26. https://doi.org/10.1248/bpb.30.21
297. Yoon H-S, Kang J-I, Kim SM, Ko A, Koh Y-S, Hyun J-W, Yoon S-P, Ahn MJ, Kim YH, Kang J-H, Yoo E-S, Kang H-K (2019) Norgalanthamine stimulates proliferation of dermal papilla cells via anagen-activating signaling pathways. Biol Pharm Bull 42:139–143. https://doi.org/10.1248/bpb.b18-00226
298. Yoon S-Y, Kim K-T, Jo SJ, Cho A-R, Jeon S-I, Choi H-D, Kim KH, Park G-S, Pack J-K, Kwon OS, Park W-Y (2011) Induction of hair growth by insulin-like growth factor-1 in 1,763 MHz radiofrequency-irradiated hair follicle cells. PloS One 6:e28474.https://doi.org/10.1371/journal.pone.0028474
299. Yoshida Y, Soma T, Matsuzaki T, Kishimoto J (2019) Wnt activator CHIR99021-stimulated human dermal papilla spheroids contribute to hair follicle formation and production of reconstituted follicle-enriched human skin. Biochem Biophys Res Commun 516:599–605. https://doi.org/10.1016/j.bbrc.2019.06.038
300. Young T-H, Lee C-Y, Chiu H-C, Hsu C-J, Lin S-J (2008) Self-assembly of dermal papilla cells into inductive spheroidal microtissues on poly(ethylene-co-vinyl alcohol) membranes for hair follicle regeneration. Biomaterials 29:3521–3530. https://doi.org/10.1016/j.biomaterials.2008.05.013
301. Young T-H, Tu H-R, Chan C-C, Huang Y-C, Yen M-H, Cheng N-C, Chiu H-C, Lin S-J (2009) The enhancement of dermal papilla cell aggregation by extracellular matrix proteins through effects on cell-substratum adhesivity and cell motility. Biomaterials 30:5031–5040. https://doi.org/10.1016/j.biomaterials.2009.05.065

302. Yu M, Kissling S, Freyschmidt-Paul P, Hoffmann R, Shapiro J, McElwee KJ (2008) Interleukin-6 cytokine family member oncostatin M is a hair-follicle-expressed factor with hair growth inhibitory properties. Exp Dermatol 17:12–19. https://doi.org/10.1111/j.1600-0625.2007.00643.x
303. Yue X, Chen F, He X, Song H, Wang T, Guo X, Qian Z (2016) [Effect of PDGF-C on biological characters of human dermal papilla cells in vitro]. Zhonghua Zheng Xing Wai Ke Za Zhi Zhonghua Zhengxing Waike Zazhi Chin. J Plast Surg 32:215–220
304. Zhang H, Su Y, Wang J, Gao Y, Yang F, Li G, Shi Q (2019) Ginsenoside Rb1 promotes the growth of mink hair follicle via PI3K/AKT/GSK-3β signaling pathway. Life Sci 229:210–218. https://doi.org/10.1016/j.lfs.2019.05.033
305. Zhang H, Zhu N-X, Huang K, Cai B-Z, Zeng Y, Xu Y-M, Liu Y, Yuan Y-P, Lin C-M (2016) iTRAQ-based quantitative proteomic comparison of early- and late-passage human dermal papilla cell secretome in relation to inducing hair follicle regeneration. PLoS One 11:e0167474.https://doi.org/10.1371/journal.pone.0167474
306. Zhang X, Xiao S, Liu B, Miao Y, Hu Z (2019) Use of extracellular matrix hydrogel from human placenta to restore hair-inductive potential of dermal papilla cells. Regen Med 14:741–751. https://doi.org/10.2217/rme-2018-0112
307. Zheng M, Jang Y, Choi N, Kim DY, Han TW, Yeo JH, Lee J, Sung J-H (2019) Hypoxia improves hair inductivity of dermal papilla cells via nuclear NADPH oxidase 4-mediated reactive oxygen species generation'. Br J Dermatol 181:523–534. https://doi.org/10.1111/bjd.17706
308. Zhou L, Yang K, Xu M, Andl T, Millar SE, Boyce S, Zhang Y (2016) Activating β-catenin signaling in CD133-positive dermal papilla cells increases hair inductivity. FEBS J 283:2823–2835. https://doi.org/10.1111/febs.13784
309. Zhou L, Zhang X, Paus R, Lu Z (2018) The renaissance of human skin organ culture: a critical reappraisal. Differ Res Biol Divers 104:22–35. https://doi.org/10.1016/j.diff.2018.10.002
310. Zhou N, Fan W, Li M (2009) Angiogenin is expressed in human dermal papilla cells and stimulates hair growth. Arch Dermatol Res 301:139–149. https://doi.org/10.1007/s00403-008-0907-5
311. Zhou Q, Song Y, Zheng Q, Han R, Cheng H (2020) Expression profile analysis of dermal papilla cells mRNA in response to WNT10B treatment. Exp Ther Med 19:1017–1023. https://doi.org/10.3892/etm.2019.8287

Chapter 9
Extracellular Vesicles Including Exosomes for Hair Follicle Regeneration

Edith Aberdam, Alizée Le Riche, Sylvie Bordes, Brigitte Closs, Byung-Soon Park, and Daniel Aberdam

Abstract *Introduction*: Extracellular microvesicles (EVs), including exosomes, represent an important mode of intercellular communication by serving as vehicles for the transfer of proteins, lipids, and RNA. Interest about potential roles and use of EVs on HF function is an emerging field of research that should reveal more information on the molecular interplay between cutaneous cell types and HF pathophysiology in the future. Here we review the accumulating reports from the past few years suggesting that EVs derived from different cellular sources benefit HF regeneration and thus could become candidates for stimulating hair growth in humans. *Methods*: For the selection of literature cited we used Pubmed database and Google Scholar. The keywords used in the Pubmed and Google Scholar research were: plant-derived extracellular vesicles (EVs); lactobacillus-derived EVs; stem cell-derived EVs; adipose-derived exosomes; stem cell conditioned media; clinical trials; alopecia, hair loss; hair growth. *Results*: Most clinical studies of EVs related to hair loss or hair growth have used stem cell-derived exosomes (especially derived from adipose) and assessed their effect on hair density and hair thickness. Recently, plant-derived EVs have also been studied—specifically the hair growth efficacy of asparagus and ginseng-derived exosomes has been reported. *Conclusion*:

E. Aberdam · A. Le Riche · D. Aberdam (✉)
INSERM U976, Paris, France
e-mail: daniel.aberdam@inserm.fr

Université de Paris, Paris, France

A. Le Riche
Monasterium Laboratory Skin & Hair Research Solutions GmbH, Muenster, Germany

S. Bordes · B. Closs
SILAB, Brives la Gaillarde, France

B.-S. Park (✉)
Cellpark Dermatology Clinic, Seoul, South Korea
e-mail: skin-md@hanmail.net

D. Aberdam
Centre de Recherche Des Cordeliers INSERM U1138, Physiopathology of Ocular Diseases Team, Paris, France

© The Author(s), under exclusive license to Springer Nature Switzerland AG 2022
F. Jimenez and C. Higgins (eds.), *Hair Follicle Regeneration*, Stem Cell Biology and Regenerative Medicine 72, https://doi.org/10.1007/978-3-030-98331-4_9

EVs derived from different cellular sources (MSC, DPC, DF, MAC, Deer Antlerogenic MSC, plant, *Lactobacillus*) could be excellent candidates for stimulating hair growth in humans. EVs therapy suggests impressive hair growth with no reported adverse effects but, due to its novelty, there is a need for clinical trials to confirm its efficacy and safety.

Keywords Plant-derived extracellular vesicles (EVs) · *Lactobacillus*-derived EVs · Stem cell-derived EVs · Adipose-derived exososomes · Stem cell conditioned media · Clinical trials · Alopecia · Hair loss · Hair growth

Summary

Secreted extracellular microvesicles (EVs), including exosomes, represent an important mode of intercellular communication by serving as vehicles for transfer of proteins, lipids, and RNA between cells. Interest on potential roles and use of EVs on HF function is a starting field of research and should reveal in the future more information on the molecular interplay between cutaneous cell types and HF pathophysiology. Here we review the accumulating reports of the last years suggesting that EVs derived from different cellular sources benefit on HF regeneration and thus could become candidates for stimulating hair growth in humans.

Key Points

1. Most if not all cells located in the dermis produce and secrete extracellular vesicles, including exosomes, which represent an important mode of intercellular communication.
2. Extracellular vesicles therapy appears to increase hair growth with no reported adverse effects but, due to its novelty, there is a need for clinical trials to confirm its efficacy and safety.
3. Due to their small size and ability of crossing biological barriers, extracellular vesicles could become a suitable system for drug delivery.

Introduction

Hair follicles (HFs) are epidermal appendages that contain both epithelial and mesenchymal compartments. Interactions between the epithelial and mesenchymal cells are important not only for differentiation during embryogenesis, but also for the regulation of hair cell proliferation and migration in adults [70] as well as hair growth cycling [3, 53]. Dermal papilla cells (DPCs) located at the bottom of HFs are mesenchymal key regulators of HF development, growth and cycling [46, 79] and are directly involved in pathological hair loss [19]. DPCs signal to the epithelial matrix keratinocyte cells to regulate their behavior into the different epithelial layers of the hair follicle mainly through paracrine mechanisms [53, 84]. DPCs release molecular signals of several pathways including the Wnt, Sonic Hedgehog, keratinocyte growth factor (KGF), fibroblast growth factor (FGF) and bone morphogenetic protein (BMP) to stimulate follicular epithelium proliferation and differentiation and modulate mesenchymal-epithelial interactions [2, 16, 24, 36, 47, 52, 57, 67] In addition to the microenvironmental interactions, HF cycling is also regulated by surrounding cells through molecular signals and extracellular materials. Dermal cells such as fibroblasts, adipocytes, endothelial cells, and immune cells are present all around the bulb of the HFs and therefore could interact with follicular cells [17]. In particular, cells release into the extracellular environment diverse types of membrane vesicles of endosomal and plasma membrane origin called exosomes and microvesicles, respectively. These extracellular microvesicles (EVs) represent an important mode of intercellular communication by serving as vehicles for transfer of proteins, lipids, and RNA between cells [81]. Due to these properties, numerous functions have been attributed to EVs in physiology and pathology. Although some studies focusing on EVs and skin wound healing have been nicely reviewed elsewhere [8], the impact of EVs in the HF micro and macroenvironment is still at its infancy. Here, we review the limited studies focused on EVs and HFs and discuss their relevance to HF regeneration.

Extracellular Vesicles and Exosomes Properties

Cells release nanometer-sized membrane-derived lipid bilayer vesicles into the extracellular environment, which are collectively termed as extracellular vesicles (EVs) with a size of 40–1000 nm [50]. These EVs are important communicators between cells and their microenvironment and can travel to short and far distances to mediate the horizontal transfer of molecular information to recipient cells. They were found to influence physiological and pathological conditions, such as immune homeostasis, pregnancy, infectious diseases, cancer, and neurological disorders [29, 50]. Small EVs (40–120 nm) with specific markers are termed exosomes. Exosomes are of endocytic origin while EVs (ectosomes, shedding vesicles) are formed by a direct outward budding of the cell's plasma membrane. They all transport proteins, nucleic acids (mRNAs and miRNAs), lipids, metabolites, and even organelles from donor cells [49] and enter recipient cells via endocytosis and release their bioactive contents into the cytosolic space [65, 95]. The composition of exosomes is influenced by a variety of factors, especially dependent on the origin of the cells from which they are released. They carry specific markers like tetraspanins (CD9, CD63, CD81, CD82), multivesicular bodies (MVBs) formation proteins (Alix, TSG101), but all present also on EVs [93]. Therefore, although exosomes differ from other EVs in terms of their biogenesis, most of the published studies on EVs, in particular about HF growth, do not discriminate between the EVs nature by generally using exosome terminology [50].

The most widely used purification method of EVs is differential centrifugation, a highly reliable method which encompasses sequential centrifugations, increasing in speed and time, to pellet particles decreasing in size [83]. However, it is a time-consuming procedure, requiring large sample volumes of cell culture to obtain low EV yields. Moreover, recent studies have shown that repeated ultracentrifugation of cell culture conditioned media alters the integrity of exosomes and thus can be detrimental to achieve the highest recovery of particles. Therefore, it is now believed that adoption of concentrating protocols will provide improved analysis of exosomes. Modern ultrafiltration devices provide a more rapid and overall higher yield of exosomes when compared to ultracentrifugation. Ultrafiltration coupled with size exclusion chromatography becomes a method that provides particle purity comparable to density gradient purification and is applicable to isolating a high yield of exosomes from conditioned media and human plasma in an efficient time frame [48].

Extracellular Vesicles from Different Cellular Sources Stimulate HF Growth

Dermal papilla cells (DPCs): The dermal papilla (DP) of the hair follicle is a central activator of epithelial progenitor cells that regenerate the cycling portion of the hair follicle and generate the hair shaft [53]. Early studies have strongly suggested that molecules secreted by DPCs modulate follicular keratinocyte proliferation during hair follicle cycling [47, 68, 84]. BMP and Wnt actors are major inhibitors and activators respectively and their balance regulates the telogen-anagen transition of hair follicles [16, 67].

Human DPC-EVs have been shown to stimulate in vitro the proliferation and differentiation of outer root sheath cells (ORSCs). When injected onto depilated dorsal skin of C56Bl/6 mice at telogen phase, DPC-EVs initiated in vivo the anagen stage while delaying the catagen stage of HF growth [98]. Human DPC-EVs cultivated as spheres (3D DPC-EVs) may promote the proliferation of DPCs and ORSCs and increase hair shaft elongation in human hair follicle organ culture. Injection of 3D DP-EVs onto depilated mice induced anagen from telogen while prolonging anagen. Furthermore, 3D DPC-EVs-treated spheres co-implanted with fresh mouse epidermal cells (hair reconstitution assay) induced hair follicle neogenesis [39].

Another study further identified miR-22-5p to be produced and secreted within EVs by DPCs. This miRNA inhibits Wnt signaling expression, regulate HFSC proliferation through targeting LEF1, a key transcriptional factor of the Wnt signaling pathway that directs HF patterning and stem cell fate [90].

A recent improvement for clinical application of EVs has been the encapsulation of human DPC-EVs in partially oxidized sodium alginate hydrogels (OSA-EVs) [11]. Such hydrogel encapsulation allows the DPC-EVs to be slowly and progressively released while stabilizing the EVs proteins. Moreover, OSA-EVs appeared to accelerate re-growth of back hair in mice after depilation.

Mesenchymal stem cells (MSCs): MSCs are present in most tissues, including the dermis. They have the capacity to support other cells by paracrine factors and enhance cutaneous wound healing [69]. When murine bone marrow (BM) MSC-EVs were injected in vivo onto dorsal skin, they promoted telogen to anagen conversion through exosomal wnt3a and wnt5a secretion in situ [63]. A hydrogel-based microneedle device made from hair-derived keratin has been designed for successful transdermal co-delivery of MSC-EVs. This microneedle system efficiently transported the MSC-EVs to the hair follicle niche and the combined administration induced efficient hair regrowth and de novo pigmentation [91].

Deer antler is an organ that undergoes subsequent regrowth, and because of its specificity, it is used as a regeneration model [10, 44]. Conditioned media isolated from deer antlers MSC (DaMSC-CM) release growth factors such as PDGF, Vascular Endothelial Growth Factor (VEGF) and bFGF which are key molecule for hair regeneration [75, 76]. Round vesicles derived from DaMSC-CM, with an average diameter of ~120 nm, have been identified, which are involved in the regulation of the expression of Wnt-3a, Wnt-10b and in hair regeneration [2, 30, 75, 76].

Dermal fibroblasts (DF): DF are located in different compartments of the dermis, some in close contact to HFs and DPCs. DF can be activated by different growth factors secreted by immune and adipocytes cells [80]. We recently investigated the impact of EVs isolated from DFs stimulated by bFGF and PDGF-AA (stDF-EVs) on human HF growth ex vivo and showed that these specific stDF-EVs were able to promote HFs length. Through transcriptomic and functional analyses, we identified Norrin, a nonconventional ligand of the Wnt/β-catenin pathway, as a non-cell-autonomous modulatory player of human hair follicle keratinocyte (HHFK) activation by DPCs in vitro and ex vivo [40]. We demonstrated that Norrin is secreted by DPCs treated with stDF-EVs [40]. The conditioned medium of these cells was sufficient to activate the Wnt/beta catenin pathway of HHFKs. Norrin specifically binds to the Frizzled-4 (Fzd4) receptor to mediate activation of β-catenin [89]. We observed that the Fzd4 receptor is poorly expressed in HHFKs but on the other hand it is expressed in DFs (stimulated and non-stimulated). Remarkably, we detected the presence of Fzd4 in the DF-EVs suggesting the incorporation of Fzd4 during the formation of EVs. We have thus hypothesized that stDF-EVs activate the expression and secretion of Norrin by DPCs while supplying HHFKs with the Fzd4 receptor, enabling β-catenin signaling (Fig. 9.1). It has been reported that Norrin acts as an ocular angiogenic ligand [74]. Since DP is highly vascular and Norrin activates the beta-catenin pathway in endothelial cells, it is likely that Norrin might act as a proangiogenic agent on HFs [12, 92, 96].

Macrophage-derived EVs (MAC-EVs): In Salamander, macrophages are required for limb regeneration after amputation, suggesting that macrophage-derived molecules create a regenerative environment in the injured tissue [22]. In fact, skin-resident macrophages contribute to the activation of skin epithelial stem cells [9] and induce HF growth through tumor necrosis factor (TNF)-induced AKT/beta-catenin signaling and Wnt protein production of HFSC, respectively [26, 82]. More recently, EVs isolated from the murine macrophage cell line Raw 264.7 were found to be able to significantly enhanced DPCs proliferation and migration in vitro. In vivo, MAC-EVs promoted HF growth and increased hair shaft size [64].

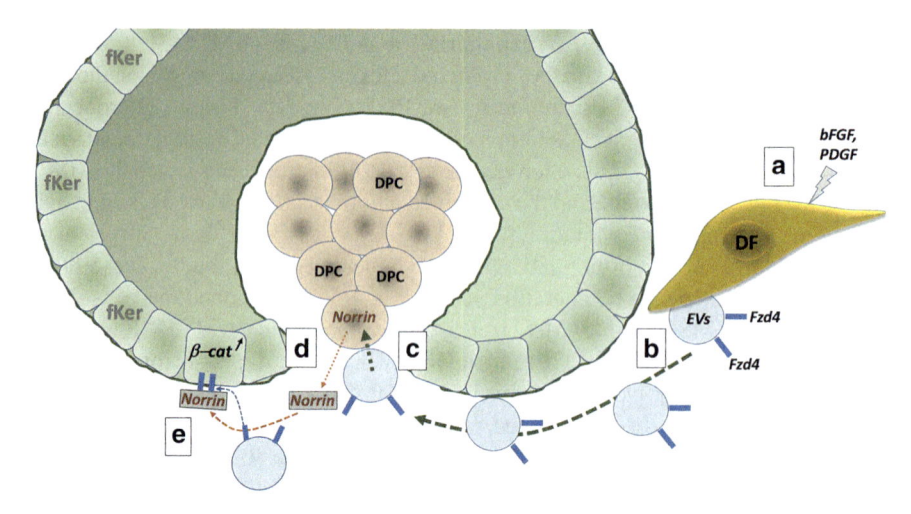

Fig. 9.1 Bottom view of a hair follicle and mechanism of growth activation through EVs. Dermal fibroblasts (DF) stimulated by bFGF and PDGF (**a**) secrete EVs (**b**) which activate the secretion of Norrin by the dermal papilla cells (**c**). In turn, soluble Norrin (**d**) activates follicular keratinocytes via its Fzd4 receptor provided by EVs (E) to eventually stimulate the beta-catenin pathway and thus follicular growth. NDP gene encodes for Norrin

Plant-derived EVs: Endosomal multivesicular body-derived exosome-like nanoparticles in plant cells may be involved in plant cell–cell communication as means to regulate plant innate immunity [1, 54, 66]. Interspecies communication of the plant EVs via crossing mammalian barriers without inducing inflammation or necrosis enable them as excellent candidates for therapy or delivery tools [15, 71]. Ginseng and its major bioactive constituents, ginsenosides, promote hair growth by enhancing proliferation of dermal papilla and preventing hair loss via modulation of various hair related cell-signaling pathways (WNT/DKK1, SHH, VEGF, TGF-β, ERK, JAK) [32, 34, 59]. Cho et al. [13] verified the spontaneous release of EVs from ginseng cells in ginseng roots and culture cells as well as their functional effects on the recovery of cellular senescence in human skin cells.

Lactobacillus-derived EVs: Several researchers have conducted studies on the skin through immunomodulation of Lactobacillus-derived exosomes. Lactobacillus plantarum-derived EVs protect atopic dermatitis (AD) by decreased interleukin (IL)-6 and IL-4 level in Staphylococus aureus-induced keratinocyte and mouse AD models [35]. Seo et al. showed an immunomodulatory effect of kefir grain-derived EVs (Lactobacillus kefirgranum, Lactobacillus kefiranofaciens and Lactobacillus kefir). Lactobacillus kefirgranum-derived EV decreased the expression level of TNF-α and increased proliferation of skin cells such as HaCaT and human DFs, which are important for hair growth [62, 75, 76]. The immunomodulatory efficacy of Lactobacillus-derived EVs including kefir grain could be applied as a new alternative to inflammation-related hair loss treatments.

Clinical Application of EVs from Different Cellular Sources for Hair growth

Given the pathogenesis of androgenetic alopecia, replenishing the signals related to hair follicle stem cell activation is promising for hair regeneration [85]. Recently, there have been diverse reports on the hair regeneration in the clinical setting using paracrine factors derived from mesenchymal stem cells [18, 77, 78, 94]. Stem cell conditioned media (CM) is obtained from stem cell culture process, and contains diverse array of paracrine factors (growth factors, angiogenic factors, hormones, cytokines, extracellular matrix proteins) and EVs (microvesicles and exosomes) [31, 33]. Besides, stem cell-derived EVs, which are purified from the CM as message entities, can be also promising for hair regrowth [63]. Human hair growth efficacy has been studied in several exosome-related products commercialized through case studies. In this review, we would like to further elaborate on studies using EVs from diverse cellular sources for human hair growth (Tables 9.1 and 9.2).

Clinical Trial Using Adipose-Derived MSC exosomes

Huh et al. conducted a study to evaluate the efficacy of adipose-derived MSC exosomes [28]. This study consisted of 39 patients (27 men and 12 women), aged 20–66 years, who received a commercialized adipose-derived MSC exosomes (AAPE® version 2.0, Prostemics. Korea) treatment with micro-needle roller. The scalp area was gently cleansed and AAPE™ version 2.0 was directly applied or sprayed close to the scalp, followed by micro-needling. Interval for treatment was every 1~2 weeks for 6~24 sessions. Hair density increased from 121.7 to 146.6 hairs/cm2 (P < 0.001), while mean hair thickness increased from 52.6 to 61.4 m (P < 0.001) (Fig. 9.2). None of the patients reported severe adverse reactions. The clinical response was maintained after discontinuation of therapy.

Clinical Trial Using Plant-Derived Exosomes

Wroński et al. have studied the application of asparagus and ginseng derived exosomes (Asprout Stemvesicle and Ginseng Stemvesicle) and its potential hair growth effect within a test period of 16 weeks [86]. The subjects used daily before bedtime, rubbed gently onto the scalp. After 8 and 16 weeks, each volunteer was examined. After using the Asprout Stemvesicle, the total amount of hair increased by 7.5% compared to 0.7% in placebo. After using Ginseng Stemvesicle the total amount of hair increased almost by 4% compared to 0.5% in placebo. During the application of Asprout Stemvesicle and Ginseng Stemvesicle, the numbers of new

Table 9.1 Lists the published studies reporting the use of EVs for hair regeneration in vitro and in vivo

Cell type origin	Treatment	Validation in vitro	Validation in vivo	Signaling pathway	Refs.
hu-DPC	–	Enhance ORS cell proliferation and migration	Accelerate HF anagen onset in mice	WNT/beta-catenin	[98]
hu-DPC	spheres	Enhance ORS cell proliferation	Accelerate HF anagen onset and prolong anagen in mice	IGF1, KGF, HGF in DPC	[39]
mu-MSC	–	Enhance ORS cell proliferation and migration	Conversion from telogen to anagen	AKT and ERK pathways Increased expression of wnts and versican	[63]
hu-DPC	EVs encapsulated in partially oxidized sodium alginate (OSA) hydrogels	OSA-EVs significantly facilitated proliferation of hair matrix cells	Accelerate hair regrowth in mice	Wnt3a and beta-catenin. Inhibition of BMP2	[11]
mu-Macrophage	–	Enhanced DPC proliferation and migration	Promoted HF growth and increased hair shaft size	Wnt3a and Wnt7b AKT pathway, VEGF, KGF	[64]
DF-EVs	bFGF/PDGF	Promote HFs length	–	Secretion of Norrin by DPC. EVs provide Fzd4 receptor for beta-cat activation	[40]
Deer antlerogenic MSCs		Enhanced DPC, HDF and keratinocyte proliferation	–	Wnt3a, Wnt10b and LEF-1	[75, 76]

Hu: human, mu: murine; DPC: dermal papilla cells; MSC: mesenchymal stem cells, DF: dermal fibroblasts; EVs: extracellular vesicles, BM: bone marrow, HF: hair follicles; ORS: outer root sheath

Table 9.2 Currently reported clinical trials with exosomes or EVs for androgenetic alopecia

Type of EVs	Treatment	Conditions	Subjects	Refs.
Adipose-derived MSC exosomes	Apply directly and micro-needling every 1~2 weeks for 6~24 sessions	AGA	39 subjects (27 males, 12 females), 24–61 years old, with AGA	[28]
Asparagus EVs/Ginseng EVs	Apply directly daily for 16 weeks	AGA	30 subjects (14 males, 16 females), 21–56 years old, with AGA	[66]

Fig. 9.2 Representative cases showing a marked response in hair growth before (**a**, **c**) and after (**b**, **d**) treatment with adipose-derived MSC exosome (AAPE® version 2.0, Prostemics, Korea). Fifteen treatment sessions were performed in 20 weeks of case (**a**, **c**), and 12 treatment sessions in 16 weeks of case (**b**, **d**), respectively

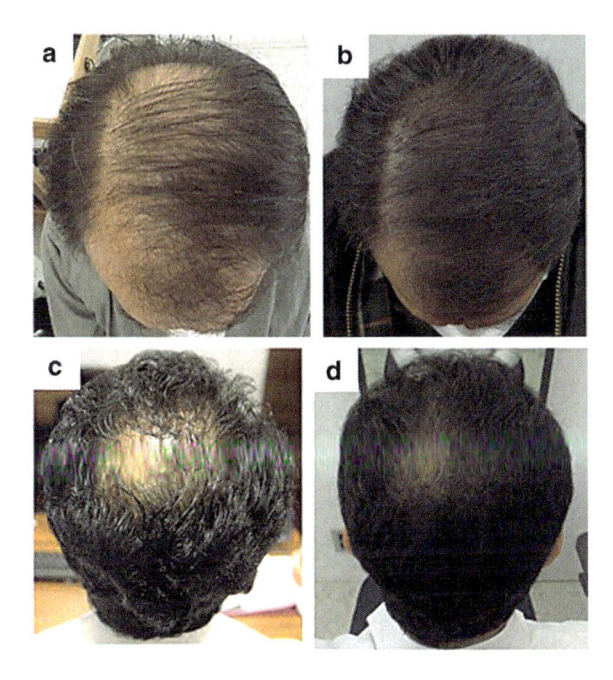

thin hairs increased by 20.7% and 10.8% respectively, while the overall thickness of the hair decreased as a result of the increase in the amount of new thin hair.

Conclusions

EVs derived from different cellular sources could be excellent candidates for stimulating hair growth in humans. EVs purified from different cell type origin (MSC, DPC, DF, MAC, Deer Antlerogenic MSC, plant, Lactobacillus) strongly suggested a potential benefit on HF cycling and wound repair. Due to their small size, their apparent non-toxic characteristics and ability of crossing the various biological barriers, they may serve as excellent delivery system for drugs. According to the ISEV (International Society for Extracellular Vesicles) position paper [43] and considering the relevant regulatory requirements, it is advised for safety purposes to systematically consider the following topics: (a) characterization of the EV source, (b) EV isolation, (c) characterization and storage strategies, (d) pharmaceutical quality control requirements, and (e) in vivo analyses of EVs. EVs display great stability on body fluids and could be embedded with specific drugs to target cells in alopecia. Specific hydrogels that could allow the slow release and stability of EVs in vivo will undoubtedly further improve their use in clinics.

Although the limited studies done in mice and humans have been promising, more robust clinical trials (double blind with control placebo) are needed to better evaluate the efficacy, safety, and limitations of EVs as therapeutic agents for the treatment of hair loss.

Acknowledgements This work was supported by grants from Fondation pour la Recherche Médicale (FRM team 2014), INSERM, Société Française de Dermatologie (SFD) and SILAB R&D.

Conflicts of Interest Byung-Soon Park is the founder and stockholder of PROSTEMICS Co., Ltd.

Bibliography

1. An Q, Hückelhoven R, Kogel KH, van Bel AJE (2006) Multivesicular bodies participate in a cell wall-associated defence response in barley leaves attacked by the pathogenic powdery mildew fungus. Cell Microbiol 8(6);1009–1019
2. Andl T, Reddy ST, Gaddapara T et al (2002) WNT signals are required for the initiation of hair follicle development. Dev Cell 2:643–653
3. Aoi N, Inoue K, Kato H, Suga H, Higashino T, Eto H et al (2012) Clinically applicable transplantation procedure of dermal papilla cells for hair follicle regeneration. J Tissue Eng Regen Med 6:85–95
4. Berger W, de Pol D, Warburg M et al (1992) Mutations in the candidate gene for Norrie disease. Hum Mol Genet 1:461–465; Familial exudative vitreoretinopathy and related retinopathies. Eye 2015; 29:1–14
5. Bian S, Zhang L, Duan L, Wang X, Min Y, Yu H (2014) Extracellular vesicles derived from human bone marrow mesenchymal stem cells promote angiogenesis in a rat myocardial infarction model. J Mol Med 92(4):387–397
6. Brown L, Wolf JM, Prados-Rosales R, Casadevall A (2015) Through the wall: extracellular vesicles in Gram-positive bacteria, mycobacteria and fungi. Nat Rev Microbiol 13:620–630
7. Cao M, Yan H, Han X, Weng L, Wei Q, Sun X, Lu W, Wei Q, Ye J, Cai X, Hu C, Yin X, Cao P (2019) Ginseng-derived nanoparticles alter macrophage polarization to inhibit melanoma growth. J Immunother Cancer 7:326
8. Carrasco E, Soto-Heredero G, Mittelbrunn M (2019) The role of extracellular vesicles in cutaneous remodeling and hair follicle dynamics. Int J Mol Sci 20:2758–2773
9. Castellana D, Paus R, Perez-Moreno M (2014) Macrophages contribute to the cyclic activation of adult hair follicle stem cells. PLoS Biol 12:e1002002
10. Cegielski M, Izykowska I, Chmielewska M et al (2013) Characteristics of MIC-1 antlerogenic stem cells and their effect on hair growth in rabbits. In Vivo 27(1):97–106
11. Chen Y, Huang J, Chen R, Yang L, Wang J, Liu B, Du L, Yi Y, Jia J, Xu Y, Chen Q, Ngondi DG, Miao Y, Hu Z (2020) Sustained release of dermal papilla derived extracellular vesicles from injectable microgel promotes hair growth. Theranostics 10:1454–1478

12. Cheng H, Zhang J, Li J, Jia M, Wang Y, Shen H (2017) Platelet-rich plasma stimulates angiogenesis in mice which may promote hair growth. Eur J Med Res 22(1):39
13. Cho EG, Choi SY, Kim H, Choi EJ, Lee EJ, Park PJ, Baek HS (2021) Panax ginseng-derived extracellular vesicles facilitate anti-senescence effects in human skin cells: an eco-friendly and sustainable way to use ginseng substances. Cells 10(3):486
14. Choi BY (2018) Hair-growth potential of ginseng and its major metabolites: a review on its molecular mechanisms. Int J Mol Sci 19(9):2703
15. Di Gioia S, Hossain MN, Conese M (2020) Biological properties and therapeutic effects of plant-derived nanovesicles. Open Med (Wars) 9;15(1):1096–1122
16. Enshell-Seijffers D, Lindon C, Kashiwagi M et al (2010) β-Catenin activity in the dermal papilla regulates morphogenesis and regeneration of hair. Dev Cell 18:633–642
17. Festa E, Fretz J, Berry R et al (2011) Adipocyte lineage cells contribute to the skin stem cell niche to drive hair cycling. Cell 146:761–771
18. Fukuoka H, Narita K, Suga H (2017) Hair regeneration therapy: application of adipose-derived stem cells. Curr Stem Cell Res Ther 12(7):531
19. Fusenig NE, Limat A, Stark HJ et al (1994) Modulation of the differentiated phenotype of keratinocytes of the hair follicle and from epidermis. J Dermatol Sci 7 (Suppl):S142eS151
20. Gentile P, Garcovich S (2019) Advances in regenerative stem cell therapy in androgenic alopecia and hair loss: Wnt pathway, growth-factor, and mesenchymal stem cell signaling impact analysis on cell growth and hair follicle development. Cells 8(5):466
21. Gilmour DF (2015) Familial exudative vitreoretinopathy and related retinopathies. Eye 29:1–14
22. Godwin JW, Pinto AR, Rosenthal NA (2013) Macrophages are required for adult salamander limb regeneration. Proc Natl Acad Sci USA 110:9415–9420
23. González R, Moffatt G, Hagner A, Sinha S, Shin W, Rahmani W et al (2017) Platelet-derived growth factor signaling modulates adult hair follicle dermal stem cell maintenance and self-renewal. NPJ Regen Med 2:11–26
24. Greco V, Chen T, Rendl M et al (2009) A two-step mechanism for stem cell activation during hair regeneration. Cell Stem Cell 4:155–169
25. Gross JC, Chaudhary V, Bartscherer K, Boutros M (2012) Active Wnt proteins are secreted on exosomes. Nat Cell Biol 14(10):1036
26. Hardman JA, Muneeb F, Pople J, Bhogal R, Shahmalak A, Paus R (2019) Human perifollicular macrophages undergo apoptosis, express wnt ligands, and switch their polarization during catagen. J Invest Dermatol 139:2543–2546
27. Harries MJ, Jimenez F, Izeta A, Hardman J, Panicker SP, Poblet E, Paus R (2018) Lichen planopilaris and frontal fibrosing alopecia as model epithelial stem cell diseases. Trends Mol Med 24(5):435–448
28. Huh CH (2019) Exosome for hair regeneration: from bench to bedside (e-Poster). In: 2019 annual meeting of American academy of dermatology (AAD), Washington, U.S.A
29. Isola AL, Chen S (2017) Exosomes: the messengers of health and disease. Curr Neuropharmacol 15:157–165
30. Ito M, Yang Z, Andl T, Cui C et al (2007) Wnt-dependent de novo hair follicle regeneration in adult mouse skin after wounding. Nature 447:316–320
31. Katsuda T, Kosaka N, Takeshita F, Ochiya T (2013) The therapeutic potential of mesenchymal stem cell-derived extracellular vesicles. Proteomics 13:1637–1653
32. Keum DI, Pi LQ, Hwang ST, Lee WS (2016) Protective effect of Korean red ginseng against chemotherapeutic drug-induced premature catagen development assessed with human hair follicle organ culture model. J Ginseng Res 40(2):169–175
33. Khanabdali R, Rosdah AA, Dusting GJ, Lim SY (2016) Harnessing the secretome of cardiac stem cells as therapy for ischemic heart disease. Biochem Pharmacol 113:1–11
34. Kim JH, Yi SM, Choi JE, Son SW (2019) Study of the efficacy of Korean red ginseng in the treatment of androgenic alopecia. J Ginseng Res 33(3):223–228
35. Kim MH, Choi SJ, Choi HI, Choi JP, Park HK, Kim EK et al (2018) Lactobacillus plantarum-derived extracellular vesicles protect atopic dermatitis induced by Staphylococcus aureus-derived extracellular vesicles. Allergy Asthma Immunol Res 10(5):516–532

36. Kishimoto J, Burgeson RE, Morgan BA (2000) Wnt signaling maintains the hair inducing activity of the dermal papilla. Genes Dev 14:1181–1185
37. Kiso M, Hamazaki TS, Itoh M, Kikuchi S, Nakagawa H, Okochi H (2015) Synergistic effect of PDGF and FGF2 for cell proliferation and hair inductive activity in murine vibrissal dermal papilla in vitro. J Dermatol Sci 79(2):110–118
38. Kwack MH, Kang BM, Kim MK, Kim JC, Sung YK (2011) Minoxidil activates beta-catenin pathway in human dermal papilla cells: A possible explanation for its anagen prolongation effect. J Dermatol Sci 62:154–159
39. Kwack MH, Seo CH, Gangadaran P, Ahn BC, Kim MK, Kim JC et al (2019) Exosomes derived from human dermal papilla cells promote hair growth in cultured human hair follicles and augment the hair-inductive capacity of cultured dermal papilla spheres. Exp Dermatol 28:854–857
40. le Riche A, Aberdam E, Marchand L, Frank E, Jahoda C, Petit I et al (2019) Extracellular vesicles from activated dermal fibroblasts stimulate hair follicle growth through dermal papilla-secreted norrin. Stem Cells 37(9):1166–1175
41. LeGrand EK, Burke JF, Costa DE, Kiorpes TC (1993) Dose responsive effects of PDGF-BB, PDGF-AA, EGF, and bFGF on granulation tissue in a guinea pig partial thickness skin excision model. Growth Factors 8(4):307–314
42. Lei M, Yang L, Chuong CM (2017) Getting to the core of the dermal papilla. J Invest Dermatol 137(11):2250–2253
43. Lener T, Gimona M, Aigner L, Börger V, Buzas E, Camussi G et al (2015) Applying extracellular vesicles based therapeutics in clinical trials–an ISEV position paper. J Extracell Vesicles 4(1):30087
44. Li C, Harper A, Puddick J, Wang W, McMahon C (2012) Proteomes and signalling pathways of antler stem cells. PloS One 7(1)
45. Li, M, Lee K, Hsu M, Nau G, Mylonakis E, Ramratnam et al (2017) Lactobacillus-derived extracellular vesicles enhance host immune responses against vancomycin-resistant enterococci. BMC Microbiol 17(1):66
46. Limat A, Breitkreutz D, Stark HJ et al (1991) Experimental modulation of the differentiated phenotype of keratinocytes from epidermis and hair follicle outer root sheath and matrix cells. Ann N Y Acad Sci 642:125–146
47. Limat A, Hunziker T, Waelti ER et al (1993) Soluble factors from human hair papilla cells and dermal fibroblasts dramatically increase the clonal growth of outer root sheath cells. Arch Dermatol Res 285:205–210
48. Lobb RJ, Becker M, Wen SW et al (2015) Optimized exosome isolation protocol for cell culture supernatant and human plasma. J Extracell Vesicles 4:27031
49. Maia J, Caja S, Strano Moraes MC, Couto N, Costa-Silva B (2018) Exosome-based cell-cell communication in the tumor microenvironment. Front Cell Dev Biol 6:18–25
50. Mathieu M, Martin-Jaular L, Lavieu G, Théry C (2019) Specificities of secretion and uptake of exosomes and other extracellular vesicles for cell-to-cell communication. Nat Cell Biol 21(1):9–17
51. Mezouar S et al (2018) Microbiome and the immune system: from a healthy steady-state to allergy associated disruption. Human Microbiome J 10:11–20
52. Millar SE (2002) Molecular mechanisms regulating hair follicle development. J Invest Dermatol 118:216–225
53. Morgan BA (2014) The dermal papilla: an instructive niche for epithelial stem and progenitor cells in development and regeneration of the hair follicle. Cold Spring Harb Perspect Med 4:a15180
54. Nielsen ME, Feechan A, Böhlenius H, Ueda T, Thordal-Christensen H (2012) Arabidopsis ARF-GTP exchange factor, GNOM, mediates transport required for innate immunity and focal accumulation of syntaxin PEN1. Proc Natl Acad Sci USA 109(28):11443
55. Oh SJ, Kim K, Lim CJ (2015) Suppressive properties of ginsenoside Rb2, a protopanaxadiol-type ginseng saponin, on reactive oxygen species and matrix metalloproteinase-2 in UV-B-irradiated human dermal keratinocytes. Biosci Biotechnol Biochem 79(7):1075–1081

56. O'Neill CA, Monteleone G, McLaughlin JT, Paus R (2016) The gut-skin axis in health and disease: a paradigm with therapeutic implications. BioEssays 38(11):1167–1176
57. Ohyama M, Kobayashi T, Sasaki T, Shimizu A, Amagai M (2012) Restoration of the intrinsic properties of human dermal papilla in vitro. J Cell Sci 125:4114–4125
58. Park BS, Kim WS, Choi JS, Kim HK, Won JH, Ohkubo F et al (2010) Hair growth stimulated by conditioned medium of adipose-derived stem cells is enhanced by hypoxia: evidence of increased growth factor secretion. Biomed Res 31(1):27–34
59. Park GH, Park KY, Cho et al (2015) Red ginseng extract promotes the hair growth in cultured human hair follicles. J Med Food 18(3):354–362
60. Paus R, Bulfone-Paus S, Bertolini M (2018) Hair follicle immune privilege revisited: the key to alopecia areata management. J Invest Dermatol Symp Proc. Elsevier:S12–S17
61. Philpott MP, Sanders DA, Bowen J, Kealey T (1996) Effects of interleukins, colony-stimulating factor and tumour necrosis factor on human hair follicle growth in vitro: a possible role for interleukin-1 and tumour necrosis factor-α in alopecia areata. Br J Dermatol 135(6):942–8
62. Choi EW, Park BS, Kang S (2018) Prostemics Co., Ltd., K.R. Exomsome and various uses thereof. K.R. Patent No. 10-2020-0060637. November 22, 2018
63. Rajendran RL, Gangadaran P, Bak SS, Oh JM, Kalimuthu S, Lee HW et al (2017) Extracellular vesicles derived from MSCs activates dermal papilla cell in vitro and promotes hair follicle conversion from telogen to anagen in mice. Sci Rep 7:15560
64. Rajendran RL, Gangadaran P, Seo CH, Kwack MH, Oh JM, Lee HW, Gopal A, Sung YK, Jeong SY, Lee SW, Lee J, Ahn BC (2020) Macrophage-derived extracellular vesicle promotes hair growth. Cells 9(4):E856
65. Raposo G, Stoorvogel W (2013) Extracellular vesicles: exosomes, microvesicles, and friends. J Cell Biol 200:373–383
66. Regente M, Pinedo M, Elizalde M, Canal LDL (2012) Apoplastic exosome-like vesicles: a new way of protein secretion in plants? Plamt Signal Behav 7(5):544–546
67. Rendl M, Polak L, Fuchs E (2008) BMP signaling in dermal papilla cells is required for their hair follicle-inductive properties. Genes Dev 22:543–557
68. Reynolds AJ, Oliver R, Jahoda C (1991) Dermal cell populations show variable competence in epidermal cell support: stimulatory effects of hair papilla cells. J Cell Sci 98:75–83
69. Rognoni E, Watt FM (2018) Skin cell heterogeneity in development, wound healing, and cancer. Trends Cell Biol 28(9):709–722
70. Roh C, Tao Q, Lyle S (2004) Dermal papilla-induced hair differentiation of adult epithelial stem cells from human skin. Physiol Genom:19:207e217
71. Rome S (2019) Biological properties of plant-derived extracellular vesicles. Food Funct 20;10(2):529–538
72. Rossi A, Cantisani C, Melis L, Iorio A, Scali E, Calvieri S (2012) Minoxidil use in dermatology, side effects and recent patents. Recent Patents Inflamm Allergy Drug Discov 6:130–136
73. Saxena R, Mittal P, Clavaud C, Dhakan DB, Hegde P, Veeranagaiah MM, Saha S, Souverain L, Roy N, Breton L, Misra N, Sharma VK (2018) Comparison of healthy and dandruff scalp microbiome reveals the role of commensals in scalp health. Front Cell Infect Microbiol 8:346
74. Seitz R, Hackl S, Seibuchner T et al (2010) Norrin mediates neuroprotective effects on retinal ganglion cells via activation of the Wnt/β-catenin signaling pathway and the induction of neuroprotective growth factors in muller cells. J Neurosci 30:5998–6010
75. Seo M, Kim J, Kim H, Choi E, Jeong S, Ki C, Jang M (2018) A novel secretory vesicle from deer antlerogenic mesenchymal stem cell-conditioned media (DaMSC-CM) promotes tissue regeneration. Stem cells Int 2018:3891404
76. Seo M, Park E, Ko S, Choi E, Kim S (2018) Therapeutic effects of kefir grain Lactobacillus-derived extracellular vesicles in mice with 2, 4, 6-trinitrobenzene sulfonic acid-induced inflammatory bowel disease. J Dairy Sci 101(10):8662–8671
77. Shin H, Ryu H, Kwon O, Park B, Jo S (2015) Clinical use of conditioned media of adipose tissue-derived stem cells in female pattern hair loss: a retrospective case series study. Int J Dermatol 54(6):730–735

78. Shin H, Won C, Chung W, Park B (2017) Up-to-date clinical trials of hair regeneration using conditioned media of adipose-derived stem cells in male and female pattern hair loss. Curr Stem Cell Res Ther 12(7):524–530
79. Taylor M, Ashcroft AT, Westgate GE et al (1992) Glycosaminoglycan synthesis by cultured human hair follicle dermal papilla cells: comparison with nonfollicular dermal fibroblasts. Br J Dermatol 126:479e484
80. Thulabandu V, Chen D, Atit RP (2018) Dermal fibroblast in cutaneous development and healing. Wiley Interdiscip Rev Dev Biol. 2018; 7(2)
81. Valadi H, Ekström K, Bossios A, Sjöstrand M, Lee JJ, Lötvall JO (2007) Exosome-mediated transfer of mRNAs and microRNAs is a novel mechanism of genetic exchange between cells. Nature Cell Biol 9:654 e659
82. Wang X, Chen H, Tian R, Zhang Y, Drutskaya MS, Wang et al (2017) Macrophages induce AKT/beta-catenin-dependent Lgr5+ stem cell activation and hair follicle regeneration through TNF. Nat Commun 8:14091
83. Witwer KW, Buzás EI, Bemis LT, Bora A, Lässer C, Lötvall J et al (2013) Standardization of sample collection, isolation and analysis methods in extracellular vesicle research. J Extracell Vesicles. 2:12–24
84. Won CH, Kwon OS, Kang YJ et al (2012) Comparative secretome analysis of human follicular dermal papilla cells and fibroblasts using shotgun proteomics. BMB Rep 45:253e258
85. Won CH, Park GH, Wu X, Tran TN, Park KH et al (2017) The basic mechanism of hair growth stimulation by adipose-derived stem cells and their secretory factors. Curr Stem Cell Res Ther 12:535–543
86. Wroński AA, Wroński A (2017) The report of hair growth efficacy test of product A and product B. Dr Koziej Instytut Badań Kosmetyków, Report number K123/JA/01
87. Xia J, Minamino S, Kuwabara K, Arai S (2019) Stem cell secretome as a new booster for regenerative medicine. Biosci Trends 13:299–307
88. Xin H, Li Y, Cui Y, Yang JJ, Zhang ZG, Chopp M (2013) Systemic administration of exosomes released from mesenchymal stromal cells promote functional recovery and neurovascular plasticity after stroke in rats. J Cereb Blood Flow Metab 33:1711–1715
89. Xu Q, Wang Y, Dabdoub A et al (2004) Vascular development in the retina and inner ear: control by Norrin and Frizzled-4, a highaffinity ligand-receptor pair. Cell 116:883–895
90. Yan H, Gao Y, Ding Q, Liu J, Li Y, Jin M et al (2019) Exosomal micro RNAs derived from dermal papilla cells mediate hair follicle stem cell proliferation and differentiation. Int J Biol Sci 15:1368–1382
91. Yang G, Chen Q, Wen D, Chen Z, Wang J, Chen G et al (2019) A therapeutic microneedle patch made from hair-derived keratin for promoting hair regrowth. ACS Nano 13:4354–4360
92. Yano K, Brown LF, Detmar M (2001) Control of hair growth and follicle size by VEGF-mediated angiogenesis. J Clin Invest 107(4):409–417
93. Yoshioka Y, Konishi Y, Kosaka N, Katsuda T, Kato T, Ochiya T (2013) Comparative marker analysis of extracellular vesicles in di_erent human cancer types. J. Extracell. Vesicles 2:20424
94. Yuan AR, Bian Q, Gao JQ (2020) Current advances in stem cell-based therapies for hair regeneration. Eur J Pharmacol:173197. https://doi.org/10.1016/j.ejphar.2020.173197. Online ahead of print
95. Zaborowski MP, Balaj L, Breakefield XO et al (2015) Extracellular vesicles: Composition, biological relevance, and methods of study. Bioscience 65:783–797
96. Zeilbeck LF et al (2016) Norrin mediates angiogenic properties via the induction of insulin like growth factor-1. Exp Eye Res 145:317–326
97. Zhang B, Wang M, Gong A, Zhang X, Wu X et al (2015) HucMSC-exosome mediated-Wnt4 signaling is required for cutaneous wound healing. Stem Cells 33:2158–2168
98. Zhou L, Wang H, Jing J, Yu L, Wu X, Lu Z (2018) Regulation of hair follicle development by exosomes derived from dermal papilla cells. Biochem Biophys Res Commun 500:325–332

Chapter 10
Stem Cell-Based Therapies for Hair Loss: What is the Evidence from a Clinical Perspective?

Byung-Soon Park and Hye-In Choi

Abstract *Introduction*: Stem cell-based therapies, especially those based on stem cell-derived conditioned medium (CM) are being widely explored in the clinical setting as a promising therapy for hair follicle regeneration. Based upon the pathogenesis of androgenetic alopecia (AGA) where the ability of hair follicle stem cells to differentiate into progenitors is disrupted, the goal of stem cell-derived CM is to restore the signals responsible for hair follicle stem cell activation. In this chapter we provide an up-to-date review of the stem cell-based technologies being used for hair regeneration, including outcomes of the clinical studies, therapeutic potential and possible mechanisms of action. *Methods*: For the selection of literature cited we used Pubmed database and ClinicalTrials.gov. The keywords used in the Pubmed and ClinicalTrials.gov. research were: alopecia, androgenetic alopecia; hair loss; hair growth; stem cell; stem cell therapy; platelet-rich plasma (PRP); stromal vascular fraction cells (SVFs); cell-free therapy; secretome; stem cell conditioned media (CM); mesenchymal stem cell exosomes; adipose-derived exosomes; clinical trials. *Results*: The use of PRP for the treatment of AGA has received considerable attention, but the lack of standard protocols (various centrifugation methods, platelet concentrations obtained, whether or not activators are added, etc.) has resulted in conflicting clinical outcomes. Several studies using SVF or stem cells show improvement in hair density in AGA patients, but need to be validated with more rigorous randomized placebo-controlled studies. Recently, many clinical studies have shown positive effects of stem cell CM on hair regeneration. Adipose stem cell derived-CM including growth factors and cytokines, enhances hair thickness and hair density, stimulates the transition to anagen, and promotes dermal papilla cell proliferation via activation of both Erk and Akt signaling pathways. *Conclusion*: The role of stem cell-based therapies in hair loss disorders is still in its infancy. Due to the some

---

COI Statement Byung-Soon Park is the founder and stockholder of PROSTEMICS Co., Ltd. Hye-In Choi is an employee of PROSTEMICS Co., Ltd.

B.-S. Park (✉)
Cellpark Dermatology Clinic, Seoul, South Korea
e-mail: skin-md@hanmail.net

H.-I. Choi
Prostemics Research Institute, Seongdong-gu, Seoul, South Korea

© The Author(s), under exclusive license to Springer Nature Switzerland AG 2022
F. Jimenez and C. Higgins (eds.), *Hair Follicle Regeneration*, Stem Cell Biology and Regenerative Medicine 72, https://doi.org/10.1007/978-3-030-98331-4_10

limitations of stem cell therapies the use of the stem cell secretome rather than stem cells themselves is emerging as a new alternative to hair loss treatment because of its efficacy in hair growth while avoiding potential risks of cell-based therapies. Overall, more randomized controlled, double-blind studies with larger sample sizes, objective evaluation methodologies and longer follow-up periods, as well as improved delivery systems are needed to evaluate and confirm the clear-cut benefits of stem cells and stem cell derived CM in the treatment of hair loss.

Keywords Alopecia · Androgenetic alopecia · Hair loss · Hair growth · Stem cell · Stem cell therapy · Platelet-rich plasma (PRP) · Stromal vascular fraction cells (SVFs) · Cell-free therapy · Secretome · Stem cell conditioned media · Mesenchymal stem cell exososomes · Adipose-derived exososomes · Clinical trials

Key Points

- Stem cell-based approaches include either stem cell implantation or stem cell-derived conditioned medium (CM).
- Cell-free therapy using CM (derived from stem cells) and/or exosomes is getting more attention due to the demand for safer and cost-effective therapies.
- Mesenchymal Stem Cell (MSC)-derived secretome composed of paracrine factors and exosomes is one of the most frequently utilized modalities among the clinical trials on stem cell-based therapies.

Introduction

In androgenetic alopecia (AGA), hair follicles (HF) on the frontal scalp transition from a terminal to a vellus (miniaturized) state, while HF on the occipital scalp are spared. Miniaturized follicles are characterized by a decrease in the number of cells within the dermal papilla (DP), located in the mesenchyme at the base of the HF. Intriguingly, the number of hair follicle stem cells (HFSC) in miniaturized HFs is equivalent to that observed in terminal HFs, however the number of HFSC progenitors is markedly decreased [1]. Depending on if the driving force in AGA is mesenchymal or epithelial there are two different proposed mechanisms for this observation, the first is that there is an intrinsic defect in HFSC (located at the bulge region) in AGA, which leads to their inability to convert to progenitors. The second is that there is an altered, or weakened signal from the DP to the bulge HFSC, initiating this conversion [2]. Currently, two therapeutic strategies for the regeneration of HFs using stem cells have been proposed, which target both of the mechanisms mentioned above: (1) reversing the pathogenesis of hair loss by activating the DP; (2) reversing the pathogenesis of hair loss by activating bulge (HFSC) [3–6] (Fig. 10.1).

The mode action by which stem cells are proposed to reverse the pathological mechanism of AGA is thought to be a paracrine one, and includes growth factors and cytokines which activate HF cells in their native niche [7–9]. The other mode of action

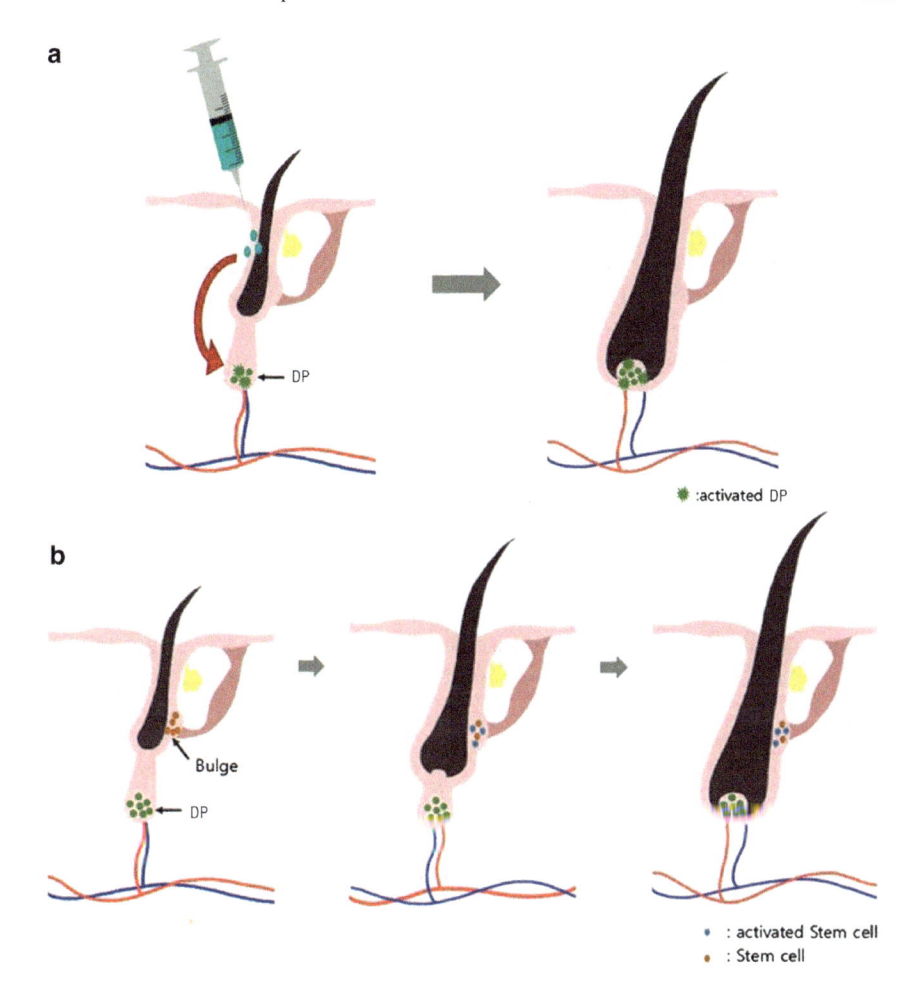

Fig. 10.1 Schematic diagram of the two possible therapeutic strategies for the regeneration of hair follicles (HFs) using stem cells: **a** reversing the pathogenesis of AGA by targeting the DP; **b** reversing the pathogenesis of AGA by targeting the HFSC

of stem cells therapies is direct differentiation. It is therefore perhaps noteworthy to mention that this type of stem cell-based tissue engineering is emerging as a thriving approach, aiming to reconstruct HFs in vitro to replace lost or damaged HFs. HF neogenesis is discussed in other chapters of this book and will therefore not be discussed within this chapter.

Current Situation of Stem Cell-Based Therapies in the Field of Hair Loss Treatments

With the advancement of regenerative medicine, stem cell-based therapies have opened new routes to cope with the challenges posed by conventional treatments for hair loss [10–13].

Multipotent stem cells found in various tissues, such as adipose, bone marrow, follicle, and umbilical cord blood, have been proposed to have the capacity to regenerate HFs in the skin [14–17]. Their high multipotent differentiation and weak immunogenic potential makes them exciting therapeutics to explore going forward [18, 19]. However, there are many limitations of stem cell therapies, not least the cost which is driven up by the short shelf life and requirement of specialized production, transportation and storage conditions [20]. Cell free conditioned medium (CM) derived from stem cells has fueled the field of hair research in recent years due to the demand for safer, more efficient and cost-effective therapies. Stem cell CM is usually obtained from cultured stem cells, and contains a diverse array of paracrine factors (growth factors, angiogenic factors, hormones, cytokines) and extracellular vesicles (microvesicles and exosomes) (see Chap. 9 by Aberdam et al.). In the following sub-sections of this chapter we will discuss both stem cell therapies, and stem cell derived CM therapies, in the context of hair regeneration.

Clinical Application of Platelet-Rich Plasma in Hair Loss

Although the use of platelet-rich plasma (PRP) in the treatment of androgenetic alopecia has gained significant attention, the clinical outcomes are conflicting and it is still considered a controversial therapy for AGA [21–25]. PRP is not in a strict sense a stem cell-based therapy; it is rather a concentrate of human platelets containing a number of growth factors stored as alpha granules [26]. The method of preparation is simply based on drawing blood from the patient, which is then centrifuged to separate out the plasma in order to obtain a high concentrate of platelets. The general consensus is that the concentration of platelets should be 4–6 times higher than the basal platelet count for optimal outcomes [27]. Once centrifuged, the PRP is injected into the reticular dermis using a 27–30 gauge needle. Systematic review articles and meta-analyses suggest that at least three injections at one month intervals should be administered in AGA patients [27–29].

There is not a standard PRP protocol due to the different methods for centrifugation, concentration of platelets obtained, the addition or not of activators such as thrombin, calcium chloride or calcium gluconate, as well as the wide biological and temporal variation even in the same person [30]. This results in variable clinical outcomes and a lack of a rigorous evaluation of the clinical trials and publications [31–33]. There are some success stories of PRP though. Patients treated with calcium gluconate-activated PRP displayed increased terminal hair thickness three months post-medical procedure [34, 35]. This can perhaps be explained by observations in cultured DP cells which show increased proliferation, improved Bcl-2 and FGF-7 levels, activated ERK and Akt proteins, and up-regulation of β-catenin when cultured in an activated PRP-enhanced growth medium [36]. Gentile et al. also reported improvements in hair density and hair count in the treated zone compared to the placebo zone [37]. They suggested that PRP injections are more effective in low and moderate-grade AGA, rather than high-grade AGA [38].

Interestingly, a lack of association between platelet counts, growth factor platelet-derived growth factor Platelet Derived Growth Factor (PDGF), EGF, and Vascular Endothelial Growth Factor (VEGF) levels, and clinical improvement has been recently reported [39], suggesting that other mechanisms or other growth factors could be involved in the hair growth promoting effects of PRP.

Clinical Applications of Stromal Vascular Fraction Cells in Hair Loss

Stromal Vascular Fraction cells (SVFs) are a heterogeneous collection of non-cultured cells isolated from adipose tissue using minimal manipulation, centrifugation, filtration, and purification (proposed by European Medicines Agency (EMA)-Committee for Advanced Therapies (CAT)), or enzymatic digestion (not proposed in the EMA-CAT suggestions). SVFs have attracted substantial attention in various fields for their potential use in regenerative medicine [40, 41]. The potential regenerative effect of SVF is proposed to occur through four possible mechanisms.

(1) Stem cell and progenitor cells within the SVF directly replace progenitor cells in the host.
(2) SVF promotes rejuvenation of tissue by reversing the cell differentiation process.
(3) Transplanted SVF contains trophic, angiogenic, or immunomodulatory factors that provide signals to nearby endogenous cells or distant cells. In the first case this refers to paracrine signals, while in the second case endocrine signaling may result in the assembly and homing of distant host cells.
(4) Cells within the SVF may combine with host cells (in a procedure known as cell combination). Heterogeneity in the condition of the cells or donors, and lack of specificity in the preparatory protocols and the type of infusion may influence clinical outcomes.

HFs are encompassed by dermal white adipose tissue which together with the dermis constitute a macro-environment that is essential for coordinating activation of both DP and HFSCs [42–44]. In the context of hair growth, adipocyte precursor cells (which are found within SVF) have been shown to activate HFSCs by releasing growth factors such as PDGF that can induce the transition of HFs from telogen to anagen in mice that genetically lack adipocyte precursors [42]. While unfractionated SVF has no effect on the telogen to anagen transition in mice [42], several groups have evaluated SVF efficacy in humans. Perez-Meza et al. reported the safety and tolerability in patients affected by AGA treated with SVF-enhanced fat graft procedures. The patients displayed a hair density improvement, with a mean increment of 31.0 hairs/cm^2, whereas those who received fat alone (without SVFs addition) showed a mean improvement of 14.0 hairs/cm^2 [45]. However, large controlled clinical studies examining the efficacy and possible side effects have not yet been reported. Recently, research trends conditioned medium using are increasing in popularity over using SVF, which will be discussed later on in this chapter.

Clinical Application of Stem Cells in Hair Loss

We conducted a search on ClinicalTrials.gov using the key word 'stem cell' and condition 'hair loss', to identify clinical trials for the treatment of hair loss. A total of 15 clinical trials were identified, however when we filtered through these to identify trials using stem cells as a therapy in the category of hair loss, just 9 were relevant (Table 10.1). Stem cells derived from adipose tissue are the most frequently used stem cell population in clinical trials, comprising 7 of 9 trials registered. The two remaining trials utilized cord-blood stem cells. Wharton's jelly has also been investigated as alternative source of stem cells although no clinical trials for hair loss have utilized this yet [19, 46]. It should be noted that despite the common misnomer that DP cells are stem cells, they do not satisfy stem cell criteria and are therefore not stem cells. Any trials containing DP cells were therefore not included here.

A method of mechanical centrifugation of punch biopsies excised from human scalp with the goal of obtaining an autologous cell suspension that could be reinjected into the patient's scalp has been described [5, 14]. A few publications show improvement in hair density of patients with AGA using this technology, but need to be validated with more rigorous randomized placebo-controlled studies [47].

Stem Cell-Free Therapy: Secretome Based Approach in Hair Loss

The secretome has been described as a complex mixture of soluble products released by the stem cells and composed of paracrine factors (constituted by growth factors and cytokines) and extracellular vesicles (microvesicles and exosomes) [48] (Fig. 10.2). The term "conditioned medium (CM)" refers to the nutrient medium where stem cells are cultured, which contains abundant secretome.

The use of a stem cell secretome rather than stem cells themselves could avoid potential risks such as immune compatibility, tumorigenicity, and transmission of infections which are observed in cell-based therapies. Gnecchi and colleagues revealed that MSCs mediated their therapeutic effects for heart disease by the

Table 10.1 Current clinical trials of stem cell therapy in hair loss (ClinicalTrials.gov as of the end of 2021 using the keywords "stem cell", and "hair loss")

Type of stem cells	Treatment	Conditions	Clinical phase	Status (Results)	NCT number
Adipose-derived tissue stromal vascular fraction	Mixture of adipose-derived SVF and high-density PRP concentrate via intravenous infusion. 60 subjects. Follow up for 12 months	AA, scarring alopecia	NA	Recruiting	NCT 03078686
Adipose-derived tissue stromal vascular fraction	Mixture of adipose-derived SVF and high-density PRP concentrate via intradermal injection of scalp. 60 subjects. Follow up for 12 months	AGA, FPHL	I/II	Recruiting	NCT 02849470
Adipose-derived tissue stromal vascular fraction	Autologous adipose-derived SVF via intradermal injection of scalp. 8 subjects. Follow up for 6 months	AGA	NA	Completed (No results posted)	NCT 02729415
Adipose-derived tissue stromal vascular fraction	Implanting a combination of SVF and human platelet rich plasma. 22 subjects. Follow up for 6 months	AGA	II	Completed	NCT 02865421

(continued)

release of soluble factors (secretome) [49]. Indeed, cell tracking analysis has revealed that when transplanted, the secretome is considered the primary attribute of MSC-mediated repair and regeneration rather than MSCs (and stem cells in general) being integrated into the injured sites [48].

ADSC derived-CM seem to have positive effects on hair regeneration [50, 51] (Fig. 10.3). Stem cell derived CM includes growth factors and cytokines that have been correlated with hair regrowth such as vascular endothelial growth factor

Table 10.1 (continued)

Type of stem cells	Treatment	Conditions	Clinical phase	Status (Results)	NCT number
Adipose-derived tissue stromal vascular fraction	Group 1-lipoaspiration and transplantation of ADSVCs (primary fresh cells without culture) Group 2-lipoaspiration and transplantation of ADSCs (after culture) 20 subjects. Follow up for 6 months	Alopecia, Hair Loss, Baldness	NA	Unknown	NCT 03427905
Adipose-derived stem cells	Mixture of adipose-derived stem cells and PRP via intradermal injection of scalp. 40 subjects. Follow up for 12 months	AGA	IV	Unknown	NCT 03388840
Adipose-derived stem cells	Apply 1.2 g of allogeneic human ADSC component extract on their scalp. 38 subjects. Follow up for 16 weeks	AGA	NA	Completed	NCT 02594046
Cord blood-derived stem cells	Patient's lymphocytes co-culture with Cord blood stem cells via intravenous infusion. 30 subjects. Follow up for 54 weeks	AA	I/II	Unknown	NCT 01673789

(continued)

10 Stem Cell-Based Therapies for Hair Loss: What is the Evidence ... 227

Table 10.1 (continued)

Type of stem cells	Treatment	Conditions	Clinical phase	Status (Results)	NCT number
Cord blood-derived stem cells	Patient's lymphocytes co-culture with Cord blood stem cells via intravenous infusion combined with oral minoxidil. 20 subjects. Follow up for 12 months	AA	II	Not yet recruiting	NCT 04011748

* ADSVCs Adipose-Derived Stromal Vascular Fraction Cells

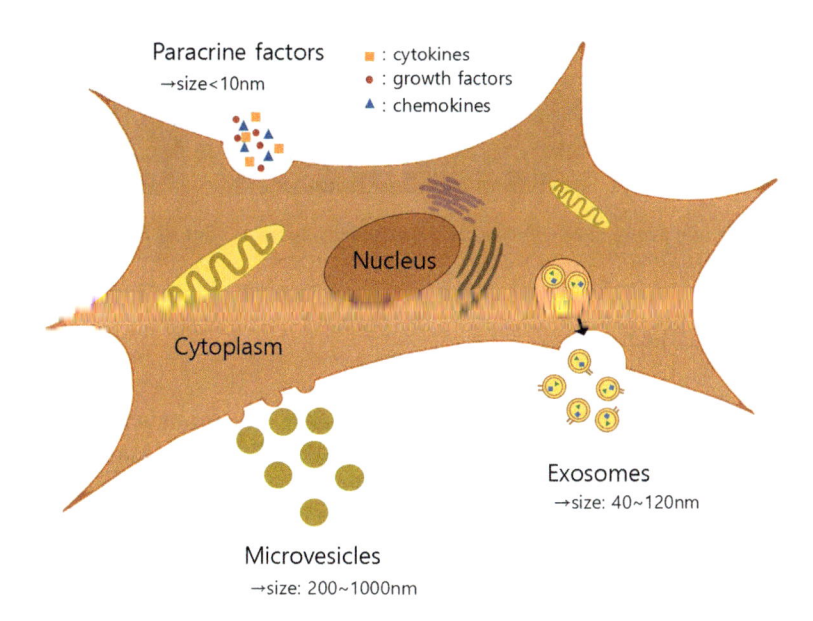

Fig. 10.2 The secretome is composed of paracrine factors and extracellular vesicles; exosomes and microvesicles

(VEGF), insulin-like growth factor (IGF), hepatocyte growth factor (HGF), platelet-derived growth factor (PDGF), bone morphogenetic proteins (BMPs), interleukin-6 (IL-6) and macrophage colony-stimulating factor (M-CSF) [52–54]. Considering that the natural response related to hair regeneration is a complex process involving the activation and differentiation of HFSCs, it seems likely that a group of paracrine factors rather than one single paracrine factor in the CM cocktail could be responsible for enhancing hair growth by increasing the number of hair follicles in anagen [20].

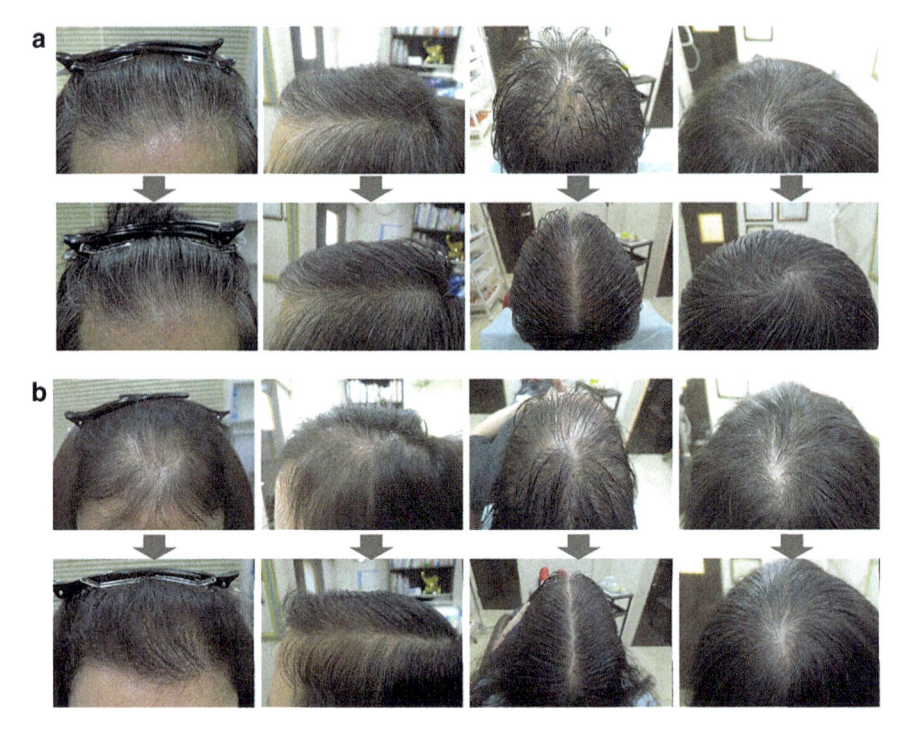

Fig. 10.3 Representative cases with marked response in hair growth with ADSC-CM. **a** 46 years old male with Hamilton-Norwood scale IV received 6 treatment sessions in 5 months. **b** 26 years old female received 4 treatment sessions in 4 months. The orders of photos are frontal view, side view of frontal scalp, vertex and side view of the vertex, respectively (unpublished data, Courtesy of Dr. Yoshizawa, Japan)

Diverse strategies have been employed to change the composition and abundance of paracrine factors, thereby influencing the regulatory effects of CM on HFs. As stem cells are usually thought to reside in hypoxic areas of the body, hypoxia serves an essential role in maintaining the stem cell niche [55–57]. It has been demonstrated that ADSCs under hypoxic conditions in vitro increase the secretion of several growth factors and when hypoxia induced ADSC-CM was subcutaneously injected into mice in vivo, it induces anagen [54, 58]. Several stimulators, such as vitamin C, LL-37 (a naturally occurring antimicrobial peptide found in the wound bed that assists in wound repair) enhance the hair growth potential of ADSC-CM by increasing secretion of hair growth-promoting factors [59, 60].

The hair growth promoting effect of ADSC derived-CM is thought to be via the enlargement and protection of the DP. In cultured DP cells, ADSC-CM promotes proliferation via activation of both Erk and Akt signaling pathways. It also protects DP cells from damage caused by androgens and reactive oxygen species [61, 62]. In addition to the injection method, a micro-needle roller in combination with an active transdermal delivery method would suffice for the application of ADSC-CM to achieve hair regeneration [51].

To date and to our knowledge, there have been 11 studies, either registered or published, that use stem cells-derived CM for the treatment of hair loss, which were searched through Clinical Trials.gov and/or Pubmed at the end of 2021 (Table 10.2). Of these, 5 studies used CM collected from hypoxia induced ADSC, while 2 used CM collected from hypoxia induced cells that were not ADSC (multipotent cells and dermal cells). One study used non-stimulated ADSC-CM, while the final study used cord blood derived MSCs as the source of CM. Four of the studies investigated the effects of AAPE® (Advanced Adipose-derived stem cell Protein & Exosome, Prostemics, Seoul, Korea), a commercially available product containing various growth factors (HGF, FGF-I, granulocyte colony-stimulating factor (G-CSF), granulocyte macrophage-colony-stimulating factor (GM-CSF), VEGF) and exosomes. In a clinical study with 22 AGA patients the mean augmentation in the hair number (hairs/0.65 cm^2) was 29 ± 4.1 in men and 15.6 ± 4.2 in women [1]. Shin et al. also used AAPE® in an observational examination of 27 women with female pattern hair loss (FPHL). They concluded that AAPE was effective in treating FPHL following 12 weeks of treatment, citing improvements in hair thickness and hair density without serious side effects [51, 63].

Concluding Remarks

The role of stem cell-based therapies in hair loss disorders is still in its infancy. While the use of stem cells in hair regeneration has got high expectations, concerns about both its convenience and cost as well as biosafety have retarded the wide use for clinical applications. To increase the likelihood of clinical translation, it is inevitable to standardize the methods of stem cell cultures, collection, preservation and validation of CM, as well as the isolation of exosomes. In addition, combination of CM and/or exosomes with non-invasive delivery systems that do not compromise their efficacy will lead to better patient compliance.

Overall, more randomized controlled, double-blind studies with larger sample sizes, objective evaluation methodologies and long follow-up periods, as well as improved delivery systems are needed to evaluate and confirm the clear-cut benefits of stem cells and stem cell derived CM in the treatment of hair loss.

Table 10.2 Currently reported clinical studies with CM derived from stem cells for hair loss (PubMed and ClinicalTrials.gov as of the end of 2021 using the keywords "hair loss" and "stem cell conditioned media")

Type of MSCs	Treatment	Conditions	Subjects	Results	NCT number or Ref.
Hypoxia-induced dermal cells	Hair Stimulating Complex (HSC) is CM of hypoxia-induced dermal cells. 0.8 ml of HSC injected intradermally into the scalp at 2 timepoints separated by 6 weeks	AGA	N = 56	Unknown	NCT 01501617
Hypoxia-induced multipotent cells	Hair Stimulating Complex (HSC) is CM of hypoxia-induced multipotent cells. 4 ml of HSC injected intradermally into the scalp at 2 timepoints separated by 6 weeks	FPHL	N = 27	Unknown	NCT 03662854
Umbilical Cord-blood	Hair serum with 5% CM of human umbilical cord blood-derived MSCs. Apply on hair and scalp twice a day for 24 weeks	AGA	N = 84	Completed (No results posted)	NCT 03676400
Hypoxia-induced ADSCs	Four weekly sessions of ADSC-CM (AAPE®) intradermal injection with 3-4 ml (0.02-0.05 ml/cm^2	AGA	N = 25	• Visual Analog Scale (mean): 2.52 to 4.20	[64]

(continued)

Table 10.2 (continued)

Type of MSCs	Treatment	Conditions	Subjects	Results	NCT number or Ref.
Hypoxia-induced ADSCs	Topical application of ADSC-CM (AAPE®) with a micro-needle stamp after non-ablative fractional laser. Once a week for 12 consecutive weeks	FPHL	N = 27	• Hair density (mean): 105.4 to 122.7 hairs/cm² • Hair thickness (mean): 57.5 to 64.0 μm	[51]
Hypoxia-induced ADSCs	Six sessions of ADSC-CM (AAPE®) intradermal injection with 3-4 ml (0.02 ml/cm²)	Alopecia	N = 11 (M) N = 11 (F)	• Increase in number of hairs (mean ± SD): – Male 29 ± 4.1 hairs/95mm² – Female 15.6 ± 4.2 hairs/95 mm²	[50]
Hypoxia-induced ADSCs	Topical application of ADSC-CM (AAPE®) with a micro-needle roller. Once a week for 12 consecutive weeks	AGA or FPHL	N = 25 (M) N = 27 (F)	• Hair density (mean): – Male 97.7 to 108.1 hairs/cm² – Female 105.4 to 122.7 hairs/cm² • Hair thickness (mean): – Male 65.4 to 71.8 μm – Female 57.5 to 64.0 μm • Split scalp study (mean ± SD): 6 Male 125.4 ± 9.5 to 137.5 ± 11.2	[63]

(continued)

Table 10.2 (continued)

Type of MSCs	Treatment	Conditions	Subjects	Results	NCT number or Ref.
Hypoxia-induced ADSCs	Six or eight sessions of ADSC-CM (AAPE®) intradermal injection with 3-4 ml (0.02-0.05 ml/cm^2	AGA or FPHL	N = 16 (M) N = 5 (F)	• Increase in number of hairs (mean ± SD):-Male 141.3 ± 31.4 hairs/95 mm^2 – Female 109.8 ± 43.5 hairs/9mm^2	[65]
Hypoxia-induced ADSCs	Six sessions of ADSC-CM (AAPE®) intradermal injection (1 vial of AAPE® in each session)	AGA or FPHL	N = 40	• Hair density and anagen hair rate increased significantly	[66]
ADSCs	Topical application of ADSC-CM with a micro-needle stamp after non-ablative fractional laser. Once a week for 12 consecutive weeks	AGA or FPHL	N = 30	• Hair density (mean ± SD): 88.5 ± 5.18 to 102.1 ± 4.09 hairs/cm^2	[67]
Umbilical Cord-blood	Topical agent with 5% CM of human umbilical cord blood-derived MSCs. Apply on scalp twice a day for 16 weeks	AA	N = 30	• Hair density (mean ± SD): 96.69 ± 16.989 to 110.06 ± 17.726 hairs/cm^2 • Hair thickness (mean ± SD): 0.074 ± 0.009 to 0.094 ± 0.010 mm • Hair growth (mean ± SD): 0.262 ± 0.039 to 0.312 ± 0.045mmday	[68]

Declaration of Competing Interest Byung-Soon Park is the founder and stockholder of PROSTEMICS Co., Ltd. Hye-In Choi is an employee of PROSTEMICS Co., Ltd.

References

1. Garza LA, Yang C, Zhao T, Blatt HB, Lee M, He H et al (2011) Bald scalp in men with androgenetic alopecia retains hair follicle stem cells but lacks CD200-rich and CD34-positive hair follicle progenitor cells. J Clin Invest 121(2):613–622
2. Pantelireis N, Higgins CA (2018) A bald statement—current approaches to manipulate miniaturisation focus only on promoting hair growth. Exp Dermatol 27(9):959–965
3. Asakawa K, Toyoshima KE, Ishibashi N, Tobe H, Iwadate A, Kanayama T, Hasegawa T, Nakao K, Toki H, Noguchi S, Ogawa M, Sato A, Tsuji T (2012) Hair organ regeneration via the bioengineered hair follicular unit transplantation. Sci Rep 2:424
4. Balañá ME, Charreau HE (2015) Leirós GJ Epidermal stem cells and skin tissue engineering in hair follicle regeneration. Word J Stem Cells 7:711–727
5. Gentile P, Scioli MG, Bielli A, Orlandi A, Cervelli V (2017b) Stem cells from human hair follicles: first mechanical isolation for immediate autologous clinical use in androgenetic alopecia and hair loss. Stem Cell Investig 4:58
6. Yuan AR, Bian Q, Gao JQ (2020) Current advances in stem cell-based therapies for hair regeneration. Eur J Pharmacol 881:173197
7. Bak DH, Choi MJ, Kim SR, Lee BC, Kim JM, Jeon ES, Oh W, Lim ES, Park BC, Kim MJ, Na J, Kim BJ (2018) Human umbilical cord blood mesenchymal stem cells engineered to overexpress growth factors accelerate outcomes in hair growth. Korean J Physiol Pharmacol. 22:555–566
8. Inukai T, Katagiri W, Yoshimi R, Osugi M, Kawai T, Hibi H, Ueda M (2013) Novel application of stem cell-derived factors for periodontal regeneration. Biochem Biophys Res Commun 430:763–768
9. Katagiri W, Osugi M, Kawai T, Ueda M (2013) Novel cell-free regeneration of bone using stem cell–derived growth factors. Int J Oral Maxillofac Implants 28:1009–1016
10. Bacakova L, Zarubova J, Travnickova M, Musilkova J, Pajorova J, Slepicka P, Kasalkova NS, Svorcik V, Kolska Z, Motarjemi H, Molitor M (2018) Stem cells: their source, potency and use in regenerative therapies with focus on adipose-derived stem cells–a review. Biotechnol Adv 36:1111–1126
11. Egger A, Tomic-Canic M, Tosti A (2020) Advances in stem cell-based therapy for hair loss. Cell R4 Repair Replace Regen Reprogram 8:e2894
12. Moore KA, Lemischka IR (2006) Stem cells and their niches. Science 311:1880–1885
13. Weissman IL (2000) Stem cells: Units of development, units of regeneration, and units in evolution. Cell 100:157–168
14. Gentile P, Cole JP, Cole MA, Garcovich S, Bielli A, Scioli MG, Orlandi A, Insalaco C, Cervelli V (2017a) Evaluation of not-activated and activated PRP in hair loss treatment: role of growth factor and cytokine concentrations obtained by different collection systems. Int J Mol Sci 18:1–16
15. Iman Hamed E, Basma Mourad M, Zeinab Abel Samad I, Said Mohamed A, Yasmina Ahmed EA, Amira Y, Maha Mostafa S, Atef T, Hala Gabr M, Mohamed MEA, Mohamed Labib S (2016) Stem cell therapy as a novel therapeutic intervention for resistant cases of alopecia areata and androgenetic alopecia. J Dermatol Treat 29:431–440
16. Yoo BY, Shin YH, Yoon HH, Seo YK, Song KY, Park JK (2010) Optimization of the reconstruction of dermal papilla like tissues employing umbilical cord mesenchymal stem cells. Biotechnol Bioprocess Eng 15:182–190
17. Zanzottera F, Lavezzari E, Trovato L, Icardi A, Graziano A (2014) Adipose derived stem cells and growth factors applied on hair transplantation: follow-up of clinical outcome. J Cosmet Dermatol Sci 4:268–274

18. Falto-Aizpurua L, Choudhary S, Tosti A (2014) Emerging treatments in alopecia. Expert Opin Emerg Drugs 19:545–556
19. Richardson SM, Kalamegam G, Pushparaj PN, Matta C, Memic A, Khademhosseini A, Mobasheri R, Poletti FL, Hoyland JA, Mobasheri A (2016) Mesenchymal stem cells in regenerative medicine: Focus on articular cartilage and intervertebral disc regeneration. Methods 99:69–80
20. Gunawardena TNA, Rahman MT, Abdullah BJJ, Abu Kasim NH (2019) Conditioned media derived from mesenchymal stem cell cultures: the next generation for regenerative medicine. J Tissue Eng Regen Med 13(4):569–586
21. Badran KW, Sand JP (2018) Platelet-rich plasma for hair loss: Review of methods and results. Facial Plast Surg Clin North Am 26:469–485
22. Gentile P, Garcovich S (2019) Advances in regenerative stem cell therapy in androgenic alopecia and hair loss: Wnt pathway, growth-factor, and mesenchymal stem cell signaling impact analysis on cell growth and hair follicle development. Cells 8:466
23. Jha AK, Udayan UK, Roy PK, Amar AKJ, Chaudhary RKP (2018) Original article: Platelet-rich plasma with microneedling in androgenetic alopecia along with dermoscopic pre- and post-treatment evaluation. J Cosmet Dermatol 17:313–318
24. Singh S (2015) Role of platelet-rich plasma in chronic alopecia areata: Our centre experience. Indian J Plast Surg. 48:7–59
25. York K, Meah N, Bhoyrul B, Sinclair R (2020) Treatment review for male pattern hair-loss. Expert Opin Pharmacother 21:603–612
26. Giordano S, Romeo M, Lankinen P (2017) Platelet-rich plasma for androgenetic alopecia: does it work? evidence from meta analysis. J Cosmet Dermatol 16:374–381
27. Stevens J, Khetarpal S (2019) Platelet-rich plasma for androgenetic alopecia: A review of the literature and proposed treatment protocol. Int J Womens Dermatol 5:46–51
28. Girijala RL, Riahi RR, Cohen PR, Girijala RL (2018) Platelet-rich plasma for androgenic alopecia treatment: a comprehensive review. Dermatol Online J 24:1–13
29. Gupta AK, Cole J, Deutsch DP et al (2019) Platelet rich plasma as treatment for androgenetic alopecia. Dermatol Surg 45:1262–1273
30. Oh JH, Kim W, Roh YH, Park KU (2015) Comparison of the cellular composition and cytokine release kinetics of various platelet-rich plasma preparations. Am J Sports Med 43:3062–3070
31. Gawdat HI, Hegazy RA, Fawzy MM, Fathy M (2014) Autologous platelet rich plasma: topical versus intradermal after fractional ablative carbon dioxide laser treatment of atrophic acne scars. Dermatol Surg 40:152–161
32. Klosova H, Stetinsky J, Bryjová I, Hledík S, Klein L (2013) Objective evaluation of the effect of autologous platelet concentrate on post-operative scarring in deep burns. Burns 39:1263–1276
33. Motolese A, Vignati F, Antelmi A, Satumi V (2015) Effectiveness of platelet-rich plasma in healing necrobiosis lipoidica diabeticorum ulcers. Clin Exp Dermatol 40:39–41
34. Cervelli V, Garcovich S, Bielli A, Cervelli G, Curcio BC, Scioli MG, Orlandi A, Gentile P (2014) The effect of autologous activated platelet rich plasma (AA-PRP) injection on pattern hair loss: clinical and histomorphometric evaluation. BioMed Res Int 2014:760709
35. Gkini MA, Kouskoukis AE, Tripsianis G, Rigopoulos D, Kouskoukis K (2014) Study of platelet-rich plasma injections in the treatment of androgenetic alopecia through a one-year period. J Cutan Aesthet Surg 7:213–219
36. Li ZJ, Choi HI, Choi DK, Sohn KC, Im M, Seo YJ, Lee YH, Lee JH, Lee Y (2012) Autologous platelet-rich plasma: A potential therapeutic tool for promoting hair growth. Dermatol Surg 38:1040–1046
37. Gentile P, Garcovich S, Scioli MG, Bielli A, Orlandi A, CervelliV (2018) Mechanical and controlled PRP injections in patients affected by androgenetic alopecia. J Vis Exp 27:131
38. Gentile P, Garcovich S (2020) Systematic review of platelet-rich plasma use in androgenetic alopecia compared with Minoxidil®, Finasteride®, and adult stem cell-based therapy. Int J Mol Sci 21:2702
39. Rodrigues B, Montalvão S, Cancela R, Silva F, Urban A, Huber S, Júnior JL, Lana JM, Annichinno-Bizzacchi JM (2019) Treatment of male pattern alopecia with platelet-rich plasma:

a double-blind controlled study with analysis of platelet number and growth factor levels. J Am Acad Dermatol 80:694–700
40. Bourin P, Bunnell BA, Casteilla L, Dominici M, Katz AJ, March KL, Gimble JM (2013) Stromal cells from the adipose tissue-derived stromal vascular fraction and ulture expanded adipose tissue-derived stromal/stem cells: a joint statement of the International Federation for Adipose Therapeutics and Science (IFATS) and the International Society for Cellular Therapy (ISCT). Cytotherapy 15(6):641–648
41. Cohen SR, Hewett S, Ross L, Delaunay F, Goodacre A, Ramos C, Saad A (2017) Regenerative cells for facial surgery: biofilling and biocontouring. Aesthetic Surg J 37(suppl_3):S16-S32
42. Festa E, Fretz J, Berry R, Schmidt B, Rodeheffer M, Horowitz M, Horsley V (2011) Adipocyte lineage cells contribute to the skin stem cell niche to drive hair cycling. Cell 146:761–771
43. Lee P, Sadick NS, Diwan AH, Zhang PS, Liu JS, Prieto VG, Zhu CC, Zhu C (2005) Expression of androgen receptor coactivator ARA70/ELE1 in androgenic alopecia. J Cutan Pathol 32:567–571
44. Zhang P, Kling RE, Ravuri SK, Kokai LE, Rubin JP, Chai JK, Marra KG (2014) A review of adipocyte lineage cells and dermal papilla cells in hair follicle regeneration. J Tissue Eng 5:1–10
45. Perez-Meza D, Ziering C, Sforza M, Krishna G, Ball E, Daniels E (2017) Hair follicle growth by stromal vascular fraction-enhanced adipose transplantation in baldness. Stem Cells Cloning Adv Appl:10
46. Sabapathy V, Sundaram B, Sreelakshmi V M, Mankuzhy P, Kumar S (2014) Human Wharton's jelly mesenchymal stem cells plasticity augments scar-free skin wound healing with hair growth. PloS one 9:e93726
47. Ruiz RG, Resell JMC, Cecarelli G et al (2019) Progenitor cell-enriched micrografts as a novel option for the management of androgenetic alopecia. J Cell Physiol:1–7
48. Teixeira FG, Carvalho MM, Sousa N, Salgado AJ (2013) Mesenchymal stem cells secretome: a new paradigm for central nervous system regeneration? Cell Mol Life Sci 70(20):3871–3882
49. Gnecchi M, He H, Liang OD, Melo LG, Morello F, Mu H, Dzau VJ et al (2005) Paracrine action accounts for marked protection of ischemic heart by Akt-modified mesenchymal stem cells. Nat Med 11(4):367–368
50. Fukuoka H, Suga H (2015) Hair regeneration treatment using adipose-derived stem cell conditioned medium: follow-up with trichograms. Eplasty 15:e10
51. Shin H, Ryu HH, Kwon O, Park BS, Jo SJ (2015) Clinical use of conditioned media of adipose tissue-derived stem cells in female pattern hair loss: a retrospective case series study. Int J Dermatol 54:730–735
52. Kinnaird T, Stabile E, Burnett MS, Lee CW, Barr S, Fuchs S, Epstein SE (2004) Marrow-derived stromal cells express genes encoding a broad spectrum of arteriogenic cytokines and promote in vitro and in vivo arteriogenesis through paracrine mechanisms. Circ Res 94:678–685
53. Kruglikov IL, Scherer PE (2016) Dermal adipocytes and hair cycling: Is spatial heterogeneity a characteristic feature of the dermal adipose tissue depot? Exp Dermato 25:258–262
54. Pawitan JA (2014) Prospect of stem cell conditioned medium in regenerative medicine. BioMed Res Int:965849
55. Haque N, Rahman MT, Abu Kasim NH, Alabsi AM (2013) Hypoxic culture conditions as a solution for mesenchymal stem cell sased regenerative therapy. Sci World J:632972
56. Hawkins KE, Sharp TV (2013) The role of hypoxia in stem cell potency and differentiation. Regen Med 8:771–782
57. Vizoso FJ, Eiro N, Cid S, Schneider J, Perez-Fernandez R (2017) Mesenchymal stem cell secretome: Toward cell-free therapeutic strategies in regenerative medicine. Int J Mol Sci 18:1852
58. Park BS, Kim WS, Choi JS, Kim HK, Won JH, Ohkubo F, Fukuoka H (2010) Hair growth stimulated by conditioned medium of adipose-derived stem cells is enhanced by hypoxia: evidence of increased growth factor secretion. Biomed Res 31:27–34
59. Kim JH, Kim WK, Sung YK, Kwack MH, Song SY, Choi JS, Park SG, Yi T, Lee HJ, Kim DD, Seo HM, Song SU, Sung JH (2014) The molecular mechanism underlying the proliferating and

preconditioning effect of vitamin C on adipose-derived stem cells. Stem Cells Dev 23:1364–1376
60. Yang Y, Choi H, Seon M, Cho D, Bang SI (2016) LL-37 stimulates the functions of adipose-derived stromal/stem cells via early growth response 1 and the MAPK pathway. Stem Cell Res Ther 7:58
61. Won CH, Yoo HG, Kwon OS, Sung MY, Kang YJ, Chung JH, Park BS, Sung JH, Kim WS, Kim KH (2010) Hair growth promoting effects of adipose tissue-derived stem cells. J Dermatol Sci 57:134–137
62. Won CH, Park GH, Wu X, Tran TN, Park KY, Park BS, Kim DY, Kwon O, Kim KH (2017) The basic mechanism of hair growth stimulation by adipose-derived stem cells and their secretory factors. Curr Stem Cell Res Ther 12:535–543
63. Shin H, Won CH, Chung WK, Park BS (2017) Up-to-date clinical trials of hair regeneration using conditioned media of adipose-derived stem cells in male and female pattern hair loss. Curr Stem Cell Res Ther 12(7):524–530
64. Fukuoka H, Suga H, Narita K, Watanabe R, Shintani S (2012) The latest advance in hair regeneration therapy using proteins secreted by adipose-derived stem cells. Am J Cosmetic Surg 29(4):273–282
65. Fukuoka H, Narita K, Suga H (2017) Hair regeneration therapy: application of adipose-derived stem cells. Curr Stem Cell Res Ther 12(7):531–534
66. Narita K, Fukuoka H, Sekiyama T, Suga H, Harii K (2020) Sequential scalp assessment in hair regeneration therapy using an adipose-derived stem cell-conditioned medium. Dermatol Surg 46(6):819–825
67. Lee YI, Kim J, Kim J, Park S, Lee JH (2020) The effect of conditioned media from human adipocyte-derived mesenchymal stem cells on androgenetic alopecia after nonablative fractional laser treatment. Dermatol Surg 46(12):1698–1704
68. Oh HA, Kwak J, Kim BJ, Jin HJ, Park WS, Choi SJ, Um S et al (2020) Migration inhibitory factor in conditioned medium from human umbilical cord blood-derived mesenchymal stromal cells stimulates hair growth. Cells 9(6):1344

Chapter 11
Induced Pluripotent Stem Cell Approach to Hair Follicle Regeneration

Antonella Pinto and Alexey V. Terskikh

Abstract *Introduction*: Current surgical hair restoration procedures redistribute existing hair follicles from one area of the scalp to another without any net increase in follicle number. The discovery of induced pluripotent stem cells (iPSCs) by Shinya Yamanaka (Nobel Prize in Physiology or Medicine 2012) ushered in a new approach to regenerative medicine with relevance to hair loss. Here we focus on dermal papilla cells, which provide cues for hair follicle growth and cycling. We describe generation of human dermal papilla cells from iPSCs via a mesenchymal route or a neural crest intermediate and cover the spontaneous eruption of hair follicles in skin organoids. *Methods*: For the selection of literature covered in this chapter we used PubMed database and our own unpublished work. We used keywords such as hair follicle development, hair follicle regeneration, induced pluripotent stem cells, neural crest, 3D cultures, and hair organoids. *Results*: Functional dermal papilla cells can be generated from human pluripotent stem cells. Combination of pluripotent stem cell derived dermal papilla cells and epithelial stem cells (which can also be generated from human pluripotent stem cells) results in formation of 3 dimensional organoids, which elongate in the dish and form hair follicles upon transplantation into Nude mice. Spontaneous eruption of hair follicles from skin organoids in vitro represents an alternative approach to generate hair follicles from human pluripotent stem cells. *Conclusions*: The iPSC-based approach provides virtually unlimited source of folliculogenic cells for de novo formation of hair follicles. However, this approach to hair regeneration must still overcome several challenges that are common to other regenerative medicine protocols based on pluripotent stem cells.

Keywords Hair follicle development · Hair follicle regeneration · Induced pluripotent stem cells · Neural crest · Fibroblast · 3D cultures · Hair organoids · Wnt · BMP · FGF

A. Pinto · A. V. Terskikh (✉)
Stemson Therapeutics, 3550 General Atomics Ct., San Diego, CA 92121, USA
e-mail: terskikh@sbp.edu

A. V. Terskikh
Sanford Burnham Prebys Medical Discovery Institute, La Jolla, CA 92037, USA

© The Author(s), under exclusive license to Springer Nature Switzerland AG 2022
F. Jimenez and C. Higgins (eds.), *Hair Follicle Regeneration*, Stem Cell Biology and Regenerative Medicine 72, https://doi.org/10.1007/978-3-030-98331-4_11

Summary

In this chapter we will review different approaches to generate hair follicles using human induced pluripotent stem cells. After a general introduction, we will begin by explaining how dermal papilla cells are generated via a mesenchymal route. Then we will explain how dermal papilla cells could be generated via a neural crest intermediate and present the validation of that route using 3 dimensional organoids that include epithelial cells. Finally, we cover the spontaneous eruption of hair follicles in skin organoids followed by the concluding remarks.

Key Points

1. Functional dermal papilla cells can be generated from human pluripotent stem cells.
2. Combination of pluripotent stem cell derived dermal papilla cells and epithelial stem cells (which can also be generated from pluripotent stem cells) results in formation of 3 dimensional organoids, which elongate in the dish and form hair follicles upon transplantation into Nude mice.
3. Spontaneous eruption of hair follicles from skin organoids in vitro represents an alternative approach to generate hair follicles from human pluripotent stem cells.

Introduction

The discovery by Shinya Yamanaka (Nobel Prize in Physiology or Medicine 2012) that mature somatic cells can be induced to acquire pluripotency ushered in a new approach to regenerative medicine with relevance to virtually all human diseases and conditions, including hair loss [1]. Somatic cells from a person experiencing hair loss can now be reprogramed in vitro to generate autologous induced pluripotent stem cells (iPSCs), which can be amplified, cryopreserved, and subsequently differentiated into various cell types associated with a functioning hair follicle. One benefit of iPSCs is that they can be obtained in large numbers and therefore provide a virtually unlimited source of folliculogenic cells for de novo formation of hair follicles. This is a major advantage compared with current surgical hair restoration procedures, which merely redistribute existing hair follicles from one area of the scalp to another without any net increase in follicle number.

Key Cell Types Required to Make a Hair Follicle

Work over the past several years has begun to reveal the remarkable complexity of the hair follicle, each of which can be considered as a mini-organ uniquely characterized by cycles of cell loss followed by full reconstitution [2–5]. This complexity adds to the challenge of deciding which follicle cells the iPSC should be differentiated into, for formation of a functional hair follicle. Cycling of hair follicle growth is controlled by dedicated stem cell niches that become active or remain quiescent in response to cues from dermal papilla (DP) cells within the hair follicle as well as cells surrounding the hair follicle [6]. These events occur through a highly-coordinated series of bidirectional epithelial–mesenchymal interactions that also control the growth activity of hair follicles starting at a very early stage of embryonic development [7, 8]. The first morphologically detectable step in hair follicle development is the formation of an epidermal placode [9]. Recent marker-independent long-term three-dimensional (3D) live imaging and single-cell transcriptomics analyses revealed that the precursors of different epithelial lineages were aligned in a 2D concentric manner in the basal layer of the hair placode [10]. Each concentric ring acquired unique transcriptomes and extended to form longitudinally aligned 3D cylindrical compartments. Prospective bulge stem cells were derived from the peripheral ring of the placode basal layer. Remarkably, this direct imaging of cell fates revealed that the fate of placode cells is determined by the cell position, rather than by the orientation of cell division [10].

The epidermal placode recruits cells from the underlying dermis to form aggregates of mesenchymal cells known as dermal condensates, which mark the location of the new hair follicle and become the DP as they further compact [11]. During this growth period, epithelial cells in physical apposition to the DP begin to differentiate into concentric cylinders to form the central hair shaft that emerges from the skin surface. Melanocytes reside above the DP within the epithelial compartment and provide pigmentation (melanin) to the hair shaft [12]. However, melanocyte stem cells are located in the bulge and maintain close contact with resident bulge epithelial stem cells (EpSCs) [13].

Once morphogenesis is complete, the follicles are prompted to enter the first hair growth cycle by a stimulus presumed to emanate from the DP; the continuous interaction of the DP with the epithelial compartment governs the transition between distinct hair cycle stages [11]. The hair cycle of growth (anagen), degeneration (catagen), and rest (telogen) is subsequently coordinated by the balance between stimulators and inhibitors and continues throughout the life of the follicle [7]. Human hair follicles on the scalp transition through these cycle phases approximately once every 4–7 years.

Strategies for Engineering a Hair Follicle

Although there are many cell types present in hair follicles [5, 14, 15], it is the interaction between DP cells and epithelial stem cells which leads to development of the follicle. It is therefore the DP and hair follicle epithelial stem cells that should be considered as the functional core of the follicle. More than half a century ago, two research groups demonstrated that transplantation of intact DPs isolated from adult donor rats and guinea pigs, into rat and guinea pig recipient skin respectively, induced de novo follicle development and hair growth [16, 17]. Rodent whisker DP expanded by number in culture are also capable of folliculogenesis, however this instructive capacity is diminished at higher passage numbers, the loss of potential is especially notable for human DP (hDP) cells at low passage numbers, where attempts to improve outcomes through modification of culture conditions has been met with limited success [18, 19]. Nevertheless, 3D culture conditions have provided a significant improvement over 2D monolayers for propagation of hDP cells in vitro [20]. Such growth conditions restored the expression of many genes in freshly isolated occipital scalp follicle-derived hDP cells and enabled the formation of some hair follicle structures upon transplantation into glabous neonatal human skin grafted on immunocompromised mice [20]. This is discussed further in Chap. 4 of this book.

Hair follicles have also been successfully deconstructed then reengineered by Tsuji and colleagues using cell types present in adult hair follicles as the starting point [21]. In this approach, primary hair follicles from the dorsal skin of mice were dissociated and epithelial and DP cells were then isolated, expanded, aggregated around a nylon thread, and transplanted onto Nude mice [21]. Recently, Tsuji and colleagues further uncovered unsuspected heterogeneity among bulge epithelial stem cells and demonstrated that CD34+ CD49f+ integrin β5 (Itgβ5)+ (triple-positive) cells are functionally important for hair regeneration [22]. These triple-positive bulge stem cells reside in the uppermost area of the bulge region, which is surrounded by tenascin in mice and humans, and are thought to be responsible for long-term hair cycling of bioengineered hair follicles [22]. Christiano and colleagues used an alternative biomimetic approach that combined folliculogenic cells obtained from dissociated adult human hair follicles within a 3D-printed collagen matrix to engineer human hair follicles growing inside a vascularized construct on the back of a mouse [23].

As highlighted above, hair follicle epithelial stem cells and DP cells are the two major cell types required to regenerate or construct hair follicles. Epithelial stem cells (EpSC) isolated from human skin were the first adult stem cells to be amplified in vitro to generate holoclones, meroclones, and paraclones with distinct clonogenic potential [24, 25]. Remarkably, in vitro-amplified human EpSCs (hEpSCs) maintain their repopulation potential [26–28] and supported autologous skin grafts [29]. Thirty years after this discovery, autologous keratinocytes (KCs) derived from such in vitro-amplified hEpSCs were used to regenerate all layers of the human epidermis [30]. Importantly, EpSCs capable of contributing to both the skin epidermis and hair follicles have been derived from hiPSCs [31]. In addition, hiPSC-derived EpSCs were recently used to repair full-thickness skin defects in Nude mice [32], paving the way for their future use in therapeutic applications.

Pluripotent Stem Cell Derived DP Cells

Fate mapping in rodents has shown that DP cells and fibroblasts in different regions of the body originate from distinct developmental lineages [33–35]. The DP cells in whiskers are derived from neural crest (NC) ectoderm while the dorsal and ventral dermis are derived from the paraxial and lateral plate mesoderm, respectively. While translating this to humans is difficult, not least because an inability to perform lineage tracing, the origin of DP in follicles on different sites of the body is accepted to be broadly correlated with androgen sensitivity. Within the beard and mustache hair (facial hair) fibroblasts and consequently DP are believed to be of NC origin. Follicles on the occipital scalp (that are protected from male pattern hair loss) are believed to have a paraxial mesoderm origin. In humans, scalp hair contains ~100,000 follicles and its epithelial compartment is distinct from beard and mustache hair both morphologically and in gene expression profile [36]. As most analysis in mice is on dorsal skin, the mechanisms of DP cell specification from dermal fibroblasts and their maturation during development from mesoderm have been elucidated in a series of elegant papers [8, 37–40].

Based on their distinct developmental origins, there are likely to be at least two routes of iPSC differentiation into DP cells: through a NC intermediate and directly from the mesoderm lineage. More recently, spontaneous formation of hair follicles within iPSC-derived skin-containing organoids has also been described [41, 42]. In the following sections, we will discuss and compare each of these approaches to hair regeneration.

Generation of DP Cells via the Mesoderm Route: Induced DP-Substituting Cells (iDPSCs)

Human and other animal hair follicles across the majority of body sites contain DP cells derived from the mesoderm germ layer. As mesenchymal stem cells (MSC) are also derived from the mesoderm, recent work from Okano and Ohyama and colleagues explored the possibility of deriving folliculogenic DP cells from iPSCs via a mesenchymal intermediate [43]. Human bone marrow contains a small (0.1%) subset of self-renewing cells characterized by high expression levels of LNGFR (CD271), THY-1 (CD90), and VCAM-1 (CD106),these cells have robust multilineage differentiation potential [44], raising the intriguing possibility that they could be directed cells towards the DP fate. Veraitch et al. exploited a novel protocol to drive LNGFR+ THY-1+ cells from hiPSCs using embryoid body formation and culture in serum-free medium containing PDGF, TGF-β, and FGF, which are conditions reported to promote MSC specification [45]. Indeed, iPSC derived MSCs were able to differentiate into the osteoblast, adipocyte, and chondrocyte lineages,these lineages have been shown to originate from the human bone marrow LNGFR+ THY-1+ subset [43]. To promote the DP fate, Veraitch et al. treated MSCs with retinoic acid and transferred them to DP cell-activating culture medium containing WNT, BMP, and FGF,a medium that had been previously developed to restore expression of DP genes in serially passaged hDP cells [18]. This treatment resulted in downregulation of multipotency-related MSC signature genes (NANOG, ZSCAN10, FZD5, BMP7, ZFP64) and upregulation of hDP signature genes (RGS2, BMP4, LEF1,

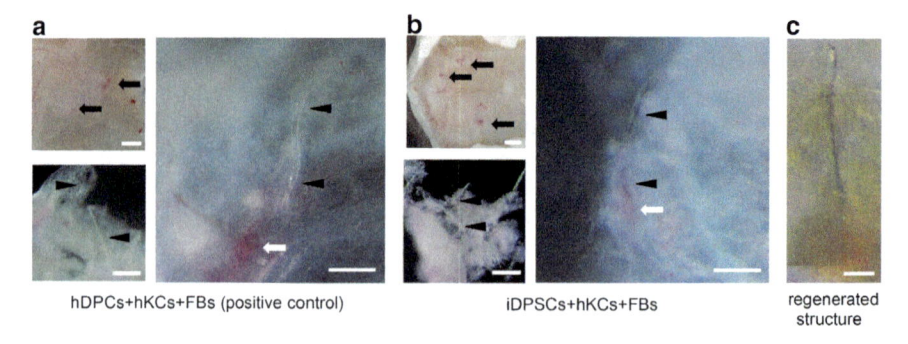

Fig. 11.1 iDPSCs exhibit functional properties of DP cells in vivo. Co-grafting of hKCs with either hDP cells (**a**) or iDPSCs (**b**) covered with FBs gave rise to cystic structures with focal aggregates (arrows), which contained fine HF-like structures (arrowheads), suggesting that iDPSCs had DP cell properties. **c** Representative structure regenerated using a combination of iDPSCs, hKCs, and FBs. hDP cells or iDPSCs were stained red with CellBrite Orange Cytoplasmic Membrane Dye. Reproduced from [43]

BAMBI, DIO2, LPL, SNCAIP7) [18] and produced cells which they termed induced DP-substituting cells (iDPSCs) [43] (Fig. 11.1).

iDPSCs are able to undergo bidirectional epithelial–mesenchymal interactions with KCs. Thus, co-culture of iDPSCs with hKCs resulted in upregulation of hDP signature genes (APL, LEF1, BMP4, IGF1) and of KC signature genes (LEF1, MSX2, KRT75), respectively. The presence of LEF1 in both signatures reflects a well-recognized key role of Wnt signaling in both the dermal and epithelial compartments of the hair follicle. Further, incubation of iDPSCs with minoxidil resulted in upregulation of the hDP gene signature, thereby mimicking the pharmacological response of hDP cells to minoxidil. Signature expression was further upregulated by co-culture of iDPSCs with hKCs plus minoxidil. Most importantly, iDPSCs contributed to the formation of hair-like structures when co-transplanted into Nude mice with primary hKCs and human fibroblasts [43]. These proof-of-principle experiments have established the feasibility of deriving hDP cells from iPSCs via the mesoderm route, which is the natural developmental pathway of DP cells in the majority of the ~5 million hair follicles throughout human skin. Further optimization of the protocol will be required to improve the efficiency of iPSC differentiation into iDPSCs and to enhance the efficiency of human follicle induction by iDPSCs in vivo. In addition to their therapeutic application, these cells will undoubtedly prove extremely useful for studying the molecular and cellular mechanisms that guide hair follicle development and cycling.

Generation of DP Cells via a Neural Crest Intermediate

On the face (which includes mustaches and beard in humans, and whiskers in rodents), both DP cells and dermal fibroblasts have cranial NC ectoderm as their developmental origin [46, 47]. Thus, DP cells can theoretically be generated via an NC intermediate. The first protocol for derivation of human embryonic stem cells (hESCs) into NC cells was developed by Studer and colleagues and was based on monolayer cultures [48]. In parallel, we developed a simplified protocol based on forming small clusters of human embryonic stem cell (hESC) colonies in suspension cultures [49]. Because hESCs produce little BMP2/4 [50], no BMP inhibitors are required to enable conversion of hESCs into NC cells in suspension cultures [49]. In these cultures, neural spheres form within 5–6 days of hESC cluster formation, attach to the culture plate plastic and produce large numbers of emigrating cells that lack SOX2 but express SOX10 and other NC markers [49]. Subsequently, we characterized the molecular signature of these NC cells and serendipitously discovered many features of cranial NC cells derived from hESCs in suspension [50]. Indeed, this protocol has subsequently been used to generate and characterize multiple human cranial cell types [51–54].

The knowledge that facial DP are derived from a NC lineage prompted us to collaborate with Prof. V. V. Terskikh (Koltsov Institute, Russian Academy of Science, Moscow, Russia) to attempt to generate hDP cells via an NC intermediate. A known property of MSCs is their preferential adherence to tissue culture plastic [55]. As DP show similarities with MSCs, expressing many similar markers [56], we enriched mesenchymal cells within our in vitro-differentiated hESC-derived NC cell cultures, by adherence to the culture dish. In our hands, about 20% of hESC-NC cells adhered to plastic, with ~50% of these expressing DP cell markers such as versican and alkaline phosphatase. We prepared epidermal cells from C57BL/6 mice on embryonic day 18.5 (E18.5) and then injected them together with hESC-derived DP (hESC-DP) cells, prepared as described above, into the skin of Nude mice using the classical patch technique [57]. Notably, we observed successful induction of hair follicles, some of which penetrated the skin, and the efficiency of hair follicle induction by hESC-DP cells was comparable to that of dermal cells from neonatal mouse skin (Fig. 11.2) [58]. The dark coloration of the hair shafts was most likely due to the presence of melanocyte progenitors within the epidermal cells prepared from dark-haired C57BL/6 mouse embryos. To ensure that hair growth was not a result of a bystander effect, we performed the same experiments with hESCs genetically engineered to express EGFP. Using confocal microscopy, we documented that EGFP+ hESC-DP cells were present within the DP, in fact, the majority of DP cells expressed EGFP [58]. Mechanistically, we have found that derivation of folliculogenic hESC-DP cells requires BMP signaling, consistent with the observation that BMP is required for the hair follicle-inductive properties of mouse DP cells in vivo [39].

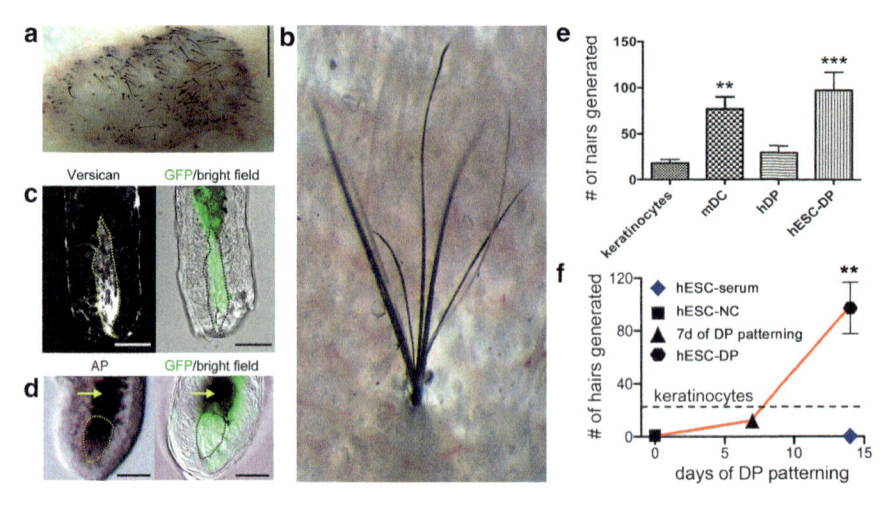

Fig. 11.2 Folliculogenic capacity of hESC-derived DP cells. **a** Stereoscopic image of whole mount skin of Nude mice transplanted with mKC + hESC-DP cells. **b** Representative hair shaft from the skin shown in (**a**). **c, d** GFP-positive DPs of newly formed hairs (GFP/bright field, confocal microscopy) express versican and alkaline phosphatase (AP). **e** Quantification of hairs induced by mKCs transplanted alone or in combination with murine dermal cells (mDC), hDP, or hESC-DP cells. **f** Dynamics of hair inductive capability of hESC-DP cells at day 0, 7, and 14 of hESC-NC differentiation, shown as the number of hairs per transplantation (red line) compared with hESCs differentiated in the presence of serum (blue diamond) or KCs alone (dashed line). Mean ± SEM. *P < 0.05, **P < 0.001 by one-way ANOVA (Kruskal–Wallis test, Dunn's multiple comparisons). Scale bars: 1 mm (**a**), 50 μM (**c, d**). Reproduced from Gnedeva et al. (2015)

Optimizing the Protocol for Clinical Translation

There are two major limitations of using hESCs for clinical application in hair restoration; namely, the lack of autologous hESCs as a source of pluripotent cells and the use of fetal bovine serum, which is a poorly defined reagent. Consequently, we refined and improved our original protocol to enable generation of DP cells from hiPSCs (as appose to hESCs) via the NC intermediate using FBS-free combination of growth factors selected to recapitulate the physiological steps of hair follicle development. Specifically, we employed Wnt10b, R-spondin, FGF20, and BMP6, which were previously implicated in mouse hair development. Of note, Wnt signaling is the key pathway for hair follicle morphogenesis [59], R-spondin is a potent physiological enhancer of Wnt signaling [60, 61], FGF20 governs the formation of primary and secondary dermal condensations in developing hair follicles [62], and BMP6 provides important signals to DP cells and is required for their hair follicle-inducing properties [39].

Cultured rodent DP cells have been shown to spontaneously aggregate after intradermal injection and form papillae-like clumps that produce large amounts of extracellular matrix [63]. However, such spontaneous aggregation has not been observed after injection of cultured hDP cells into skin [20]. To overcome this problem, we employed the hanging drop method and AggreWell™ plates (StemCell Technologies) to aggregate iPSC-derived DP cells (iPSC-DPCs) and observed that the aggregates formed translucent halos indicative of extracellular matrix deposition. Remarkably, aggregation of hiPSC-DPCs into spheroid-like structures resulted in upregulation of several genes (e.g., syndecan-1 and $\alpha 9$ integrin) compared with cells cultured in monolayers. We employed single-cell sequencing to compare the gene signatures of freshly isolated hDP cells with iPSC-DPCs both grown as monolayers and upon aggregation. As expected, monolayer grown iPSC-DPCs and hDP cells had distinct signatures, however, both cell populations dramatically altered their expression profiles upon aggregation into spheres (Fig. 11.3) and upregulated many genes characteristic of growth in the 3D environment compared with monolayer-grown hDP or hiPSC-DP cells [18, 20]. Moreover, we have demonstrated that hiPSC-DPCs are capable of inducing hair follicles upon transplantation together with mouse or hEpSCs into Nude mice (Pinto et al., manuscript in revision).

Interaction between functionally competent DP cells and epithelial cells leads to characteristic morphological and gene expression changes in these cells [43, 64]. Therefore, we postulated that if iPSC-DP cells are functionally competent to interact with epithelial cells, they should undergo similar changes upon co-culture with

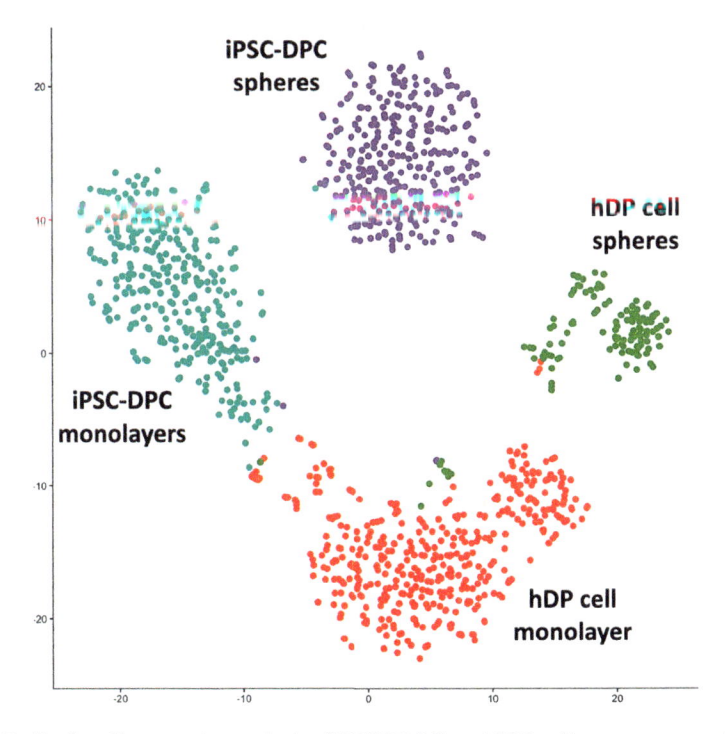

Fig. 11.3 Single cell sequencing analysis of hiPSC-DPCs and hDP cells grown as monolayers or aggregated in spheres. Note a compact nature of hiPSC-DPCs clusters suggesting a homogenous cell population of iPSC-derived DPCs

human or mouse epithelial cells. In fact, such changes could be useful for validating that cell–cell interactions had indeed taken place.

Lindner et al. showed that human hair follicle-derived DP and epithelial cells could form 3D organoids, which they termed "neopapillae", resembling developing hair follicles [65]. Over several days in culture, ~13% of these neopapillae generated visible emerging fibers similar in appearance to human vellus (unpigmented) hair, albeit lacking cuticular scales [65]. This observation suggested that at least some elements of hair follicle neogenesis and growth cycling could be recapitulated in vitro, and the requirement for epithelial cells and time course also underscored the importance of sustained crosstalk between the epithelial and mesenchymal compartments. Although intriguing, this approach remains dependent on isolation of primary cells from human hair follicles and thus has limited scalability as well as high sample-to-sample variability, likely precluding its development for clinical application.

We recently developed a 3D organoid approach based on combining hiPSC-DP cells and epithelial cells of both human (neonatal foreskin or iPSC-derived) and mouse (E18.5) origin (Pinto et al., manuscript in revision). hiPSC-epithelial cells were derived by modifying and optimizing a previously published protocol [31] to yield cultures in which ~70% of differentiating cells were CD200+ ITGA6+, a marker combination present on human hair follicle bulge stem cells [66]. qPCR analysis demonstrated that the cells progressed from pluripotent stem cells (OCT4+ and NANOG+) first to ectoderm lineages characterized by early differentiation markers (LGR5, TCF4, LEF),then to cells expressing epithelial progenitor markers (keratin [KRT] 5, 8, 14 and 15) and finally to cells expressing markers of more differentiated follicular epithelial cells (KRT71 and 74, markers of the inner root sheath).

Organoid formation was initiated by combining equal numbers of iPSC-DP cells and epithelial cells. After 24 h in culture, the cells formed spheroids and then engaged in self-directed organization to form polarized clusters that developed further into more complex and organized structures mimicking the structural features of hair follicles. The organoids can be developed in suspension cultures or in Matrigel-encapsulated cultures; the latter method permits cell growth and remodeling within an environment without imposed constraints (such as a plastic surface) on spatial cell patterning. We observed robust elongation of the epithelial compartment and cessation of proliferation of hiPSC-DP cells within such 3D organoids (Fig. 11.4). We have also adapted the conditions for organoid growth in 384-well plates and developed an automated imaging and analysis pipeline enabling us to screen small molecules and media components that modify the crosstalk between the DP and epithelial compartments in these 3D organoids.

Despite these advances, several challenges remain before we can reconstitute properly organized functional hair follicle organoids. For example, the human follicular unit includes stem cell compartment/niches and sebaceous glands, and, upon transplantation, engages a responsive sensory nerve fiber network. The possibility that hair follicles can be developed in vitro to mimic properties of a healthy or diseased organ is an exciting research area that has the potential to contribute to our

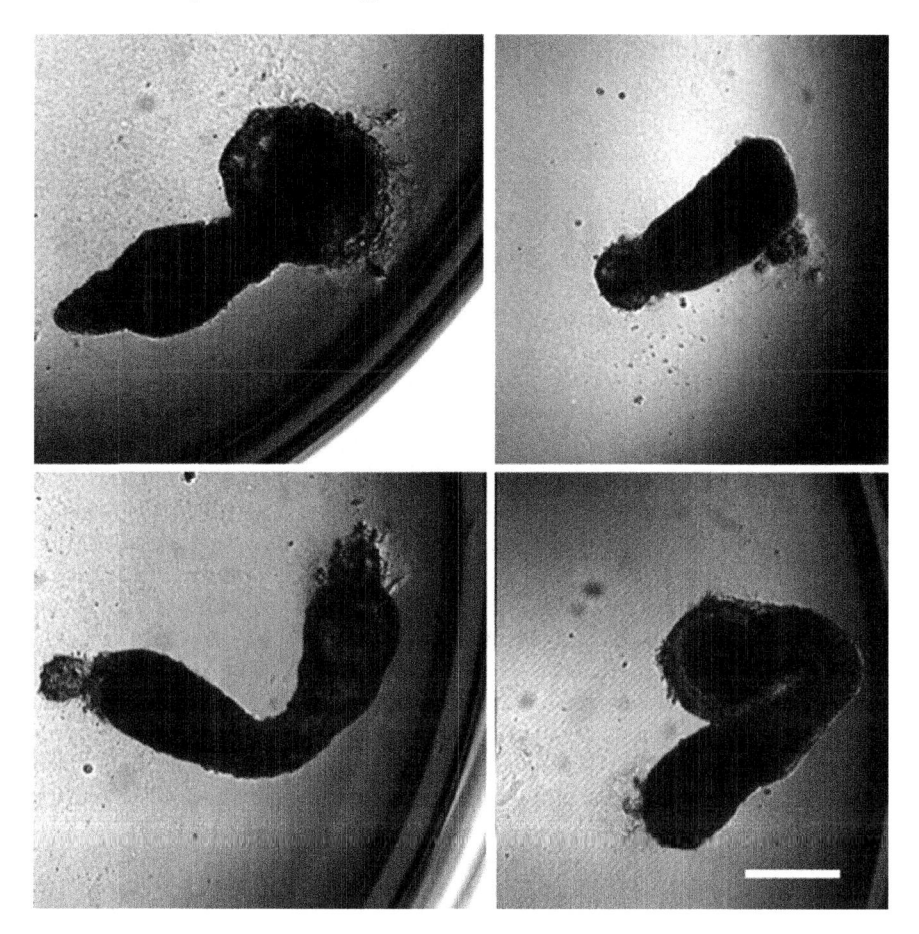

Fig. 11.4 hiPSC-DP support the formation of 3D organoids. Representative examples of day 27 organoids formed by co-culture of hiPSC-DP cells and human foreskin KCs. Organoids were generated in the agarose microwell (12 k cells total) and moved on day 3 to a low attachment, optical 96 well plate, with 2% of Matrigel in the medium. The organoids were kept in these conditions for the rest of the elongation assay. Scale bar, 200 μm

understanding of hair loss. Further optimization is required to fully harness the power of the high-throughput screening using 3D hair organoids and explore the differences between the DP of mesoderm origin vs DP of neural crest origin. In the future, such high throughput automated approach could lead to the discovery of small molecules that can modulate hair follicle function (e.g. by modulating the signaling between the DP compartment and epithelial compartment) to either promote or prevent hair growth.

Spontaneous Generation of Hair Follicles Within Skin Organoids

Another approach to generating hair follicles from iPSCs has been recently reported by Koehler and colleagues. After optimizing a mouse system for use with human cells, the authors demonstrated the spontaneous generation of hair follicles within skin organoids [41, 42]. Further, while optimizing a protocol for inner ear organoid production using mouse pluripotent stem cells (PSCs), the authors occasionally observed formation of protruding bulb-like structures extending from the KRT5+ epidermis associated with SOX2+ DP-like cells, reminiscent of nascent guard, awl, or auchene hair follicles [41]. In light of these unexpected findings, the authors further characterized the hair folliculogenesis process and found that a 3D cyst composed of surface ectodermal cells encapsulating mesenchymal cells could self-organize and mimic cell-to-cell signaling events. They initially treated day 3 stem cell aggregates with SB431542, a TGF-β signaling inhibitor, and recombinant BMP4 to induce surface ectoderm formation at the outer layer of the spheroid cell aggregates. On day 4, treatment with FGF-2 and LDN-193189, a BMP4 inhibitor, promoted induction of placodal epithelium. By day 8, the aggregate comprised an undifferentiated stem cell core, an intermediate layer containing mesoderm and neuroectoderm cells, and an outer layer of surface ectoderm. Thereafter, the aggregates underwent inside-out transformation as the intermediate layer erupted to cover the surface ectoderm, and finally an epidermal cyst was formed bearing radial-growing hair follicle structures. The newly generated hair follicles continued to grow throughout the experiment (32 days) albeit accompanied by a decline in growth rate with time. These findings suggested that the hair follicles underwent catagen, although the shafts were not shed as they typically would do during a typical growth cycle in vivo. More recently, Koehler's group employed a similar protocol with hiPSCs in which the timing of FGF-2 and BMP inhibitor treatment was optimized to promote induction of cranial NC cells that could generate mesenchymal cells, including DP cells [42]. This differentiation strategy produced uniform epithelial cysts that gave rise to stratified epidermis, fat-rich dermis, and pigmented hair follicles after 4–5 months (Fig. 11.5). Clearly visible from day 18, the organoids gradually became bipolar, with the epidermal cyst partitioned to one pole and an opaque cell mass at the opposite pole (referred to by the authors as the head and tail, respectively). The skin organoids reached a hair-bearing stage only after 70 days in culture, when hair-germ-like buds began to extend radially outward from the organoid surface. After more than 100 days in culture, the skin organoids were visually comparable to the structure of 18-week human fetal skin viewed from the dermal side, with melanocytes distributed evenly throughout the epithelium and concentrated in the matrix region of hair follicles (Fig. 11.5). Using immunostaining and electron microscopy, the authors could detect most of the unique cellular layers of hair follicles, except the medulla layer characteristic of adult terminal hair follicles, suggesting that the organoid follicles are similar to the vellus hairs found in the skin of the cheek and outer ear.

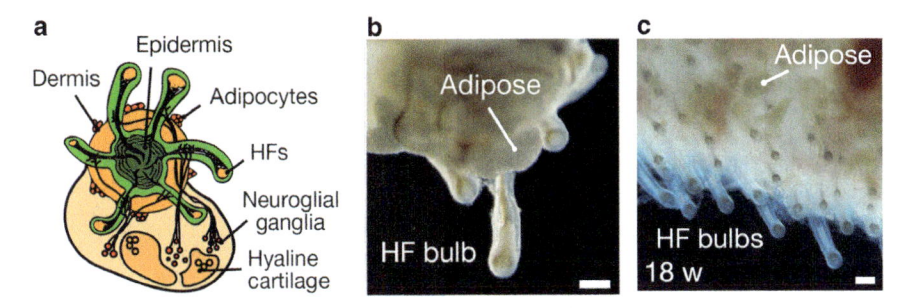

Fig. 11.5 Structure of hair follicles in skin organoids. **a** Schematic of a typical skin organoid. **b, c** Dark-field images of day-140 skin organoids (**b**) and 18-week human fetal forehead skin (**c**). HF, hair follicle. Scale bars, 250 μm (**b**) and 100 μm (**c**). Reproduced from [42] with permission from Nature

Koehler and colleagues performed single-cell RNA sequencing followed by unbiased clustering analysis on organoids at day 6, 29, and 48 [42]. The results showed a clear transition from a predominantly epithelial phenotype at the early stages to a more mesenchymal phenotype with definitive markers of the dermal lineage at later stages. Moreover, the neuroglial cells identified after the early stages matured into a network of sensory neurons, Schwann cells, and Merkel cells that formed nerve-like bundles, thus mimicking the neural circuitry associated with the human touch.

Finally, after about 150 days in culture, the skin organoids displayed accumulation of squamous cells in the core and abnormal hair follicle morphologies, suggesting that this time point could be the upper limit for skin organoid culture [42]. The authors also tested whether skin organoids could integrate into mouse skin and sustain hair growth. Notably, 55% of the xenografts integrated with the host epidermis and supported the development of follicular units composed of sebaceous glands and bulge stem cells [42]. These results suggested that the hair follicles of skin organoids reached a level of maturity roughly equivalent to that of second-trimester human fetal hair and also contained the cellular components required for further maturation. Providing the improved robustness and reproducibility of derivation of skin organoids, this approach has a potential for a translational applications such as high throughput screening for small molecules that modifies the growth of hair follicles without disturbing the skin structure and function.

Concluding Remarks

The iPSC-based approach provides a unique opportunity to apply the wealth of knowledge acquired in the field of pluripotent stem cells to the treatment of hair loss. In general, there are two distinct opportunities to use iPSC technology: (1) as a source of cells for transplantation (including autologous or patient specific approach and allogenic or universal cells approach) and (2) as a source of cells and 3D organoids

Fig. 11.6 Applications of the human iPSC-derived cells. iPSC-derived cells could be transplanted back to the donor (1, autologous transplantation) or 2, used to screen for personalized drugs (personalized medicine, green arrows. Universal cell approach (blue arrows) includes screening for the drugs that could be useful for many unrelated people (3, universal pill) and developing cells that could be transplanted into many donors (4, universal cell)

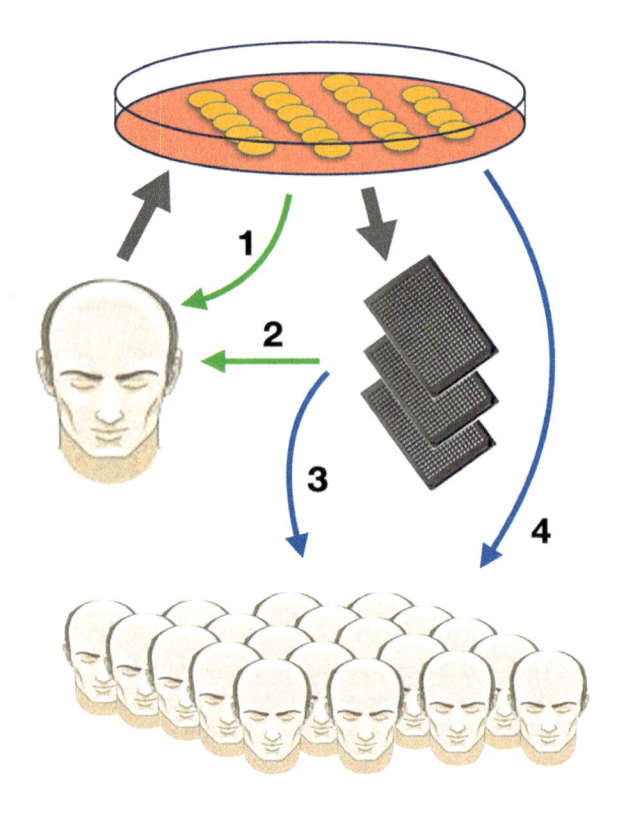

for drug screening (which could be subdivided into patient specific and universal drug screening) (Fig. 11.6). The first clinical trial of patient-derived iPSC therapy to replace and repair dying cells in the retina has recently been launched by the National Eye Institute at the United States National Institutes of Health (NCT04339764). This and other iPSC-based therapies are currently under consideration by the Food and Drug Administration, and they will undoubtedly pave the way for the development of iPSC-based therapies for other diseases and disorders, including hair loss, in the near future. The advantages of iPSC-based therapies lie in the unique characteristics of iPSCs. First, they can be massively amplified in vitro while preserving their genomic and epigenetic integrity, pluripotency, and differentiation potential [67]. Such robust amplification provides a virtually endless supply of differentiated cells for therapeutic applications. Second, it is likely that iPSC-derived dermal and epithelial cells represent immature fetal-like cellular states, as has been observed with other iPSC-derived cells [68], which form hair follicles more efficiently than adult-derived cells upon transplantation. This expectation is in line with the *bone fide* developmental potential of fetal cells used for classical transplantation experiments. Third, iPSCs have the potential to generate all of the cellular fates required to regenerate hair follicles, making this an attractive 'one-stop-shop' for production of all necessary components/cell lineages of hair follicle.

Methods for the generation and characterization of patient-specific iPSCs are being refined by many investigators [69]. Such efforts are aimed at ensuring robust manufacturing processes that adhere to GMP guidelines [67] and can be made cost-effective through the use of automation and robotics [70]. The iPSC-based approach to hair regeneration must still overcome some technical challenges that are common to other regenerative medicine protocols based on pluripotent stem cells. One of the key problems is the development of robust and reproducible differentiation protocols that enable homogeneous cell populations to be generated at an appropriate maturation stage. In addition, current iPSC-based approaches employ transplantation of autologous cells, which ensures perfect histocompatibility but is extremely expensive and severely restricts the number of potential patients. Several strategies have been proposed to enable the use of allogeneic iPSC-based approach in the near future, these include generating "universal" iPSC lines through HLA editing [71] and the development of large iPSC banks to enable optimal HLA matching between donor and recipient [72]. An allogeneic cell-based approach will inevitably bring down the cost of iPSC-based hair restoration, eventually making it an affordable option for all people experiencing hair loss.

References

1. Takahashi K, Yamanaka S (2006) Induction of pluripotent stem cells from mouse embryonic and adult fibroblast cultures by defined factors. Cell 126:663–676
2. Castro AR, Logarinho E (2020) Tissue engineering strategies for human hair follicle regeneration: how far from a hairy goal? Stem Cells Transl Med 9:342–350
3. Oh JW, Kloepper J, Langan EA, Kim Y, Yeo J, Kim MJ, Hsi TC, Rose C, Yoon GS, Lee SJ et al (2016) A guide to studying human hair follicle cycling in vivo. J Invest Dermatol 136:34–44
4. Paus R, Cotsarelis G (1999) The biology of hair follicles. N Engl J Med 341:491–497
5. Sasaki GH (2019) Review of human hair follicle biology: dynamics of niches and stem cell regulation for possible therapeutic hair stimulation for plastic surgeons. Aesthet Plast Surg 43:253–266
6. Hsu YC, Li L, Fuchs E (2014) Emerging interactions between skin stem cells and their niches. Nat Med 20:847–856
7. Muller-Rover S, Handjiski B, van der Veen C, Eichmuller S, Foitzik K, McKay IA, Stenn KS, Paus R (2001) A comprehensive guide for the accurate classification of murine hair follicles in distinct hair cycle stages. J Invest Dermatol 117:3–15
8. Rendl M, Lewis L, Fuchs E (2005) Molecular dissection of mesenchymal-epithelial interactions in the hair follicle. PLoS Biol 3:e331
9. Pispa J, Thesleff I (2003) Mechanisms of ectodermal organogenesis. Dev Biol 262:195–205
10. Morita R, Sanzen N, Sasaki H, Hayashi T, Umeda M, Yoshimura M, Yamamoto T, Shibata T, Abe T, Kiyonari H et al (2021) Tracing the origin of hair follicle stem cells. Nature 594:547–552
11. Sennett R, Rendl M (2012) Mesenchymal-epithelial interactions during hair follicle morphogenesis and cycling. Semin Cell Dev Biol 23:917–927
12. Slominski A, Wortsman J, Plonka PM, Schallreuter KU, Paus R, Tobin DJ (2005) Hair follicle pigmentation. J Invest Dermatol 124:13–21
13. Tanimura S, Tadokoro Y, Inomata K, Binh NT, Nishie W, Yamazaki S, Nakauchi H, Tanaka Y, McMillan JR, Sawamura D et al (2011) Hair follicle stem cells provide a functional niche for melanocyte stem cells. Cell Stem Cell 8:177–187

14. Chen CL, Huang WY, Wang EHC, Tai KY, Lin SJ (2020) Functional complexity of hair follicle stem cell niche and therapeutic targeting of niche dysfunction for hair regeneration. J Biomed Sci 27:43
15. Schneider MR, Schmidt-Ullrich R, Paus R (2009) The hair follicle as a dynamic miniorgan. Curr Biol 19:R132–R142
16. Cohen J (1961) The transplantation of individual rat and guineapig whisker papillae. J Embryol Exp Morphol 9:117–127
17. Oliver RF (1970) The induction of hair follicle formation in the adult hooded rat by vibrissa dermal papillae. J Embryol Exp Morphol 23:219–236
18. Ohyama M, Kobayashi T, Sasaki T, Shimizu A, Amagai M (2012) Restoration of the intrinsic properties of human dermal papilla in vitro. J Cell Sci 125:4114–4125
19. Qiao J, Zawadzka A, Philips E, Turetsky A, Batchelor S, Peacock J, Durrant S, Garlick D, Kemp P, Teumer J (2009) Hair follicle neogenesis induced by cultured human scalp dermal papilla cells. Regen Med 4:667–676
20. Higgins CA, Chen JC, Cerise JE, Jahoda CA, Christiano AM (2013) Microenvironmental reprogramming by three-dimensional culture enables dermal papilla cells to induce de novo human hair-follicle growth. Proc Natl Acad Sci U S A 110:19679–19688
21. Toyoshima K-E, Asakawa K, Ishibashi N, Toki H, Ogawa M, Hasegawa T, Irié T, Tachikawa T, Sato A, Takeda A et al (2012) Fully functional hair follicle regeneration through the rearrangement of stem cells and their niches. Nat Commun 3:784
22. Takeo M, Asakawa K, Toyoshima KE, Ogawa M, Tong J, Irie T, Yanagisawa M, Sato A, Tsuji T (2021) Expansion and characterization of epithelial stem cells with potential for cyclical hair regeneration. Sci Rep 11:1173
23. Abaci HE, Coffman A, Doucet Y, Chen J, Jackow J, Wang E, Guo Z, Shin JU, Jahoda CA, Christiano AM (2018) Tissue engineering of human hair follicles using a biomimetic developmental approach. Nat Commun 9:5301
24. Barrandon Y, Green H (1987) Three clonal types of keratinocyte with different capacities for multiplication. Proc Natl Acad Sci U S A 84:2302–2306
25. Rheinwald JG, Green H (1975) Serial cultivation of strains of human epidermal keratinocytes: the formation of keratinizing colonies from single cells. Cell 6:331–343
26. Barrandon Y, Green H (1987) Cell migration is essential for sustained growth of keratinocyte colonies: the roles of transforming growth factor-alpha and epidermal growth factor. Cell 50:1131–1137
27. Barrandon Y, Li V, Green H (1988) New techniques for the grafting of cultured human epidermal cells onto athymic animals. J Invest Dermatol 91:315–318
28. Morgan JR, Barrandon Y, Green H, Mulligan RC (1987) Expression of an exogenous growth hormone gene by transplantable human epidermal cells. Science 237:1476–1479
29. O'Connor N, Mulliken J, Banks-Schlegel S, Kehinde O, Green H (1981) Grafting of burns with cultured epithelium prepared from autologous epidermal cells. The Lancet 317:75–78
30. Hirsch T, Rothoeft T, Teig N, Bauer JW, Pellegrini G, De Rosa L, Scaglione D, Reichelt J, Klausegger A, Kneisz D et al (2017) Regeneration of the entire human epidermis using transgenic stem cells. Nature 551:327–332
31. Yang R, Zheng Y, Burrows M, Liu S, Wei Z, Nace A, Guo W, Kumar S, Cotsarelis G, Xu X (2014) Generation of folliculogenic human epithelial stem cells from induced pluripotent stem cells. Nat Commun 5:3071
32. Zhou H, Wang L, Zhang C, Hu J, Chen J, Du W, Liu F, Ren W, Wang J, Quan R (2019) Feasibility of repairing full-thickness skin defects by iPSC-derived epithelial stem cells seeded on a human acellular amniotic membrane. Stem Cell Res Ther 10:155
33. Driskell RR, Clavel C, Rendl M, Watt FM (2011) Hair follicle dermal papilla cells at a glance. J Cell Sci 124:1179–1182
34. Fernandes KJ, McKenzie IA, Mill P, Smith KM, Akhavan M, Barnabe-Heider F, Biernaskie J, Junek A, Kobayashi NR, Toma JG et al (2004) A dermal niche for multipotent adult skin-derived precursor cells. Nat Cell Biol 6:1082–1093

35. Wong CE, Paratore C, Dours-Zimmermann MT, Rochat A, Pietri T, Suter U, Zimmermann DR, Dufour S, Thiery JP, Meijer D et al (2006) Neural crest-derived cells with stem cell features can be traced back to multiple lineages in the adult skin. J Cell Biol 175:1005–1015
36. Rutberg SE, Kolpak ML, Gourley JA, Tan G, Henry JP, Shander D (2006) Differences in expression of specific biomarkers distinguish human beard from scalp dermal papilla cells. J Invest Dermatol 126:2583–2595
37. Driskell RR, Lichtenberger BM, Hoste E, Kretzschmar K, Simons BD, Charalambous M, Ferron SR, Herault Y, Pavlovic G, Ferguson-Smith AC et al (2013) Distinct fibroblast lineages determine dermal architecture in skin development and repair. Nature 504:277–281
38. Mok KW, Saxena N, Heitman N, Grisanti L, Srivastava D, Muraro MJ, Jacob T, Sennett R, Wang Z, Su Y et al (2019) Dermal condensate niche fate specification occurs prior to formation and is placode progenitor dependent. Dev Cell 48:32–48 e35
39. Rendl M, Polak L, Fuchs E (2008) BMP signaling in dermal papilla cells is required for their hair follicle-inductive properties. Genes Dev 22:543–557
40. Zhang Y, Tomann P, Andl T, Gallant NM, Huelsken J, Jerchow B, Birchmeier W, Paus R, Piccolo S, Mikkola ML et al (2009) Reciprocal requirements for EDA/EDAR/NF-kappaB and Wnt/beta-catenin signaling pathways in hair follicle induction. Dev Cell 17:49–61
41. Lee J, Bscke R, Tang PC, Hartman BH, Heller S, Koehler KR (2018) Hair follicle development in mouse pluripotent stem cell-derived skin organoids. Cell Rep 22:242–254
42. Lee J, Rabbani CC, Gao H, Steinhart MR, Woodruff BM, Pflum ZE, Kim A, Heller S, Liu Y, Shipchandler TZ et al (2020) Hair-bearing human skin generated entirely from pluripotent stem cells. Nature 582:399–404
43. Veraitch O, Mabuchi Y, Matsuzaki Y, Sasaki T, Okuno H, Tsukashima A, Amagai M, Okano H, Ohyama M (2017) Induction of hair follicle dermal papilla cell properties in human induced pluripotent stem cell-derived multipotent LNGFR(+)THY-1(+) mesenchymal cells. Sci Rep 7:42777
44. Mabuchi Y, Morikawa S, Harada S, Niibe K, Suzuki S, Renault-Mihara F, Houlihan DD, Akazawa C, Okano H, Matsuzaki Y (2013) LNGFR(+)THY-1(+)VCAM-1(hi+) cells reveal functionally distinct subpopulations in mesenchymal stem cells. Stem Cell Reports 1:152–165
45. Ng F, Boucher S, Koh S, Sastry KS, Chase L, Lakshmipathy U, Choong C, Yang Z, Vemuri MC, Rao MS et al (2008) PDGF, TGF-beta, and FGF signaling is important for differentiation and growth of mesenchymal stem cells (MSCs): transcriptional profiling can identify markers and signaling pathways important in differentiation of MSCs into adipogenic, chondrogenic, and osteogenic lineages. Blood 112:295–307
46. Nagoshi N, Shibata S, Kubota Y, Nakamura M, Nagai Y, Satoh E, Morikawa S, Okada Y, Mabuchi Y, Katoh H et al (2008) Ontogeny and multipotency of neural crest-derived stem cells in mouse bone marrow, dorsal root ganglia, and whisker pad. Cell Stem Cell 2:392–403
47. Shakhova O, Sommer L (2008) Neural crest-derived stem cells. In: StemBook (Cambridge, MA)
48. Lee G, Kim H, Elkabetz Y, Al Shamy G, Panagiotakos G, Barberi T, Tabar V, Studer L (2007) Isolation and directed differentiation of neural crest stem cells derived from human embryonic stem cells. Nat Biotechnol 25:1468–1475
49. Curchoe CL, Maurer J, McKeown SJ, Cattarossi G, Cimadamore F, Nilbratt M, Snyder EY, Bronner-Fraser M, Terskikh AV (2010) Early acquisition of neural crest competence during hESCs neuralization. PLoS ONE 5:e13890
50. Bajpai R, Coppola G, Kaul M, Talantova M, Cimadamore F, Nilbratt M, Geschwind DH, Lipton SA, Terskikh AV (2009) Molecular stages of rapid and uniform neuralization of human embryonic stem cells. Cell Death Differ 16:807–825
51. Bajpai R, Chen DA, Rada-Iglesias A, Zhang J, Xiong Y, Helms J, Chang CP, Zhao Y, Swigut T, Wysocka J (2010) CHD7 cooperates with PBAF to control multipotent neural crest formation. Nature 463:958–962
52. Bayless NL, Greenberg RS, Swigut T, Wysocka J, Blish CA (2016) Zika virus infection induces cranial neural crest cells to produce cytokines at levels detrimental for neurogenesis. Cell Host Microbe

53. Rada-Iglesias A, Bajpai R, Prescott S, Brugmann SA, Swigut T, Wysocka J (2012) Epigenomic annotation of enhancers predicts transcriptional regulators of human neural crest. Cell Stem Cell 11:633–648
54. Rada-Iglesias A, Prescott SL, Wysocka J (2013) Human genetic variation within neural crest enhancers: molecular and phenotypic implications. Philos Trans R Soc Lond B Biol Sci 368:20120360
55. Pittenger MF, Mackay AM, Beck SC, Jaiswal RK, Douglas R, Mosca JD, Moorman MA, Simonetti DW, Craig S, Marshak DR (1999) Multilineage potential of adult human mesenchymal stem cells. Science 284:143–147
56. Hoogduijn MJ, Gorjup E, Genever PG (2006) Comparative characterization of hair follicle dermal stem cells and bone marrow mesenchymal stem cells. Stem Cells Dev 15:49–60
57. Zheng Y, Du X, Wang W, Boucher M, Parimoo S, Stenn K (2005) Organogenesis from dissociated cells: generation of mature cycling hair follicles from skin-derived cells. J Invest Dermatol 124:867–876
16. Gnedeva K, Vorotelyak E, Cimadamore F, Cattarossi G, Giusto E, Terskikh VV, Terskikh AV (2015) Derivation of hair-inducing cell from human pluripotent stem cells. PLOS ONE 10(1):e0116892. https://doi.org/10.1371/journal.pone.0116892
59. Rabbani P, Takeo M, Chou W, Myung P, Bosenberg M, Chin L, Taketo MM, Ito M (2011) Coordinated activation of Wnt in epithelial and melanocyte stem cells initiates pigmented hair regeneration. Cell 145:941–955
60. Hagner A, Shin W, Sinha S, Alpaugh W, Workentine M, Abbasi S, Rahmani W, Agabalyan N, Sharma N, Sparks H et al (2020) Transcriptional profiling of the adult hair follicle mesenchyme reveals R-spondin as a novel regulator of dermal progenitor function. iScience 23:101019
61. Schuijers J, Junker JP, Mokry M, Hatzis P, Koo BK, Sasselli V, van der Flier LG, Cuppen E, van Oudenaarden A, Clevers H (2015) Ascl2 acts as an R-spondin/Wnt-responsive switch to control stemness in intestinal crypts. Cell Stem Cell 16:158–170
62. Huh SH, Narhi K, Lindfors PH, Haara O, Yang L, Ornitz DM, Mikkola ML (2013) Fgf20 governs formation of primary and secondary dermal condensations in developing hair follicles. Genes Dev 27:450–458
63. Jahoda CA, Oliver RF (1984) Vibrissa dermal papilla cell aggregative behaviour in vivo and in vitro. J Embryol Exp Morphol 79:211–224
64. Lei M, Schumacher LJ, Lai YC, Juan WT, Yeh CY, Wu P, Jiang TX, Baker RE, Widelitz RB, Yang L et al (2017) Self-organization process in newborn skin organoid formation inspires strategy to restore hair regeneration of adult cells. Proc Natl Acad Sci U S A 114:E7101–E7110
65. Lindner G, Horland R, Wagner I, Atac B, Lauster R (2011) De novo formation and ultrastructural characterization of a fiber-producing human hair follicle equivalent in vitro. J Biotechnol 152:108–112
66. Ohyama M, Terunuma A, Tock CL, Radonovich MF, Pise-Masison CA, Hopping SB, Brady JN, Udey MC, Vogel JC (2006) Characterization and isolation of stem cell-enriched human hair follicle bulge cells. J Clin Invest 116:249–260
67. Shafa M, Walsh T, Panchalingam KM, Richardson T, Menendez L, Tian X, Suresh Babu S, Dadgar S, Beller J, Yang F et al (2019) Long-term stability and differentiation potential of cryopreserved cGMP-compliant human induced pluripotent stem cells. Int J Mol Sci 21
68. Patterson M, Chan DN, Ha I, Case D, Cui Y, Van Handel B, Mikkola HK, Lowry WE (2012) Defining the nature of human pluripotent stem cell progeny. Cell Res 22:178–193
69. Chen Y, Tristan CA, Chen L, Jovanovic VM, Malley C, Chu PH, Ryu S, Deng T, Ormanoglu P, Tao D et al (2021) A versatile polypharmacology platform promotes cytoprotection and viability of human pluripotent and differentiated cells. Nat Methods 18:528–541
70. Konagaya S, Ando T, Yamauchi T, Suemori H, Iwata H (2015) Long-term maintenance of human induced pluripotent stem cells by automated cell culture system. Sci Rep 5:16647
71. Koga K, Wang B, Kaneko S (2020) Current status and future perspectives of HLA-edited induced pluripotent stem cells. Inflamm Regen 40:23
72. Lee S, Huh JY, Turner DM, Lee S, Robinson J, Stein JE, Shim SH, Hong CP, Kang MS, Nakagawa M et al (2018) Repurposing the cord blood bank for haplobanking of HLA-homozygous iPSCs and their usefulness to multiple populations. Stem Cells 36:1552–1566

Chapter 12
Biofabrication Technologies in Hair Neoformation

Carla M. Abreu, Luca Gasperini, and Alexandra P. Marques

Abstract *Introduction* The hair follicle (HF) is an exclusive adnexal structure with important cosmetic and physiological value, embodying stem cells that contribute to skin homeostasis and response to injury. Therefore, sustaining de novo hair formation in damaged skin would be a clinical breakthrough both in the management of critical wounds and in irreversible hair disorders, such as common androgenetic alopecia. However, the complex nature of hair development—hierarchically controlled by multiple cellular compartments—implicates not only the use of appropriate cells but also advanced biotechnologies and instructive matrices capable of recreating the HF specific microarchitecture and biochemical profile. *Methods* This chapter starts by exploring the cellular and biomaterial elements commonly considered when bioengineering the HF. Next, we reviewed strategies previously used to elicit HF neoformation, from simpler cell-based approaches to the ones assisted by microfabrication technologies, closing with a discussion on future directions. *Results* Earlier efforts to stimulate HF neoformation have mostly attempted to re-establish epithelial and mesenchymal interactions and respective cellular organization, while disregarding the specific nature of the HF extracellular matrix (ECM) and remaining regulatory cells. Nevertheless, the use of microfabrication technologies to mimic the spatial organization of these compartments or most recently, to regenerate HFs within skin models denote increasing efforts towards the recapitulation of the HF 3D architecture and milieu. *Conclusions* Despite the ever-growing number of strategies and emergent biofabrication platforms arising from tissue engineering these have just started to be considered towards HF regeneration, holding the promise of providing innovative and clinically useful solutions to promote hair neoformation in patients.

Keywords Hair follicle · Bioengineering · Dermal papilla (DP) cells · Epithelial cells · Adipose tissue · Cell signalling · Extracellular matrix (ECM) ·

C. M. Abreu · L. Gasperini · A. P. Marques (✉)
3B's Research Group, I3Bs—Research Institute on Biomaterials, Biodegradables and Biomimetics, Headquarters of the European Institute of Excellence on Tissue Engineering and Regenerative Medicine, AvePark–Parque de Ciência e Tecnologia, University of Minho, Barco, Portugal
e-mail: apmarques@i3bs.uminho.pt

ICVS/3B's–PT Government Associate Laboratory, Guimarães, Portugal

© The Author(s), under exclusive license to Springer Nature Switzerland AG 2022
F. Jimenez and C. Higgins (eds.), *Hair Follicle Regeneration*, Stem Cell Biology and Regenerative Medicine 72, https://doi.org/10.1007/978-3-030-98331-4_12

Biomaterials · Hair follicle germ (HFG) · Organotypic substitutes · Microfabrication · Moulding · Bioprinting · Laser ablation · Vasculature

Key points:

- Strategies to promote hair neogenesis include the implantation of trichogenic cells, the bioengineering of HF units or the reformation of follicular appendages in skin grafts. However, the few successful human-based strategies required dermal papilla cell in combination with newborn epithelial cells, preventing autologous approaches.
- Materials used to fabricate or regenerate HF have been mostly used to produce extracellular matrix-bioinspired hydrogels to encapsulate dermal papilla cells or promote their homotypic/ heterotypic self-assembly with epithelial cells. Some of the most developed strategies include the production of hair germ-like structures and dermal papilla cell inclusion in organotypic skin models.
- Although the recapitulation of the 3D architecture, milieu and interactions of the HF could benefit from advanced biofabrication techniques, these have been modestly explored in HF bioengineering. In the future, the development of biomaterials that specifically emulate the HF extracellular environment and a deeper and targeted exploration of these technologies towards hair neoformation will certainly improve regenerative outcomes.

Introduction

The hair follicle (HF), besides being of paramount importance for the individual's aesthetic appearance, also contributes to the main biological functions of the skin, including physical protection, thermoregulation, immunosurveillance and aiding sensory perception. The presence of distinct stem cell niches and instructive cell populations confer a high regenerative capacity to this skin appendage, as naturally demonstrated by its capacity to cycle over the entire human lifetime. However, new HFs cannot be naturally formed post-natal. After extensive skin damage, such as burning or traumatic wounding, the repaired skin lacks HFs. Moreover, in hair loss disorders, the hair-growth machinery becomes dysfunctional compromising its natural regeneration. While suitable solutions are still missing to promote de novo HF formation in wounded skin, autologous hair transplantation for hair loss disorders is limited by the availability of sufficient transplantable follicles in the donor occipital scalp. Thus, innovative approaches are needed to address these needs.

Strategies that rely on providing cellular cues capable of triggering the reformation of new HFs in situ, and bioengineering hair-bearing skin grafts or transplantable HFs, have been proposed. Like in other mammal tissues, HF development and growth are controlled by reciprocal interactions between epithelial and mesenchymal cells [1]. Pioneering studies demonstrated that both intact dermal papilla (DP) [2] and the adjacent dermal sheath (DS) tissue [3], the two mesenchymal compartments existing within the HF, can regenerate hairs by interacting with the recipient epithelium upon transplantation. Therefore, isolating and obtaining a high number of inductive mesenchymal cells that could then be transplanted back into patients, alone or in combination with competent epithelial cells, represent an important step when aiming to generate numerous new HFs in patients (Fig. 12.1). Moreover, the HF is a highly complex skin appendage with a particular structure composed of multiple epithelial layers (at different degrees of differentiation) that periodically self-renews under well-guided cellular interplay. Given this complexity, tissue engineering multidisciplinary and integrated approaches may provide the best solution to create a naturalized microenvironment capable of replicating the HF development process and subsequent neoformation. For that, the recreation of this tissue in bioartificial substitutes will necessarily require the development of complex 3D microenvironments that specifically replicate the tissue's cellular organization and extracellular matrix (ECM) nature, to support the reestablishment of native interactions between the different cell types and between cells and the neighbouring ECM. Altogether, these implicate that more advanced 3D microfabrication technologies and biomimetic matrices are needed to recapitulate the HFs organization, interactions and milieu, thereby improving the chances of successfully regenerating new HFs.

Cellular Players to Bioengineer the Hair Follicle

Reconstitution of the Epithelial-Mesenchymal Crosstalk

The development and continuous remodelling of the HF growth machinery after each cycle are dependent on highly coordinated epithelial-mesenchymal interactions (EMIs). Hence, the success of HF regenerative strategies is intimately interconnected with the efficiency in combining inductive mesenchymal cells and responsive epithelial cells as cellular blocks to elicit EMIs [4]. The hair mesenchyme comprises both the DP and DS compartments, but the highly specialized DP cells are considered the primary inductive cells [5, 6]. Unfortunately, the hair inductive ability of human DP (hDP) cells is seriously compromised when these cells are grown in culture [7], therefore requiring changes in the culture conditions for its preservation or recovery, as explored in Chap. 4 (Patelireis, Goh and Clavel) of this book.

Epithelial cells, as the responsive component of HF development, need to be instructed by DP cells to self-organize, proliferate and differentiate in the multiple epithelial cell layers that constitute the HF sophisticated structure. Independent studies demonstrate that both follicular and interfollicular cell sources can be used [8], nonetheless, the literature strongly suggests a positive correlation between the low differentiation level of the epithelial cells used and their capacity to sustain HF neoformation. This was visibly demonstrated in co-grafting experiments in which human fetal dermal cells were combined with epithelial cells collected from donors of different ages [9]. The combination of cells from fetal origin enabled the formation

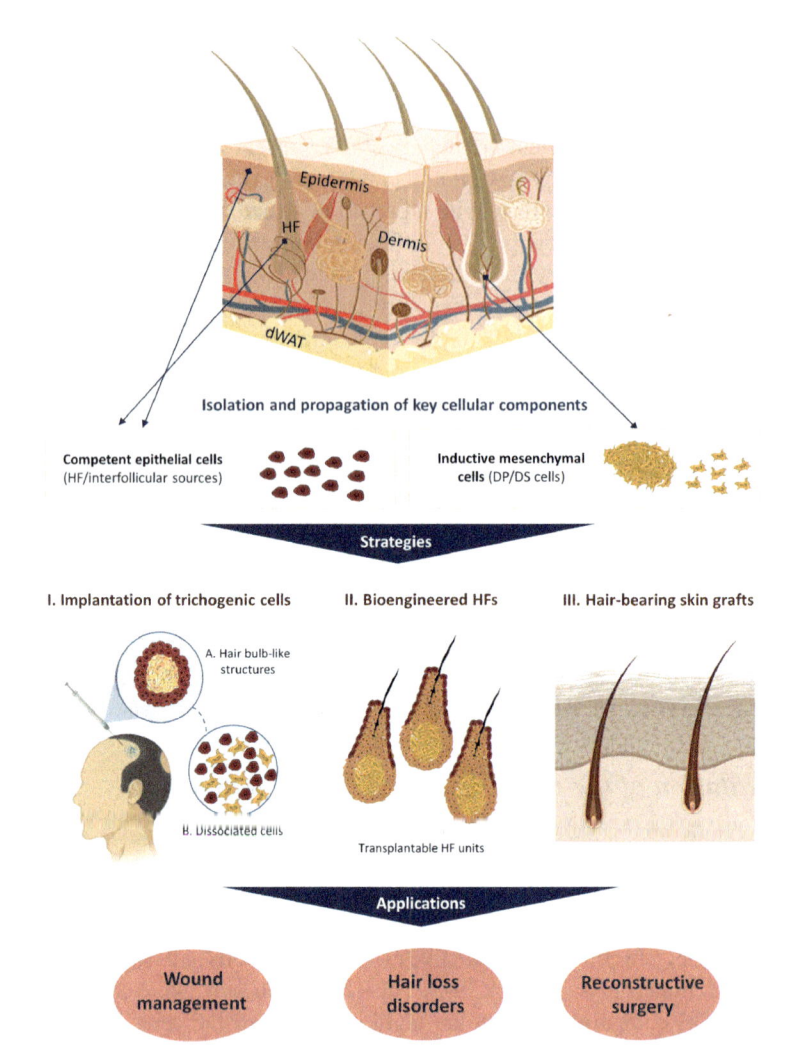

Fig. 12.1 Schematic representation of different strategies to promote hair neogenesis. Inductive mesenchymal cells and epithelial cells receptive to their signals can be obtained from patients and expanded in culture to obtain a higher number of cells for regenerative applications. Afterwards, heterotypic dissociated cell combinations can be directly transplanted into patients (I) or used to produce ready-to-use bioartificial HFs (II) or follicular appendages in tissue-engineered skin equivalents (III). These could then be used either as products for hair restoration or reconstructive applications or as improved skin substitutes for more effective wound management of extensive skin defects. DP-dermal papilla; DS- dermal sheath; dWAT- dermal white adipose tissue; HF-hair follicle

of an ample number of human-origin HFs, however, the efficiency dropped more than 85% when adult epithelial cells were used. Even when neonatal cells are grafted, the number of HFs formed and their diameter significantly decreases with the passage of the cells [10, 11]. This explains why studies successfully reporting human HF reformation adopted the use of epidermal cells isolated from either fetal [9] or newborn [7, 9, 12] tissues. Despite their higher responsiveness to mesenchymal cell signals, ethical and availability issues (especially when considering autologous applications) hinder their use for bioengineering applications. Ideally, the selection of adult epithelial cells with a low differentiation degree and high clonogenic capacity should be preferred. We recently showed that interfollicular epithelial cells with stem-like characteristics ($\alpha 6$ integrin^{high}/CD71^{low} expression) co-grafted with passaged adult hDP cells supported the formation of immature HFs and sebaceous glands [13]. Moreover, the selection and use of human bulge-enriched stem cells, previously profiled with a CD200^{high}CD24^{low}CD34^{low}CD71^{low}CD146^{low} phenotype [14], could also represent a viable epithelial fraction for autologous strategies.

Considering the challenges associated with the recovery of the trichogenic properties of both adult hDP cells and epithelial cells, technologies to promote hair neoformation should also consider other cell types that reside within the HF and contribute to the homeostasis and physiological functions of this tissue.

Role of Other Key Components

Hair colour and the photoprotection it provides depend on melanin production by bulbar melanocytes and its transfer to the keratinocytes of the growing hair shaft at each anagen phase [15]. Therefore, the production of aesthetically suitable hairs will depend on the inclusion of functionally active melanocytes in HF analogues, or their capacity to recruit them from the surrounding tissue.

During anagen, the proximal HF epithelium and bulb benefit from an immunologically privileged state due to a low expression of the major histocompatibility complex class (MHC) Ia antigen and a localized expression of immunosuppressive molecules [16]. Notwithstanding, immune cells residing in the remaining portions of the HF contribute to skin immunity. Those cells are mainly located in the distal HF epithelium (Langerhans cells and CD4 + /CD8 + T cells) and perifollicular DS (macrophages and mast cells) [17]. Besides contributing to the skin defence system, immune cells have also a role in controlling the HF stem cell niche during hair cycling [18, 19], rendering them interesting elements to consider and further study in HF regenerative strategies. Additionally, hair growth depends on the reestablishment of the DP and perifollicular vascular network upon each anagen onset, which rely respectively on DP cell and outer root sheath cell capacity to produce vascular endothelial growth factor (VEGF) [20, 21]. Thereby, using cells with improved capacity to produce VEGF or including this angiogenic mediator when designing HF or hair-bearing skin substitutes may represent an additional advantage to support HF formation.

Additionally, it would add to the neovascularization of skin substitutes thus fostering their integration into the host tissue.

Finally, hair bulbs are deeply rooted in the dermal white adipose tissue (dWAT), a layer of dermis-derived adipose tissue underlying the reticular dermis [22] with important regulatory capacity over the HF cycle and skin homeostasis. Indeed, adipogenesis in the dWAT is coupled with the HF stem cell activity and cycle progression, in which immature adipocytes are required for stem cell activation [23] whilst mature adipocytes promote follicular differentiation [24]. The physiological role of dWAT is discussed in Chap. 6 (Ramos and Plikus), however the recreation of the HF extrafollicular macroenvironment may itself represent a strategic way to promote this appendage reformation. On the other hand, the correct placing of trichogenic cells or HF equivalents nearby or into the dWAT could also be explored to promote that crosstalk. In a previous study, mice epithelial and mesenchymal cells injected in the dWAT (close to the *panniculus carnosus*) were shown to impact the number of new HFs formed, in opposition to their injection in the deep subcutis after which almost no HFs were formed [25]. While there are currently no distinctive markers for the dWAT preventing the isolation and use of cells from this specialized adipose area, intradermal injection of adipose-derived stem cell (ASC)-conditioned medium showed HF and skin regenerative effects in patients with hair loss (see chapter by Park et al.) [26]. Moreover, the recovery of the inductivity of DP cells in vitro supported by ASCs conditioned medium [27], further validates the potential of ASCs as a promising cellular source to recreate the adipose tissue surrounding the hair bulb.

Biomaterials to Bioengineer the HF

The ultimate goal of any bioengineering approach is to recapitulate, as much as possible, the tissue microenvironment in vitro. But, if the use of relevant cell sources and the preservation of the trichogenic capacity of the cells has been comprehensively explored, the relevance of the ECM and its influence in supporting and/or intensifying EMIs have been generally overlooked (Fig. 12.2).

Rudimentary and Hair Germ-Like Structures

So far, in the context of HF bioengineering, biomaterials have been mostly used as culture surfaces to facilitate the self-assembly of mesenchymal and epithelial cells into biomaterial-free 3D structures, including the formation of DP cell spheroids and epithelial-mesenchymal aggregates, or to provide a 3D environment to dissociated cells that is still far from the ideal representation of the native ECM. As an example of the former, Matrigel™ has been widely used to produce DP cell aggregates [31], benefiting from these cells tendency to self-aggregate. Heterotypic cell aggregates were also attained; the addition of matrix keratinocyte to the hDP cell

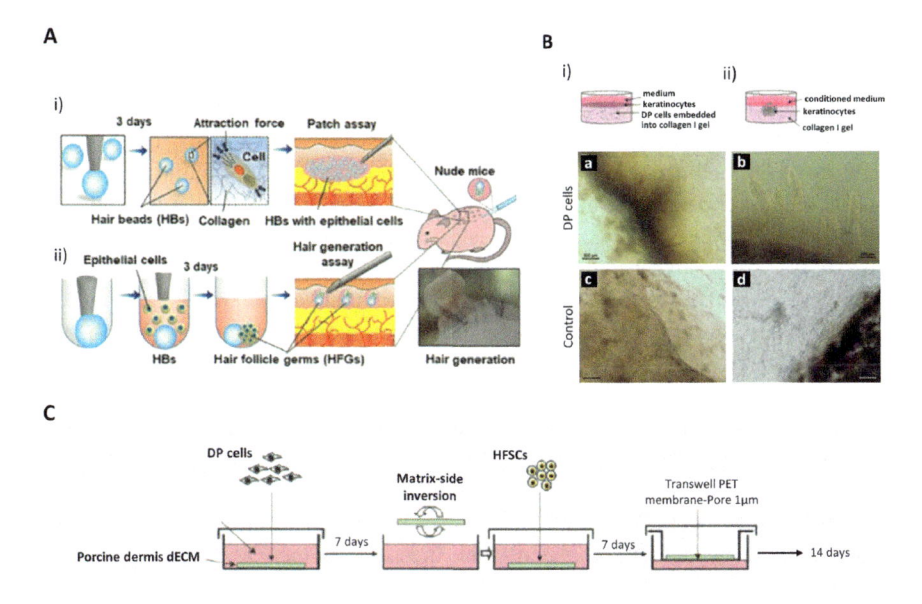

Fig. 12.2 Different biomaterial approaches used to generate follicular-like structures in vitro. **A** Preparation of hair beads (i) and hair follicle germs (ii) and respective evaluation of their trichogenic activity in vivo. Reprint (adapted) from Kageyama et al. [28] Creative Commons CC BY. **B** Inclusion of DP cells in a skin organotypic models (i) or their conditioned medium (ii) both promoted epithelial tubulogenesis (a, d), which was not observed when dermal fibroblasts (c) or their conditioned medium (d) were used instead as negative controls. Reprint (adapted) from Chermnykh et al. [29] Copyright Springer Nature 2010. **C** Schematic representation of the procedure used to generate skin constructs containing DP cells and using a porcine dermis dECM as the scaffold. Reprint (adapted) from Leirós et al. [30] Copyright John Wiley and Sons

spheroids allowed the recreation of a rudimentary hair shaft [31], and the coating of hDP cell spheroids with basement membrane ECM molecules (collagen IV, laminin, fibronectin) onto which matrix keratinocytes and melanocytes were added lead to the formation of a hair fibre-producing "microfollicle" with a vellus-like appearance [32]. Despite the promising results, which denote ongoing EMIs and subsequent guided epithelial differentiation in vitro, the capacity of these aggregates in promoting hair neoformation upon in vivo implantation was not tested.

Larger scale approaches that would allow overcoming the onerous task of handling and properly implanting micrometre-sized cell aggregates, have relied on the encapsulation of human mesenchymal and epithelial cells in Matrigel™ mixed with [33] or layered onto collagen I [34, 35]. However, independently of the spatial arrangement of the cells, or even the source of epithelial cells used, the produced constructs failed to induce a follicular-type of organization in vitro. Instead, rudimentary cellular aggregates in which epithelial cells often formed cyst-like structures, without surrounding mesenchymal cells, were obtained.

On the other hand, collagen I has been extensively used on its own with different strategies leading to different degrees of success. Mouse embryonic mesenchymal or hDP cells encapsulated within collagen 3D droplets, named hair beads, showed significantly higher alkaline phosphatase (ALP) activity and versican expression (two markers associated with DP cell inductivity) than the respective biomaterial-free 3D spheroids and monocultures, demonstrating the positive effect of the hydrogel in the recovery of DP cell properties [28]. This was confirmed by the significantly higher amount of hairs formed after hair beads transplantation with neonatal mice epithelial cells in a patch assay (Fig. 12.2a). Curiously, when epithelial cells were subsequently added to the hair beads the cells self-organized into a compartmentalized hair follicle germ (HFG) which displayed significantly higher inductive ability. This was not seen when both cells were concomitantly mixed within the collagen drops. When cells were co-cultured without the collagen matrix, self-segregation was also observed, but the number of generated new hairs was significantly lower, once again showing that the collagen hydrogel had a significant positive impact on HF neogenesis. This work was quite relevant by particularly showing an independent positive effect of the collagen hydrogels, but some of the most often referred bioengineered HFGs were firstly developed by Tsuji's group [36]. By mixing high density dissociated epithelial and mesenchymal isolated from mice embryonic whisker [36] or pelage [37] into adjacent regions of a collagen drop, cell-type compartmentalization and the formation of new HFs following subrenal capsule implantation in vivo were similarly attained. Shortly after, the same was reported for cultured DP cells and bulge-derived epithelial cells isolated from vibrissae [38]. Noteworthy, the formation of human-derived HFs was achieved but only after the implantation of follicle germs containing intact DPs.

Although rarer, the use of biomaterials unrelated to the ECM has been also considered in the context of HF regeneration. Silk fibroin denatured collagen (gelatin) hydrogels were used to encapsulate hDP cell spheroids [39]. Upon co-culture with epithelial cells, DP cells displayed higher viability, improved proliferation, ECM production and expression of DP cell signature gene markers than in the biomaterial-free aggregates. Similarly, hydrogels formed by self-assembled RAD16-I synthetic peptide [40] were able to enhance hDP cells inductive capacity (ALP activity, DP cell signature gene expression), strengthening the importance of 3D matrices. Yet, whether these hydrogels support hair reformation in vivo is not known.

Skin Substitutes

Considering that the HF arises during the early stages of skin development through EMIs, in which DP cells condensate beneath the epidermis and instruct HF development, some researchers have tried to reproduce this process by differently modifying the collagen-based organotypic skin model with the inclusion of DP cells. In one of the first attempts, dermal fibroblasts were replaced by DP cells as the dermal cellular component in the organotypic model [29] (Fig. 12.2b). The generation of epithelial tube-like structures that resembled the hair germ formation during the initial HF embryonic development was observed [29]. A similar outcome was shown when hDP cell spheroids were included in the reconstructed dermis of the model [41]. This guided an epithelial down-growth movement with the formation of a proper basement membrane, but without clearly demonstrating the reconstitution of a hair bulb or evidencing differences between the epidermis of the reconstructed skin and the one from the invaginated epidermis. The formation of epithelial invaginations and DP condensates represent the primary events required for HF development, but while those strategies seem to promote the first, maintaining DP cells in a 3D structure and their inductivity is not simple. These cells self-aggregative behaviour is increasingly lost in vitro rapidly disintegrating when on a 2D cell culture surface [31] or inside of a collagen I matrix [42, 43]. This is also related to DP cells origin as shown in a study where collagen-glycosaminoglycan hydrogels populated with murine DP (mDP) or hDP cells were used to produce skin substitutes [44]. Only the constructs containing mDP cells formed follicular structures following in vivo implantation, independently of the species origin of the keratinocytes used in the constructs. Further analysis demonstrated that upon implantation, dermal ALP activity strongly increased in grafted substitutes that contained mDP cells, whereas for hDP cells the improvement was minor. Moreover, both this and other works that built skin substitutes containing murine DP cells and human epithelial cells [45, 46] testified the formation of chimeric HFs. The most successful human-based strategy was achieved when scalp hDP cells were part of dermal-epidermal substitutes [12]. Hair neoformation was observed in animals implanted with the constructs that contained the higher percentage of hDP cells positive for ALP activity ($\geq 52\%$). Moreover, this work shows that neonatal keratinocytes are required to achieve hair reformation. This seems to cement the hypothesis that human-based HF regeneration can only be reached when DP cells inductivity is not compromised in vitro and the differentiation degree of the epithelial cells used is low and, therefore, the cells are more likely to be competent to receive and respond to DP cell signals.

Decellularized ECM (dECM) also represent a common trend for regenerative strategies since they have the advantage of preserving, to some extent, the native tissue macro- and -microarchitecture and composition [47]. Decellularized dermis has been also used to recreate skin substitutes relevant for HF regeneration. Porcine dermis dECM seeded with hDP cells and follicular keratinocytes enriched in stem cells on opposite sides (Fig. 12.2c) resulted in the formation of epithelial structures similar to hair buds upon transplantation into skin wounds [30]. The expression of k6hf in these hair bud-like structures, a specific marker of the companion layer, indicated differentiation commitment to the HF lineage. In a different study, human dermis dECM repopulated with human keratinocytes and mDP cells at a 10:1 ratio, also lead to the production of de novo HF regeneration at the wound site [46]. Curiously, the human placenta, given that its matrix components are akin to the skin, has been also explored. High passage (P8) DP cell spheroids encapsulated in placenta dECM-derived hydrogels were able to recover the gene and protein expression of inductive-related markers (ALP, β-catenin and versican) to levels similar to the ones observed in P2 cells [48]. Moreover, grafting with mouse neonatal epidermal cells resulted in newly formed hairs in a hair-inductive assay, which was not observed when same passage monocultured cells were used. Unfortunately, the lack of controls containing only DP cell spheroids (without the hydrogels), known to improve DP cell hair inductive efficiency by itself [7] and to promote chimeric hair reformation after combination with newborn epithelial cells [49], prevents assessing the contribution of the dECM to these results.

Optimization of biomaterial strategies to provide proper positional cues and mimic cellular compartmentalization symbolize an important step for the bioengineering of HFs. Yet, more than supporting the cells, the biomaterial should transduce the ECM regulatory role over cell fate and critically support the re-establishment of EMIs. So far, the importance of the ECM and its specific biochemical composition in the different HF compartments are some of the most understudied factors in HF bioengineering. In the future, advanced recreation of this non-cellular self-sufficient factor, including the incorporation of biochemical cues present in the native tissue, is expected to significantly boost HF regenerative strategies.

Biofabrication Approaches

The term "biofabrication" doesn't have a definition that is universally accepted and, as the field evolves, the demarcation between microfabrication, biofabrication, bioprinting and other technologies becomes blurry making it a matter of debate [50]. Some researchers in the field, for example, consider the use of cells as building blocks a crucial aspect of the biofabrication approach. If that is the case, there would be very few works related to hair formation that fall into this definition of the field. For this reason, we adopted a relatively wide definition and, in this subchapter, the term biofabrication will be related to the definition set forth by Mironov et al. In the inaugural issue of the journal Biofabrication: *"the production of complex living and non-living biological products from raw materials such as living cells, molecules, ECM and biomaterials"*[51]. This definition comprises an array of methodologies that includes not only 3D bioprinting but also other technologies that have been used in the HF regeneration field including micro moulding, layer-by-layer and laser-assisted fabrication of biological products.

Micro Moulding

Micro moulding is a technology used for the 2D or 3D patterning of hydrogels. It is a versatile technology that allows the use of different materials and crosslinking strategies to form the hydrogel [54]. Commonly, an elastomer such as polydimethylsiloxane (PDMS) is cast onto a patterned surface (master) and cured to form a mould. The hydrogel precursor is then poured onto the mould and crosslinked to obtain a patterned hydrogel. Micro moulding approaches have been used, together with 3D printing technologies, in the context of HF bioengineering [42, 55, 56]. In these cases, the patterning surface, or the mould, is designed and directly fabricated by 3D printing. Kageyama et al. used a 3D subtractive manufacturing approach to manufacture the initial template used for a multi-step micro moulding process that can be scaled up for the production of hundreds of HFGs [55]. The authors used a micro miller to fabricate a negative mould into an olefin plate, followed by casting an epoxy resin to form its positive counterpart. Afterwards, an elastomer was further cast onto the positive mould to obtain a chip constituted by 300 microwells functionalized with Pluronic to render it non-adhesive for cells. Each microwell was then filled with mouse embryonic epithelial and mesenchymal cells suspensions, which spontaneously self-sorted into HFGs. When transplanted intracutaneously in nude mice, these structures can form hairs at a high density. Following this line of work of self-organizing functional microtissues, the same group proposed an approach using a mixture of hDP cells, murine fetal epithelial cells and human vascular endothelial [56]. In this case, the initial single heterotypic cell aggregate rearranged into a dumbbell-shaped HFG with endothelial cells localized predominantly in the region of DP cells. Interestingly the presence of endothelial cells in these constructs significantly improved the expression of DP signature genes (including ALP, versican, BMP4 and LEF1) in vitro and the number of generated hairs in vivo. Micro moulding techniques were also used as a tool to guide epithelial cells invagination in a skin organotypic model, as occurs during the HF embryonic development. In this case, addictive 3D printing was used to fabricate a mould containing geometrical features with a high aspect ratio, namely follicular extensions with 500 μm diameter and 4 mm length [42] (Fig. 12.3a). This approach provides good flexibility in the design of the moulds, allowing the fabrication of accurate geometries at different densities. After being designed using computer-aided software, the mould was fabricated using a 3D printer that uses a jetting system similar to desktop printers to dispense, layer-by-layer, a polymeric solution of UV-curable VeroWhite. The mould was then used to form an array of microwells that acted as follicular cavities within a fibroblast-populated collagen dermal-like analogue. DP cells were then seeded on the dermal sample and spontaneously aggregate at the base of the microwells, whose dimensions (controlled by the mould) defined the size of the spheroids formed in situ. Neonatal skin keratinocytes seeded on top of the construct then engulfed the spheroids and started to proliferate upwards, initiating the HF-lineage differentiation and every so often forming unpigmented hair shafts. Furthermore, constructs with a density of 255 HF per square centimetre showed hair formation in 4 out of 7 constructs successfully grafted in nude mice.

Soft Lithography

Soft lithography is a microfabrication technique that uses an elastomeric mould to replicate the geometric features of a master. The term soft is related to the use of a suitable elastomer to facilitate the replication of a pattern, in contrast with other lithography techniques, such as photolithography, that uses light to transfer a pattern using a combination of photoresist and photosensitive materials [57]. Among the different

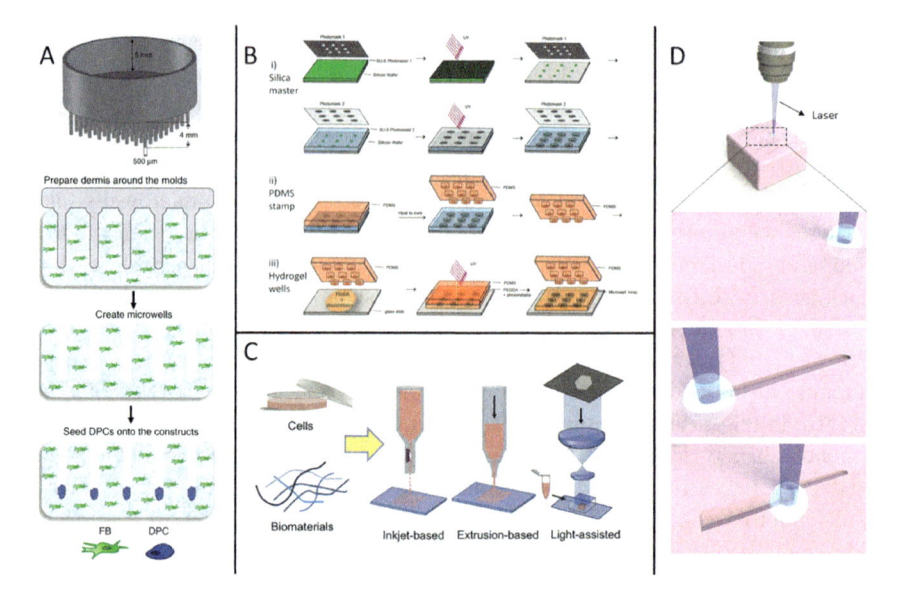

Fig. 12.3 Different advanced biofabrication approaches used in the HF regeneration field. **a** Micromoulding of collagen type-1 gel using 3D printed moulds. Reprint (adapted) from Abaci et al. [42] Creative Commons CC BY **b** Schematic representation of hair follicle-like mould fabrication, (i) silicon master manufacturing, (ii) PDMS stamp production, and (iii) hydrogel wells fabrication. Reprint (adapted) with permission from Pan et al. [52], Copyright John Wiley and Sons 2013. **c** Schematic diagram showing the use of 3D bioprinting to build in vitro constructs. Reprint (adapted) with permission from Ma et al. [53]. Copyright Elsevier 2018. **d** Microscopy-guided laser ablation where at the focus of the laser, the material is vaporized due to thermal decomposition. Reprint (adapted) from Abreu et al. [43] Creative Commons CC BY

elastomers, PDMS is one of the preferable choices for this particular technique given its low surface energy and chemical inertness that limit the interaction with the master, hence facilitating its removal [58]. This technique was combined with moulding to fabricate a 3D microwell array containing center islets (Fig. 12.3b) that mimic the architecture of human hair follicles [52]. A master was created by photolithography and used to create a PDMS mould depicting geometrical features that allow replicating the natural distribution of the DP and matrix within the HF bulb. This mould was then used to fabricate by photopolymerization poly(ethylene glycol) diacrylate (PEGDA) hydrogels encapsulating human dermal fibroblasts. These hydrogels are formed by an array of wells of controlled size, ranging from 50 to 400 µm, into which HaCaT keratinocytes were cultured. The authors showed uncompromised viability of keratinocytes inside the wells while the viability of fibroblasts decreased and stabilized to 50% during the second week of culture, probably due to the presence of unreacted photoinitiator and monomers. Although to our knowledge, further exploration of this approach was not performed so far, its ability to more accurately replicate the 3D architecture of HF and create well-arranged compartmentalized constructs holds the promise to more effectively bioengineer HF structures.

Layer-By-Layer

Layer-by-layer assembly allows fabricating polymeric multilayer thin films with tailored thicknesses around a liquid suspension of cells [59]. This is accomplished by the sequential adsorption of complementary multivalent polymers on a sacrificial core. This technique has been historically investigated for the xenotransplantation of cells to treat endocrine diseases such as diabetes by avoiding graft rejection [60]. For this purpose, cells are suspended in an alginate solution that is then ionically crosslinked in the presence of divalent cations as calcium. The alginate beads are then coated with a polycation, poly-L-lysine, forming a polyelectrolyte complex. Successive layers alginate and poly-L-lysine, following a layer-by-layer approach, are deposited up to balanced protection from immune reactivity and controlled diffusion of small molecules is achieved [61]. This complex is commonly referred to as alginate-poly-L-lysine-alginate (APA) and can be further manipulated using a sodium citrate solution to liquefy the inner core of the bead, obtaining a capsule. This technique was also used to generate DP cell loaded-microcapsules, which lead to the formation of HF- and sebaceous gland-like structures upon implantation in rat footpads [62]. Another work demonstrated that APA-encapsulated hDP cells displayed significantly higher expression of DP signature genes than conventionally cultured cells over a period of 21 days, being also able to sustain HF reformation when combined with epidermal cells of newborn mice [63]. Layer-by-layer approaches have been followed also for the nanoencapsulation of single mDP cells using gelatin as polycation to mimic the native ECM collagen [64]. The resultant DP cell spheroids depicted a native morphology and their gene and protein expression profile (ALP and versican) were improved compared to the same passage cells cultured under 2D conditions. Moreover, the hair-inductive capacity of these nanoencapsulated spheroids was demonstrated in vivo after co-grafting with freshly isolated newborn keratinocytes in nude mice.

Bioprinting

Bioprinting is a technology developed for the computer-aided layer-by-layer deposition of cells and biomaterials [65]. While many different approaches to this technique exist, in general, bioprinting can be seen as a specialized form of 3D printing where the ink or the filament is composed by a suspension of cells in a polymeric solution, often a hydrogel precursor (Fig. 12.3c). In the last few years, there has been a substantial increase in interest to use this technology for printing tissues and organ analogues to overcome some of the limitations of traditional scaffold fabrication approaches. Bioprinting allows the fabrication of precisely controlled geometrical microfeatures ultimately opening the possibility to recreate complex biological environments. On the other hand, bioprinting processing parameters need a careful selection to limit cell stress. For example, high pressure during extrusion bioprinting leads to high shear stress at the dispensing tip and to cell damage [66]. Although there have been many attempts to recapitulate the layered structure of the skin through bioprinting, there is still not much work related to hair regeneration [67–69]. A relatively recent patent discloses bioprinting methods to generate HF-like structures by injecting cells through a plurality of needles into a hydrogel [67]. Coaxial needles are used to inject DP cells (through the inner one), into a collagen I or gelatin hydrogel down to the desired depth. During the ensuing needle withdrawal, keratinocytes are delivered (through the outer portion) generating epithelial tube-like structures. In a different approach, normal needles prefilled with keratinocytes and then DP cells are used to perforate the hydrogel and, similarly to the previous method, dispense sequentially DP cells and keratinocytes during withdrawal. In another work a multi-nozzle 3D printer and fibrinogen-based bioinks were used to bioprint complex skin trilayered substitutes composed of human epidermal (keratinocytes and pigmented melanocytes), dermal (fibroblasts, hDP cells and microvascular endothelial cells) and hypodermal (pre-adipocytes) cells [68]. While this work may have not led to the reformation of structures morphologically akin to HFs it is, to our knowledge, the only work in the literature using a DP cell-loaded bioink. Still, it is interesting to underline that bioprinting has emerged as a biofabrication technology that gained lots of scientific momentum, being therefore expectable new developments in the near future.

Laser Ablation

Laser ablation is a technique that uses pulses of light of the desired wavelength to selectively and controllably remove material. The laser can be coupled with a microscope that focuses the laser beam, increasing the energy density in the focal point and allowing high accuracy and minimal thermal damage of the surroundings [70, 71]. The process is influenced by both the characteristics of the laser, such as wavelength and by the characteristics of the sample that determine light absorption. Recently, laser ablation was used to facilitate direct interaction between hDP cells and keratinocytes in an organotypic skin model. This approach followed a similar idea explored in other works that is to recreate a microwell into a collagen matrix where DP cell aggregates are placed, to allow epidermal invagination from the outside surface towards it. In this case, the DP cell aggregates were firstly formed in low-adherent wells and posteriorly cultured with keratinocytes to form compartmentalized multi-cellular aggregates. Differently from other approaches, these aggregates were incorporated into the fibroblast-populated collagen lattice that constitutes the dermal layer and the microwells were fabricated afterwards, using microscopy-guided laser ablation [43]. This was performed by successively ablating layer-by-layer sections of the collagen hydrogel (Fig. 12.3d), from the top surface of the construct down to reach the aggregates containing the DP cells. This approach allows the removal of 30 μm of collagen in each layer making it more challenging than other methods to get deeper microwells. In turn, being a laser-assisted technique this method has the advantage of not requiring contact and providing higher resolution than others, allowing the fabrication of microwells down to 50 μm in diameter. Overall, this proposed strategy enabled to recreate in the organotypic skin model immature HF-like structures with the reformation of a hair bulb and multistratified epithelia that was more complex than the epidermis from where it was derived.

Conclusion and Future Trends

Despite the reports demonstrating de novo human HF formation following the use of DP cells and epithelial cells from neonatal tissue sources, successful HF regenerative strategies promoted entirely from human adult cells have not yet been reached. Most tactics are focused on partially recovering DP cell inductivity through 3D spheroidal cultures or in multicellular aggregates combined with epithelial cells, in an attempt to prompt the EMIs that promote hair development and growth. Contrasting to what is often targeted in other tissue engineering strategies, in which scaffolds are designed to act as bioinstructive matrices, in the hair restoration field biomaterials have often been used as low-adhesive substrates to promote cellular aggregation rather than ECM-inspired matrices to guide cellular behaviour and growth in a biomimetic fashion. The majority of works have mostly used collagen I matrices nonetheless, the DP is embedded in a specialized ECM that differs from the non-follicular dermis by being richer in basement membrane components such as collagen type IV, fibronectin and laminin and by displaying a higher content of sulphated glycosaminoglycans and versican, among other proteoglycans [72–74]. Moreover, the ECM content of the DP increases during anagen and decreases afterwards, being almost unnoticeable at telogen [72], suggesting that the highly packed DP cell spheroids would benefit from a biomaterial-assisted increase in the ECM-to-cell ratio and respective cell-ECM interactions. However, the specific replication of the ECM of the DP either through the modification of natural or synthetic materials represents a considerable research gap in the field. Similarly, the rapid differentiation that epithelial cells undergo in culture and that jeopardizes their regenerative potential could also be subject of control with matrices that emulate the native skin basement membrane, regulatory integrins or even specific cell anchorage proteins known to maintain/regulate the HF stem cell pool, such as the collagen XVII (COL17A1) strongly expressed in the hair bulge [75]. The development of well-designed instructive biomaterials and their association with advanced biofabrication techniques, such as the use of bioinks for 3D bioprinting of biomimetic hydrogel-based constructs, or their casting into complexly shaped 3D printed moulds, holds the promise to significantly foster the recreation of the HF and/or skin architecture, cellular interactions and microenvironment with higher precision, therefore improving the chances of inducing HF formation with higher efficiency and reproducibility. Moreover, and as often considered by tissue engineers, biomaterials could also be explored for the controlled delivery of growth factors or other biochemical cues. The incorporation of signalling molecules capable of initiating processes reminiscent of the HF development or instead, instructing cellular behaviour in situ or in tissue constructs could represent a breakthrough in HF regenerative strategies. Regrettably, the complex nature of this appendage and the multiple signalling pathways involved in its regulation make it difficult to identify and take a translational advantage of such biomolecules without a deeper understanding of their regulatory actions in the human HF. Alternatively, the use of the cell sources of these biological signals/pathway ligands may be used to introduce the necessary complexity to the every so often simplified epithelial-mesenchymal systems commonly used. As previously mentioned, increasing proof demonstrates that both the dWAT and resident HF immune cells play a role in the control of the HF stem cell niche and hair cycle, but their potential also remains vastly unexplored. Ultimately, successful biofabrication strategies would require long-term functionality of the bioengineered HF units (or hair-bearing skin equivalents) and their integration with the hosts' vasculature and sensory networks in addition to proper activation of hair shaft pigmentation mechanisms. Vascularization strategies, to our knowledge, haven't been applied to hair neoformation but there are examples in literature applied to other fields of tissue engineering. These strategies are fundamentally conceptualized around the use of a sacrificial template to fabricate channels inside another material. The most common approach is to construct a structure that can then be removed by changing the temperature. Sacrificial materials such as poloxamer 407 (commonly known as pluronic-127) or even ice, represent some examples. Other approaches involve the use of materials that liquefy by changing environmental parameters other than the temperature, such as shellac modified hydrogels as a pH-responsive material. Alternatively, vascularization strategies that do not involve a sacrificial template include the use of laser-assisted degradation or ablation of a substrate to recapitulate the vascular-like structure. These allow a higher resolution that may be suited to recapitulate small blood vessels such as capillaries [76] and, in the order of few microns, to reproduce the in vivo transport phenomena in vascularized engineered structures [77]. These represent some of the strategies already developed that could be explored to improve the viability, integration and even functional capacity of bioengineered products containing follicular units.

Considering all these premises, challenges and the HF innate complexity, it is not surprising that even with the advent of complex biofabrication technologies the formation of new hair remains an unfulfilled need. However, the optimization of isolation and culture methods to induce or preserve trichogenic properties in adult human cells and its multidisciplinary convergence with cutting edge enabling technologies and biomimetic biomaterials will gradually bring us closer to develop clinically suitable HF bioengineering solutions.

Acknowledgements Authors would like to acknowledge the financial support from the Consolidator Grant "ECM_INK" (ERC-2016-COG-726061) and from FCT/MCTES (*Fundação para a Ciência e a Tecnologia/Ministério da Ciência, Tecnologia, e Ensino Superior*) grants PD/59/2013, PD/BD/113800/2015 (CA).

References

1. Pispa J, Thesleff I (2003) Mechanisms of ectodermal organogenesis. Dev Biol 262(2):195–205
2. Jahoda CAB, Oliver RF, Reynolds AJ, Forrester JC, Gillespie JW, Cserhalmi-Friedman PB et al (2001) Trans-species hair growth induction by human hair follicle dermal papillae. Exp Dermatol 10(4):229–237
3. Reynolds AJ, Lawrence C, Cserhalmi-Friedman PB, Christiano AM, Jahoda CA (1999) Trans-gender induction of hair follicles. Nature 402(6757):33–34
4. Ohyama M, Veraitch O (2013) Strategies to enhance epithelial–mesenchymal interactions for human hair follicle bioengineering. J Dermatol Sci 70(2):78–87
5. Rendl M, Lewis L, Fuchs E (2005) Molecular dissection of mesenchymal–epithelial interactions in the hair follicle. Hogan B (ed). PLoS Biol 3(11):e331
6. Driskell RR, Clavel C, Rendl M, Watt FM (2011) Hair follicle dermal papilla cells at a glance. J Cell Sci 124(8):1179–1182
7. Higgins CA, Chen JC, Cerise JE, Jahoda CAB, Christiano AM (2013) Microenvironmental reprogramming by three-dimensional culture enables dermal papilla cells to induce de novo human hair-follicle growth. Proc Natl Acad Sci 110(49):19679–19688
8. Yang C-C, Cotsarelis G (2010) Review of hair follicle dermal cells. J Dermatol Sci 57(1):2–11
9. Wu X, Scott L, Washenik K, Stenn K (2014) Full-thickness skin with mature hair follicles generated from tissue culture expanded human cells. Tissue Eng Part A 20(23–24):3314–3321
10. Thangapazham RL, Klover P, Li S, Wang J, Sperling L, Darling TN (2014) A model system to analyse the ability of human keratinocytes to form hair follicles. Exp Dermatol 23(6):443–446
11. Ehama R, Ishimatsu-Tsuji Y, Iriyama S, Ideta R, Soma T, Yano K et al (2007) Hair follicle regeneration using grafted rodent and human cells. J Invest Dermatol 127(9):2106–2115
12. Thangapazham RL, Klover P, Wang JA, Zheng Y, Devine A, Li S et al (2014) Dissociated human dermal papilla cells induce hair follicle neogenesis in grafted dermal-epidermal composites. J Invest Dermatol 134(2):538–540
13. Abreu CM, Pirraco RP, Reis RL, Cerqueira MT, Marques AP (2021) Interfollicular epidermal stem-like cells for the recreation of the hair follicle epithelial compartment. Stem Cell Res Ther 12(1):62
14. Ohyama M (2005) Characterization and isolation of stem cell-enriched human hair follicle bulge cells. J Clin Invest 116(1):249–260
15. Slominski A, Wortsman J, Plonka PM, Schallreuter KU, Paus R, Tobin DJ (2005) Hair follicle pigmentation. J Invest Dermatol 124(1):13–21
16. Paus R, Ito N, Takigawa M, Ito T (2003) The hair follicle and immune privilege. J Investig Dermatol Symp Proc 8(2):188–194
17. Christoph T, Müller-Röver S, Audring H, Tobin DJ, Hermes B, Cotsarelis G et al (2000) The human hair follicle immune system: cellular composition and immune privilege. Br J Dermatol 142(5):862–873
18. Ali N, Zirak B, Rodriguez RS, Pauli ML, Truong H-A, Lai K et al (2017) Regulatory T cells in skin facilitate epithelial stem cell differentiation. Cell 169(6):1119-1129.e11
19. Castellana D, Paus R, Perez-Moreno M (2014) Macrophages contribute to the cyclic activation of adult hair follicle stem cells. In Nusse R (ed). PLoS Biol 12(12):e1002002
20. Lachgar C, Gall B (1998) Minoxidil upregulates the expression of vascular endothelial growth factor in human hair dermal papilla cells. Br J Dermatol 138(3):407–411
21. Yano K, Brown LF, Detmar M (2001) Control of hair growth and follicle size by VEGF-mediated angiogenesis. J Clin Invest 107(4):409–417
22. Driskell RR, Jahoda CAB, Chuong C-M, Watt FM, Horsley V (2014) Defining dermal adipose tissue. Exp Dermatol 23(9):629–631
23. Festa E, Fretz J, Berry R, Schmidt B, Rodeheffer M, Horowitz M et al (2011) Adipocyte lineage cells contribute to the skin stem cell niche to drive hair cycling. Cell 146(5):761–771
24. Plikus MV, Mayer JA, de la Cruz D, Baker RE, Maini PK, Maxson R et al (2008) Cyclic dermal BMP signalling regulates stem cell activation during hair regeneration. Nature 451(7176):340–344

25. Zheng Y, Du X, Wang W, Boucher M, Parimoo S, Stenn K (2005) Organogenesis from dissociated cells: generation of mature cycling hair follicles from skin-derived cells. J Invest Dermatol 124(5):867–876
26. Narita K, Fukuoka H, Sekiyama T, Suga H, Harii K (2020) Sequential scalp assessment in hair regeneration therapy using an adipose-derived stem cell-conditioned medium. Dermatol Surg 46(6):819–825
27. Huang C-F, Chang Y-J, Hsueh Y-Y, Huang C-W, Wang D-H, Huang T-C et al (2016) Assembling composite dermal papilla spheres with adipose-derived stem cells to enhance hair follicle induction. Sci Rep 6(1):26436
28. Kageyama T, Yan L, Shimizu A, Maruo S, Fukuda J (2019) Preparation of hair beads and hair follicle germs for regenerative medicine. Biomaterials 212:55–63
29. Chermnykh ES, Vorotelyak EA, Gnedeva KY, Moldaver MV, Yegorov YE, Vasiliev AV et al (2010) Dermal papilla cells induce keratinocyte tubulogenesis in culture. Histochem Cell Biol 133(5):567–576
30. Leirós GJ, Kusinsky AG, Drago H, Bossi S, Sturla F, Castellanos ML et al (2014) Dermal papilla cells improve the wound healing process and generate hair bud-like structures in grafted skin substitutes using hair follicle stem cells. Stem Cells Transl Med 3(10):1209–1219
31. Miao Y, Sun Y Bin, Liu BC, Jiang JD, Hu ZQ (2014) Controllable production of transplantable adult human high-passage dermal papilla spheroids using 3D matrigel culture. Tissue Eng Part A 20(17–18):2329–2338
32. Lindner G, Horland R, Wagner I, Ataç B, Lauster R (2011) De novo formation and ultrastructural characterization of a fiber-producing human hair follicle equivalent in vitro. J Biotechnol 152(3):108–112
33. Havlickova B, Biro T, Mescalchin A, Tschirschmann M, Mollenkopf H, Bettermann A et al (2009) A human folliculoid microsphere assay for exploring epithelial- mesenchymal interactions in the human hair follicle. J Invest Dermatol 129(4):972–983
34. Limat A, Breitkreutz D, Hunziker T, Klein CE, Noser F, Fusenig NE et al (1994) Outer root sheath (ORS) cells organize into epidermoid cyst-like spheroids when cultured inside Matrigel: a light-microscopic and immunohistological comparison between human ORS cells and interfollicular keratinocytes. Cell Tissue Res 275(1):169–176
35. Havlickova B, Biro T, Mescalchin A, Arenberger P, Paus R (2004) Towards optimization of an organotypic assay system that imitates human hair follicle-like epithelial-mesenchymal interactions. Br J Dermatol 151(4):753–765
36. Nakao K, Morita R, Saji Y, Ishida K, Tomita Y, Ogawa M et al (2007) The development of a bioengineered organ germ method. Nat Methods 4(3):227–230
37. Asakawa K, Toyoshima K, Ishibashi N, Tobe H, Iwadate A, Kanayama T et al (2012) Hair organ regeneration via the bioengineered hair follicular unit transplantation. Sci Rep 2(1):424
38. Toyoshima K, Asakawa K, Ishibashi N, Toki H, Ogawa M, Hasegawa T et al (2012) Fully functional hair follicle regeneration through the rearrangement of stem cells and their niches. Nat Commun 3(1):784
39. Gupta AC, Chawla S, Hegde A, Singh D, Bandyopadhyay B, Lakshmanan CC et al (2018) Establishment of an in vitro organoid model of dermal papilla of human hair follicle. J Cell Physiol 233(11):9015–9030
40. Betriu N, Jarrosson-Moral C, Semino CE (2020) Culture and differentiation of human hair follicle dermal papilla cells in a soft 3D self-assembling peptide scaffold. Biomolecules 10(5):684
41. Vahav I, Broek LJ, Thon M, Monsuur HN, Spiekstra SW, Atac B et al (2020) Reconstructed human skin shows epidermal invagination towards integrated neopapillae indicating early hair follicle formation in vitro. J Tissue Eng Regen Med 14(6):761–773
42. Abaci HE, Coffman A, Doucet Y, Chen J, Jacków J, Wang E et al (2018) Tissue engineering of human hair follicles using a biomimetic developmental approach. Nat Commun 9(1):5301
43. Abreu CM, Gasperini L, Lago MEL, Reis RL, Marques AP (2020) Microscopy-guided laser ablation for the creation of complex skin models with folliculoid appendages. Bioeng Transl Med 6:e10195

44. Sriwiriyanont P, Lynch KA, Maier EA, Hahn JM, Supp DM, Boyce ST (2012) Morphogenesis of chimeric hair follicles in engineered skin substitutes with human keratinocytes and murine dermal papilla cells. Exp Dermatol 21(10):783–785
45. Sriwiriyanont P, Lynch KA, McFarland KL, Supp DM, Boyce ST (2013) Characterization of hair follicle development in engineered skin substitutes. In Slominski AT (ed). PLoS One 8(6):e65664
46. Zhang L, Wang W, Jin JY, Degan S, Zhang G, Erdmann D et al (2019) Induction of hair follicle neogenesis with cultured mouse dermal papilla cells in de novo regenerated skin tissues. J Tissue Eng Regen Med 13(9):1641–1650
47. Taylor DA, Sampaio LC, Ferdous Z, Gobin AS, Taite LJ (2018) Decellularized matrices in regenerative medicine. Acta Biomater 74:74–89
48. Zhang X, Xiao S, Liu B, Miao Y, Hu Z (2019) Use of extracellular matrix hydrogel from human placenta to restore hair-inductive potential of dermal papilla cells. Regen Med 14(8):741–751
49. Kang BM, Kwack MH, Kim MK, Kim JC, Sung YK (2012) Sphere formation increases the ability of cultured human dermal papilla cells to induce hair follicles from mouse epidermal cells in a reconstitution assay. J Invest Dermatol 132(1):237–239
50. Groll J, Boland T, Blunk T, Burdick JA, Cho D-W, Dalton PD et al (2016) Biofabrication: reappraising the definition of an evolving field. Biofabrication 8(1):13001
51. Mironov V, Trusk T, Kasyanov V, Little S, Swaja R, Markwald R (2009) Biofabrication: a 21st century manufacturing paradigm. Biofabrication 1(2):22001
52. Pan J, Yung Chan S, Common JEA, Amini S, Miserez A, Birgitte Lane E et al (2013) Fabrication of a 3D hair follicle-like hydrogel by soft lithography. J Biomed Mater Res Part A 101(11):3159–3169
53. Ma X, Liu J, Zhu W, Tang M, Lawrence N, Yu C et al (2018) 3D bioprinting of functional tissue models for personalized drug screening and in vitro disease modeling. Adv Drug Deliv Rev 132:235–251
54. Yanagawa F, Sugiura S, Kanamori T (2016) Hydrogel microfabrication technology toward three dimensional tissue engineering. Regen Ther 3:45–57
55. Kageyama T, Yoshimura C, Myasnikova D, Kataoka K, Nittami T, Maruo S et al (2018) Spontaneous hair follicle germ (HFG) formation in vitro, enabling the large-scale production of HFGs for regenerative medicine. Biomaterials 154:291–300
56. Kageyama T, Chun Y-S, Fukuda J (2021) Hair follicle germs containing vascular endothelial cells for hair regenerative medicine. Sci Rep 11(1):624
57. Xia Y, Whitesides GM (1998) Soft lithography. Annu Rev Mater Sci 28(12):153–184
58. Künzler JF (1996) Silicone hydrogels for contact lens application. Trends Polym Sci 4(2):52–59
59. Borges J, Mano JF (2014) Molecular interactions driving the layer-by-layer assembly of multilayers. Chem Rev 114(18):8883–8942
60. Lim F, Sun A (1980) Microencapsulated islets as bioartificial endocrine pancreas. Science (80-) 210(4472):908–910
61. Strand BL, Ryan L, Veld PI, Kulseng B, Rokstad AM, Skjåk-Braek G et al (2001) Poly-L-Lysine induces fibrosis on alginate microcapsules via the induction of cytokines. Cell Transpl 10(3):263–275
62. Lin C, Li Y, Ji Y, Keng H, Cai X, Zhang J (2008) Microencapsulated human hair dermal papilla cells: a substitute for dermal papilla? Arch Dermatol Res 300(9):531–535
63. Xie B, Chen M, Ding P, Lei L, Zhang X, Zhu D, et al (2020) Induction of dermal fibroblasts into dermal papilla cell-like cells in hydrogel microcapsules for enhanced hair follicle regeneration. Appl Mater Today 21:100805
64. Wang J, Miao Y, Huang Y, Lin B, Liu X, Xiao S et al (2018) Bottom-up nanoencapsulation from single cells to tunable and scalable cellular spheroids for hair follicle regeneration. Adv Healthc Mater. 7(3):1700447
65. Gasperini L, Mano JF, Reis RL (2014) Natural polymers for the microencapsulation of cells. J R Soc Interface 11(100):20140817
66. Nair K, Gandhi M, Khalil S, Yan KC, Marcolongo M, Barbee K et al (2009) Characterization of cell viability during bioprinting processes. Biotechnol J 4(8):1168–1177

67. WO2018089750A1—Bioprinted hair follicles and uses thereof [Internet] (2019) [cited 2021 May 5]. Available from: https://patents.google.com/patent/WO2018089750A1/en
68. Jorgensen AM, Varkey M, Gorkun A, Clouse C, Xu L, Chou Z et al (2020) Bioprinted skin recapitulates normal collagen remodeling in full-thickness wounds. Tissue Eng Part A 26(9–10):512–526
69. Wang R, Wang Y, Yao B, Hu T, Li Z, Huang S et al (2019) Beyond 2D: 3D bioprinting for skin regeneration. Int Wound J 16(1):134–138
70. McCullen SD, Gittard SD, Miller PR, Pourdeyhimi B, Narayan RJ, Loboa EG (2011) Laser ablation imparts controlled micro-scale pores in electrospun scaffolds for tissue engineering applications. Ann Biomed Eng 39(12):3021
71. Koo S, Santoni SM, Gao BZ, Grigoropoulos CP, Ma Z (2017) Laser-assisted biofabrication in tissue engineering and regenerative medicine. J Mater Res 32(1):128–142
72. Messenger AG, Elliott K, Temple A, Randall VA (1991) Expression of basement membrane proteins and interstitial collagens in dermal papillae of human hair follicles. J Invest Dermatol 96(1):93–97
73. Jahoda CA, Mauger A, Bard S, Sengel P (1992) Changes in fibronectin, laminin and type IV collagen distribution relate to basement membrane restructuring during the rat vibrissa follicle hair growth cycle. J Anat 181(Pt 1):47–60
74. Soma T, Tajima M, Kishimoto J (2005) Hair cycle-specific expression of versican in human hair follicles. J Dermatol Sci 39(3):147–154
75. Tanimura S, Tadokoro Y, Inomata K, Binh NT, Nishie W, Yamazaki S et al (2011) Hair follicle stem cells provide a functional niche for melanocyte stem cells. Cell Stem Cell 8(2):177–187
76. Heintz KA, Bregenzer ME, Mantle JL, Lee KH, West JL, Slater JH (2016) Fabrication of 3D biomimetic microfluidic networks in hydrogels. Adv Healthc Mater. 5(17):2153–2160
77. Pradhan S, Keller KA, Sperduto JL, Slater JH (2017) Fundamentals of laser-based hydrogel degradation and applications in cell and tissue engineering. Adv Healthc Mater. 6(24):1700681

Part IV
Hair Follicle Regeneration and Injury

Chapter 13
Wound Healing Induced Hair Follicle Regeneration

Yiqun Jiang and Peggy Myung

Abstract *Introduction* Adult mammals possess very limited regeneration potential. Wound-induced hair neogenesis (WIHN) is a rare regenerative event observed in mammals, in which de novo hair follicles are formed at the wound site. The phenomenon has been observed more than half a century ago, yet the mechanism remains unclear. Recently, more mechanistic studies have shed light onto the complex cellular signaling and mechanical events during WIHN. Interestingly, many of the mechanisms in WIHN are shared with those found in embryonic hair follicle development. Here we review emerging works regarding WIHN mechanisms and potential clinical applications for hair follicle regeneration and improved wound healing outcomes. *Methods* For the selection of literature covered in this chapter we used Pubmed database. We used keywords such as wound-induced hair follicle neogenesis, embryonic hair follicle development, scarless regeneration, Wnt signaling, Hedgehog signaling. *Results* Hair follicles generated in WIHN are formed from non-hair-residing epithelial and dermal cells. Extrinsic signals such as Wnt and Shh shown to promote embryonic hair follicle formation are also crucial players in WIHN. Understanding WIHN mechanisms provides a basis to achieve regenerative wound healing with appendage formation. *Conclusions* Although there has been much progress in identifying the critical factors that promote adult hair follicle neogenesis, there is a great need for exploring the different cellular behavior and signaling mechanisms involved in WIHN and how they can be applied clinically.

Keywords Wounding · Regeneration · Wound-induced hair neogenesis · Embryonic hair follicle · Wnt · Shh · Mechanic transduction · Fibroblast · Fibrosis · Macrophages · Myofibroblasts

---

Y. Jiang
Departments of Dermatology and Pathology, Yale University, New Haven, CT, USA
e-mail: yiqun.jiang@yale.edu

P. Myung (✉)
Department of Molecular, Cellular and Developmental Biology, Yale University, New Haven, CT, USA
e-mail: peggy.myung@yale.edu

© The Author(s), under exclusive license to Springer Nature Switzerland AG 2022
F. Jimenez and C. Higgins (eds.), *Hair Follicle Regeneration*, Stem Cell Biology and Regenerative Medicine 72, https://doi.org/10.1007/978-3-030-98331-4_13

Summary

Skin is composed of epidermis, dermis, and associated adnexal structures (e.g. hair follicles, nails, sweat glands). Those appendages are functionally important for environmental protection, thermoregulation as well as for cosmesis. As skin is under constant assault, wound healing is key to animal survival. Complete skin regeneration after wound healing should ideally involve regenerating the epidermis, dermis, and associated appendages to a state prior to wounding. However, wounding the skin of adult mammals usually results in scarring with no appendage regeneration. One theory speculates that mammals have evolved to increase the rate of repair for self-protection following wounding at the expense of the quality of regeneration (i.e. no appendages). Intriguingly, rare regenerative events such as wound-induced hair neogenesis (WIHN) have been reported in adult rabbits and mice but rarely in humans, yet the mechanism(s) by which neogenesis occurs is not fully understood. In this chapter, we aim to review the origin of cells that make up newly generated hair follicles in WIHN. We will then review the extrinsic signals that support WIHN, conferred by the wounding environment or transgenic approaches. This chapter will particularly focus on the comparison between WIHN and embryonic hair follicle development since several lines of evidence suggest the connection between the two. We will also discuss how uncovering the mechanisms behind WIHN can inform clinical studies for human applications.

Key Points:

- Wound healing in mammals leads to fibrosis instead of regeneration of associated appendages.
- Wound-induced hair follicle neogenesis (WIHN) has been observed in rabbits and mice but rarely in humans, the mechanism of which is not clear yet.
- Modulating development signals (Wnt, Shh) in the wound can improve outcome for WIHN.

Introduction

Skin is the largest organ in our body and the first barrier against multiple types of environmental insults. Wound healing in adult mammals usually leads to fibrosis, commonly known as scarring, characterized by aggregation of excessive extracellular matrix proteins deposited by fibroblasts in the wound bed. The type of collagen and poor organization of scar tissue renders it weaker than healthy tissue. In addition, there is typically failure to regenerate functional appendages (e.g. hair follicles, sweat glands and nails) at the healed wound site. However, many species (e.g. axolotl, *Xenopus*) and even prenatal mammals can regenerate wounded skin without a scar and with functional appendages. This full regenerative ability appears to compete with the commonly observed fibrotic, appendage-deficient, rapid-repair process designed to facilitate expedited wound closure in mammals. From an evolutionary standpoint, it is reasonable to speculate that adult mammals prioritize rapid wound closure over full regeneration of appendages because the former presumably grants the animal a greater survival advantage.

Nevertheless, to confer the human skin barrier with full regenerative capacity that achieves both appendage reconstitution and scarless healing is clinically valuable. For instance, burn wounds exhibit difficulties in restoring both the epithelium barrier and functional appendages of the skin, affecting protection and thermoregulation. In addition, pathological scarring such as hypertrophic scarring and keloids as well as scarring alopecia where hair follicles are lost and replaced with scarring tissue are a huge cosmetic and functional challenge for patients; even a small scar can impose psychological and social stress.

Despite the lack of complete skin regeneration in most scenarios, it is possible that adult mammals still preserve some regenerative ability. Wound-induced hair follicle neogenesis (WIHN) is one example that demonstrates the residual regenerative potential of adult mammals in which de novo hair follicles are observed at the wound site in adult rabbits, humans and mice [1–4]. WIHN reconstitutes both the dermal papilla (DP), a cluster of specialized mesenchymal cells that regulates the cyclical growth and differentiation of the adult hair follicle, and the epithelium downgrowth from which the new hair shaft emerges. Ito et al. [4] observed newly formed hair follicles only at the center of large excisional wounds. All regenerated hairs had similar length and bends with those produced by native hair follicles.

A major focus of current research is to regenerate hair follicles in the wound bed of adult mammals including humans. However, the mechanisms that convert scar tissue to hair follicle neogenesis still remain poorly understood. What we do know is that certain developmental signals such as Wnt and Sonic Hedgehog (SHH) can potentiate WIHN [5]. We still don't quite understand why WIHN occurs only at the wound center whereas the periphery remains devoid of new hairs in mice [4], but new research suggests that the lower tissue stiffness in the center of the wound bed could somewhat explain the higher number of hair follicles regenerated therein [6]. Thus, it is possible that the 'softness' of scarring tissue can facilitate hair follicle neogenesis. However, to turn the wound into normal healthy skin, an optimal balance needs to be achieved between reestablishing the mechanical strength of scar tissue and inducing hair follicle neogenesis within the wound bed. How to engineer the perfect mechanical environment for scarless regeneration is beyond the scope of this review, but definitely is an important future direction to look at.

Although great progress in investigating WIHN mechanisms in animal models have been achieved, we still don't understand why WIHN has only rarely been reported in humans [3, 7]. The goal of this chapter is to provide an overview of the regenerative WIHN and current perspectives for regenerative strategies, hopefully providing ideas for WIHN-based mechanistic investigation and clinical application in both experimental model organisms and humans.

Wound Healing Process and New Hair Follicles

General Wound Healing Process

Since WIHN occurs in context of wound healing, we need to first understand the complex, intricately regulated wound healing process. After the initial acute inflammatory response that usually occurs within 24 to 48 h post wounding, cells proliferate for roughly 2 weeks to compensate for skin cell loss. While the keratinocytes migrate and proliferate to bring wound edges together and regenerate the epithelial barrier [8, 9], the underlying dermal cell types such as fibroblasts and macrophages participate in the granulation process to generate new tissue and to form a wound bed to facilitate unimpaired keratinocyte migration [10, 11].

There are two distinct although overlapping mechanisms to close the wound. Re-epithelialization, or epithelial re-surfacing, refers to the active closing of the epithelium barrier by migrating keratinocytes whereas wound contraction is facilitated by the adhesion of myofibroblasts, activated fibroblasts, to extracellular matrix (ECM) components post wounding. Depending on the species, either epithelial resurfacing or wound contraction dominate the cutaneous wound closure process. Animals with loose skin such as mice close the wound mainly by contraction mediated by contractile myofibroblasts in the dermis, whereas in humans up to 80% of wound closure is accounted for by resurfacing by the epithelium [12, 13]. Re-epithelialization is still important in mice especially in larger wounds where WIHN is mostly observed.

Wound-Induced Hair Follicle Neogenesis in Animals and Humans

Out of the two wound closure mechanisms, it is re-epithelialization that seems to be directly associated with WIHN in animals. WIHN in wild-type mice has only been observed in the center of a large wound (2.25 cm^2) [4]: Wound contraction accounts for most closure, leaving a final area of 0.25 cm^2 for re-epithelialization and WIHN. The onset of re-epithelialization seems to directly trigger WIHN in rabbits too: Hair follicle neogenesis is only observed in the middle of the re-epithelialized area after wound contraction has stopped [2]. It may be speculated that hair follicle neogenesis occurs only after epidermal inductive signals are restored post re-epithelialization. The exact reason why WIHN is only observed in large wounds after contraction stops in animal models remains elusive, although newest research suggests that the mechanical 'softness' of center of the large wound may contribute to hair follicle neogenesis [6].

The occurrence of WIHN is even more scarce in humans. There has been one report of the scarless re-formation of facial vellus hair follicles in humans [3] which was published 70 years ago in 1956. The reason why the more commonly observed fibrotic wounds in humans have not been associated with WIHN remains to be investigated. It is even more odd that WIHN is hardly seen in humans when human wounds heal by re-epithelialization.

One of the best characterized WIHN examples in animal models is found in the center of large wounds in mice 2–3 days after re-epithelialization is complete [4]. Here, de novo hair follicles expressed differentiation markers similar to those found in embryonic hair follicles, such as keratin 17 (KRT17), Lef1, alkaline phosphatase (AP), Wnt10b and Shh. Hair follicle differentiation markers KRT17 and Lef1 only started to be expressed as new hair germs were formed, suggesting a true nascent nature of the new hair follicles. This study suggests that the newly regenerated hair follicles share similar features with embryonic hair follicles.

Epithelial Cell of Origin of Newly Formed Hair Follicles

It has been established that adult hair follicles regenerate from resident stem cells during the adult hair cycle. Several stem cell populations have been studied including Keratin 15 + (KRT15) keratinocytes residing in the hair follicle bulge [14, 15] and Lgr6 + cells found in the central isthmus area above the bulge [16]. One hypothesis is that the new hair follicles observed after wounding arise from those stem cells that have migrated from surrounding intact hair follicles. Ito et al. [4] demonstrated that non-bulge stem cells gave rise to new hair follicles in response to wounding. The authors lineage labeled KRT15 + progenies derived from the bulge region of intact hair follicle and revealed that most neogenic hair follicle epithelial cells did not originate from those bulge stem cells. In contrast to KRT15 + hair follicle bulge stem cells, Lgr6 + stem cells that reside above the bulge persist in the reformed epithelium and new hair follicles for more than three months [16]. This evidence suggests a hierarchy in the potential of skin-related stem cells to contribute to de novo hair follicles. In contrast to the relatively well characterized hair follicle epithelium post wounding, the origin of the neogenic dermal papilla (DP) is less understood. We will discuss this as we look at the dermal fibroblast cell fates upon wounding.

Signals from Both Dermis and Epidermis Are Essential for WIHN

Fibrosis, Remodeling and Source of HF-Related Mesenchyme

Following the proliferative re-epithelialization phase of wound healing is the remodeling phase, during which time newly deposited ECM strengthens the healing wound. During this stage, fibrosis is promoted by secretion of growth factors and cytokines of immune cells e.g. macrophages, prompting fibroblasts to deposit extracellular matrix proteins including collagen that transitions from type I to type III collagen to form the bulk of scar [17]. Remodeling can persist up to 6 months or even longer.

Besides mediating contraction and collagen deposition, dermal fibroblasts form inductive mesenchyme for new hair follicles. However, the ability to form inductive mesenchyme is heterogenuous among fibroblasts. There are two distinct fibroblast lineages that generate new mesenchyme post-wounding (Fig. 13.1). In Driskell et al. [18], the authors characterized the embryonic origin of those two distinct fibroblast populations by performing lineage tracing experiments in mice. Dlk1(Delta like homologue 1) expression persists in the lower dermis after mouse embryonic day E18.5 and Blimp1 (B-lymphocyte-induced maturation protein 1) is expressed in upper papillary dermis between E16.5 and E18.5. When lineage-traced to postnatal timepoints, Dlk1-lineage dermal fibroblasts go to lower reticular dermis and hypodermis whereas Blimp1-lineage gives rise to the upper papillary dermis that induces

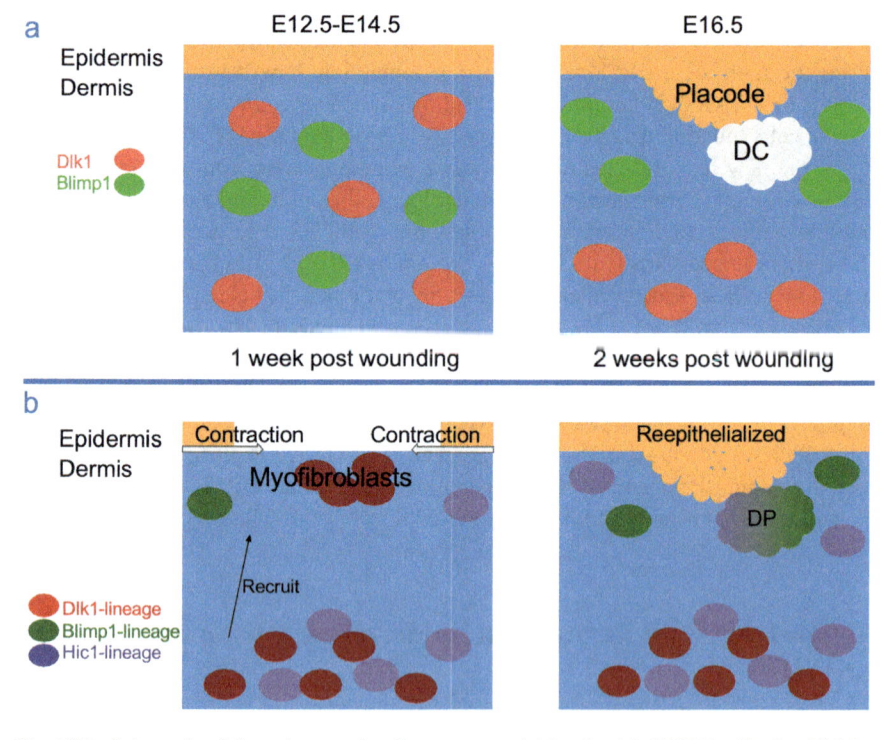

Fig. 13.1 Schematic of dermal progenitor lineage potential involved in WIHN. **a** During E12.5 to E14.5 Dlk1 and Blimp1 are expressed throughout the dermis. At E16.5, Blimp1 and Dlk1 expression start to be stratified. DC is dermal condensate. **b** Seven days post wounding in adult murine back skin, the Dlk1-lineage was largely found in myofibroblasts at the wound site while the Blimp1-lineage was not. In contrast, the Blimp1-lineage cells were exclusively found immediately underneath the reformed epidermis 17 days post wounding. 2–3 weeks post wounding, Blimp-1 lineage and Hic-1 lineage fibroblasts are found in the upper dermis as well as in dermal sheath and dermal papilla in neogenic hair follicles

hair-follicles. Then they asked whether the two lineages differentially contributed to dermal fibroblasts post wounding. They found that the Dlk1-lineage was recruited immediately after wounding to become myofibroblasts at wound site whereas the Blimp-lineage is not recruited until re-epithelialization and contributes exclusively to papillary dermis, potentially to support new hair follicle formations. In addition to the embryonic lineages of fibroblasts. a study by Abbasi et al. [19] described a postnatal, extrafollicular fibroblast progenitor population of Hic1(quiescence-associated factor hypermethylated in cancer 1), which preferentially locate to the reticular dermis and less in the papillary dermis. Interestingly, Hic1-lineage vastly contributed to the neodermis and more than 90% of the neogenic dermal papilla cells (Fig. 13.1). This collective evidence emphasizes the distinct behaviors of fibroblast lineages and their respective contributions to neogenic hair follicle mesenchyme.

On the other hand, hair follicle associated dermal stem cells (hfDSC) that reside in the dermal cup area at the tip of dermal papilla and regenerate dermal papillae during homeostatic hair cycle contribute only moderately to new HFs post wounding. The minor contribution of hfDSC to neogenic HFs mirrors the scenario in the epidermis, where the existing hair follicle bulge stem cells (K15 + ones) contribute to the new hair follicle epithelium in a limited fashion.

Regulation of WIHN by Developmental Signals in the Epidermis and Dermis

In addition to characterizing the lineage potential of dermal fibroblasts in skin repair, we want to ask if and how environmental cues might affect WIHN. Here we review several developmental signals that are critical for WIHN. Embryonic hair follicle morphogenesis is known to be a tightly regulated process involving complex signaling crosstalk between the epidermis and the dermis [20] and so is WIHN. Consistent with this, recent scRNA-seq analyses capturing ligand-receptor interactions in WIHN tissue revealed an epithelial-mesenchyme interactome reminiscent of embryonic hair follicle morphogenesis [19].

Wnt signaling is one of the earliest and most important regulators in the embryonic hair follicle development process. Around embryonic day 12.5, epidermal Wnt ligands secretion activates broad dermal Wnt activity. At E13.5, the dermal Wnt activity through an unknown downstream signaling initiates the formation of patterned preplacodes. Following this, focal epithelium thickenings, the placodes, emerge from preplacodal progenitors and signal to the dermis. At around E14.5, a morphological dermal condensate(DC), the embryonic precursor of the dermal papilla(DP), is instructed to form. The signals including Wnt from the DC further stimulates HF downgrowth and the DC matures at around E16.5 (Fig. 13.2a).

Similarly, in WIHN, Wnt signaling plays an essential role. It has been shown that inhibition of epithelial Wnt signaling via Wnt inhibitor Dkk1 prevents hair follicle neogenesis [4]. Neither the DP marker, alkaline phosphatase (AP) nor K17

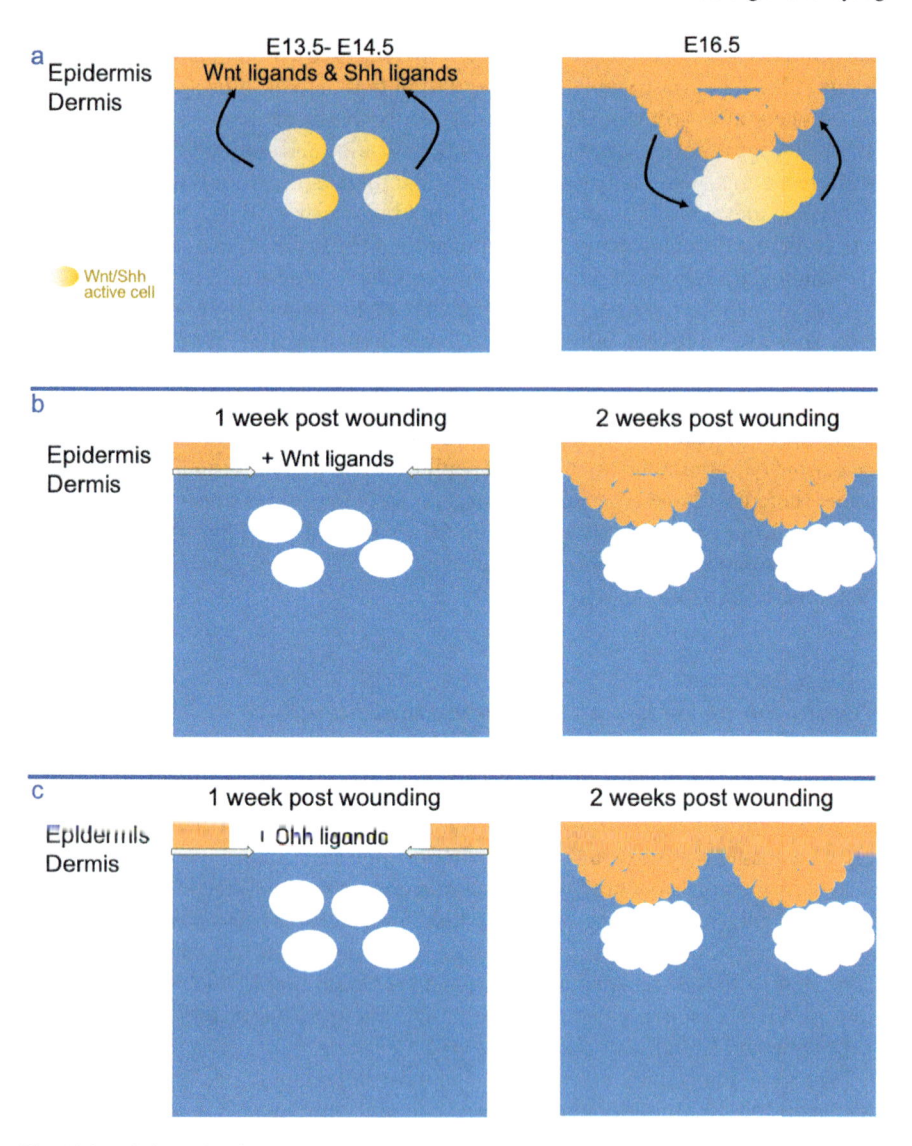

Fig. 13.2 a Schematic of progenitor lineage potential and developmental signals involved in embryonic hair development. Between E13.5 and 14.5, epidermal Wnt ligands secretion activate broad dermal Wnt activity, which through an unknown downstream signaling initiates the formation of placode, focal epithelium thickening, and dermal condensate(DC). The epidermis continues to secrete Wnt and Shh ligands to form the mature DC, which in turn signals back to the HF epithelium for further HF downgrowth. **b** Over expressing Wnt ligands in epidermis promotes hair follicle neogenesis. **c** Over activating Shh in Wnt-active myofibroblasts post wounding leads to ectopic HF formation

+ hair follicle epithelial cells could be detected in the Wntless (Wls) knockout mouse model which prevents Wnt ligand secretion [21]. In addition, overexpression of the Wnt ligand, Wnt 7a, throughout the epidermis dramatically increased the number of regenerated hair germs post wounding [4]. Increasing epidermal Wnt activity via stabilized b-catenin also enhanced ectopic hair follicle neogenesis in the wound bed [18]. These examples collectively show that epidermal Wnt signaling is essential for de novo hair follicle formation. Modulating dermal Wnt signaling also has a significant impact on WIHN. When activated through Wnt signaling, upper papillary dermal fibroblasts expanded and the number of regenerated hair follicles increased [18], suggesting that Wnt activation in the wound dermis renders it permissive for hair follicle formation. Fgf9 secreted by $\gamma\delta$ Tcells in the wound has also been shown to promote dermal Wnt signaling to induce WIHN [22]. It appears that Wnt signaling in the epidermis and the dermis are important for WIHN.

Sonic Hedgehog (Shh) is another essential signaling pathway during hair follicle morphogenesis and WIHN. During HF morphogenesis, epidermis begins secretion of Shh ligands to sustain Shh signaling in both epidermis and dermis [23–25]. The loss of Shh ligands arrests hair follicle downgrowth at an early stage [23, 26]. These studies clearly indicate that Sonic Hedgehog (Shh) signaling is required for embryonic hair follicle growth.

Lim et al. [5] demonstrated that over-activity of the Shh pathway in the wound dermis can affect WIHN, which is consistent with the theory that WIHN reutilizes many of the same mechanisms that govern embryonic hair follicle development. This work suggests that induction of specific signaling pathways could reprogram the cell fate of dermal fibroblasts into hair-follicle inductive mesenchyme. The authors expressed a constitutively active Shh receptor, Smoothened, in dermal myofibroblasts. The over-activation of Shh signalling in those fibroblasts resulted in ectopic DP formation and adult follicle neogenesis. Interestingly, Shh activation in WIHN could even overcome the requirement for a large wound size and was sufficient to induce hair follicle neogenesis in small wounds, including at the wound periphery where WIHN does not typically occur. Thus, it is possible that developmental signals like Shh can induce lineage plasticity in dermal fibroblasts to regenerate new hair follicle-associated mesenchyme(Fig. 13.2c).

Studying WIHN with a developmental focus may also shed light on the molecular roadmap of hair follicle neogenesis. Loss of Shh doesn't prohibit the initial formation of the embryonic placode and DC but only blocks further embryonic hair growth/DC formation. It is not known whether neogenic hair follicles in WIHN show a similar arrest in hair follicle formation. Dissecting the defect in Shh mutant in a WIHN context will help us understand the critical molecular steps required for adult hair follicle neogenesis.

Although Wnt and Shh are reciprocally activated in the hair follicle epithelium and DC, their function within these two compartments may be different. During embryonic development, Wnt and Shh in the follicular epithelium are antagonists. Wnt^{hi} basal cells are unresponsive to Shh signaling. Ouspenskaia et al. [27] used live imaging to show that a Wnt^{hi} basal cell divides asymmetrically to place one of its daughter cells in the suprabasal Wnt^{low} environment where the daughter cell starts to respond to Shh and proliferate symmetrically. However, how Wnt and Shh interact in the dermis is less studied. Lim et al. [5] showed that loss of either Shh or Wnt inhibited new follicle formation in adults. Over-activating Shh in Wnt-active fibroblasts promoted WIHN and the Wnt-active lineage traced to de novo DP. We could thus infer that Shh and Wnt cooperate to redirect fibroblast cell fate. It will be interesting to dissect the mechanisms by which Wnt and Shh interact within both the dermis and epidermis to generate new hair follicles, both in WIHN and in embryonic development as the insights from either field can benefit the other.

Differences Between Embryonic Hair Morphogenesis and Adult WIHN

Despite the remarkably conserved mechanisms between WIHN and embryonic hair follicle morphogenesis, there are several differences worth noting. Firstly, the upstream source to initiate a signaling cascade to promote WIHN could come from immune cells in the wound environment. For example, Fgf9, secreted by $\gamma\delta$ Tcells in the wound, is a known inducer of Wnt ligand secretion in development, and can promote dermal Wnt signaling to induce WIHN [22]. Macrophage-induced TNF signaling leads to hair follicle neogenesis in a dose-dependent manner [28]. There are also inflammatory mediators in the wound that inhibit follicle neogenesis, such as the lipid signaling molecule Prostaglandin D2 [29]. On the other hand, in embryonic development, the first dermal signal(s) to induce embryonic hair follicle morphogenesis still remain(s) poorly characterized, presumably not from the poorly developed immune system in embryos.

Secondly, the absolute requirement of signal-activating ligands differs in embryonic hair morphogenesis and adult WIHN. In Lim et al. [5], WIHN was prohibited when Wnt ligand expression was abrogated in myofibroblasts, whereas dermal Wnt ligand secretion is dispensable for embryonic hair follicle initiation [30]. This requirement of dermal Wnt ligands in adult WIHN may reveal a particular vulnerability of adult hair follicle neogenesis under the loss of epidermal signals until re-epithelialization.

Thirdly, ablation of epidermal Shh in WIHN results in no apparent hair follicle placode, whereas in embryonic Shh KO placode formation is maintained. This suggests that the epidermal and dermal components of de novo hair follicles may form in an order different from that of embryonic ones. Understanding the difference between embryonic hair morphogenesis and WIHN may help us to think about what are the absolutely critical versus dispensable steps for regenerating de novo hair follicles and what are the developmental signals that could potentially be modulated by therapeutics to promote hair follicle regeneration.

Outlook for Appendage Regeneration and Scarless Healing in Humans

One of the ultimate goals of studying WIHN in animal models is to provide therapeutic means of regenerating hair follicles in adult humans. The dermal papilla is a key to adult hair follicle regeneration. The transplantation of isolated dermal papilla can induce de novo hair follicle neogenesis and hair growth in recipient skin [31]. Now that it is established that Shh signalling can reprogram Wnt-active fibroblasts into inductive dermal papilla to promote WIHN [5], further dissecting how Shh interacts with the other signals (e.g. Wnt) in the wounding environment could help inform strategies to install a regenerative field that promotes hair follicle neogenesis in adult skin.

Establishing the epidermal competency has also been shown to be a promising candidate for hair follicle neogenesis. For instance, epithelial Msx2, an important modulator in growth and differentiation, expression is required for hair follicle neogenesis [32]. Epidermal Wnt activation also clearly has a regenerative role: Inducing epidermal Wnt signaling promotes de novo HF formation [33] and ablating it abolishes WIHN [4, 21].

Ultimately, the frontier is to leverage the signals that regulate WIHN to regenerate new hair follicles in non-wounding conditions. To this end, activating Shh in the keratinocytes and fibroblasts without wounding can generate de novo hair follicles even in the hairless paw of adult mice [34]. Nevertheless, the role of Shh overactivation in skin tumor formation is a major adverse effect that could prove challenging to address; however, possibly fine-tuning "when and where" Shh should be activated could be explored.

Understanding how WIHN occurs can also help improve clinical outcomes for wound healing. The wide array of current cutaneous wound healing therapies includes growth factor delivery, skin grafting, bioengineered skin substitutes and cell-based therapies [10]. Although those approaches are immensely beneficial for patients with life-threatening wounds, there is still a long way to go before we achieve true scarless regeneration accompanied by appendage formation. For example, dermal substitutes derived from acellular matrices or human amnion do not reduce scarring. Skin grafts that provide epidermal coverage suffer from contracture and no appendage formation. Neutralizing antibodies to TGF-b3, a contributor to fibrotic repair, reduce scarring but do not regenerate hair follicles [35]. Studying WIHN will uncover the relationship between wound healing and appendage regeneration and how to organically combine the two. For example, fibrosis has generally been thought of as the opposite of regeneration, but the study from Lim et al. [5] suggests that the interplay between fibrosis and regeneration is more complicated than that. Activation of Shh in fibroblasts can reprogram them into new dermal papilla despite the onset of fibrosis. The heterogeneous potential of dermal fibroblasts may partly explain why fibrosis and hair follicle neogenesis can co-exist: lower dermal cells contributes to fibrosis [18], and Shh mostly acts on the upper dermal Wnt-active lineage contributing to new hair follicle mesenchyme. Besides reinstalling dermal papilla, fibroblasts have long been recognized as the critical cell population to target for scar reduction: when embryonic skin containing fibroblasts are transplanted subcutaneously into adult wounds, wounds heal with collagen patterns indistinguishable from healthy skin [36]. Thus, engineering specific fibroblast populations at the wound site with proper signal cues may be a potential transdifferentiation approach that induce hair follicle neogenesis with reduced scarring.

Macrophages are another critical cell population for aiding hair neogenesis. Macrophages prune excessive extracellular matrix to reduce scarring at later stages of wound healing [37]. Depleting macrophages has also been shown to reduce adult WIHN [38]. One mechanism macrophages acts on WIHN is through activation of dermal Wnt signaling through phagocytosis of Wnt inhibitor, SFRP4 [39]. The variety of cell types and origins in adult WIHN potentially provides more combinations of targets to stimulate new follicle formation and reduce scarring at the same time. Understanding the different cell types at mouse and human wounds could also potentially explain why WIHN is rarely observed in humans. For example, in mouse wounds $\gamma\delta$ T cells have been shown to secrete Fgf9 to induce WIHN [22]; however, $\gamma\delta$ T cells are not found in humans. Ultimately, dissecting the mechanisms of how different cells are reprogrammed into regenerative sources in WIHN and elucidating the relationship between scarring and regeneration will help fulfill the goal for scarless regeneration and appendage regeneration.

Acknowledgements We'd like to thank Dr. Mayumi Ito, Dr. Denise Gay and Dr. Chae Ho Lim for their helpful feedback on the manuscript.

References

1. Breedis C (1954) Regeneration of hair follicles and sebaceous glands from the epithelium of scars in the rabbit. Cancer Res 14(8):575–579
2. Billingham RE, Russell PS (1956) Incomplete wound contracture and the phenomenon of hair neogenesis in rabbits' skin. Nature 177(4513):791–792
3. Kligman AM, Strauss JS (1956) The formation of vellus hair follicles from human adult epidermis. J Invest Dermatol 27(1):19–23
4. Ito M, Yang Z, Andl T, Cui C, Kim N, Millar SE et al (2007) Wnt-dependent de novo hair follicle regeneration in adult mouse skin after wounding. Nature 447(7142):316–320
5. Lim CH, Sun Q, Ratti K, Lee S-H, Zheng Y, Takeo M et al (2018) Hedgehog stimulates hair follicle neogenesis by creating inductive dermis during murine skin wound healing. Nat Commun 9(1):4903
6. Harn H, Wang S, Lai Y, Van Handel B, Liang Y, Tsai S et al (2021) 609 Symmetry breaking of tissue mechanics in wound induced hair follicle regeneration. J Invest Dermatol 141(5):S106
7. Wong T-W, Hughes M, Wang S-H (2018) Never too old to regenerate? Wound induced hair follicle neogenesis after secondary intention healing in a geriatric patient. J Tissue Viability 27(2):114–116
8. Park S, Gonzalez DG, Guirao B, Boucher JD, Cockburn K, Marsh ED et al (2017) Tissue-scale coordination of cellular behaviour promotes epidermal wound repair in live mice. Nat Cell Biol 19(3):155–163

9. Aragona M, Dekoninck S, Rulands S, Lenglez S, Mascré G, Simons BD et al (2017) Defining stem cell dynamics and migration during wound healing in mouse skin epidermis. Nat Commun 8:14684
10. Sun BK, Siprashvili Z, Khavari PA (2014) Advances in skin grafting and treatment of cutaneous wounds. Science 346(6212):941–945
11. Yang R, Liu F, Wang J, Chen X, Xie J, Xiong K (2019) Epidermal stem cells in wound healing and their clinical applications. Stem Cell Res Ther 10(1):229
12. Walmsley GG, Maan ZN, Wong VW, Duscher D, Hu MS, Zielins ER et al (2015) Scarless wound healing: chasing the holy grail. Plast Reconstr Surg 135(3):907–917
13. Sorg H, Tilkorn DJ, Hager S, Hauser J, Mirastschijski U (2017) Skin wound healing: an update on the current knowledge and concepts. Eur Surg Res 58(1–2):81–94
14. Cotsarelis G, Sun TT, Lavker RM (1990) Label-retaining cells reside in the bulge area of pilosebaceous unit: implications for follicular stem cells, hair cycle, and skin carcinogenesis. Cell 61(7):1329–1337
15. Liu Y, Lyle S, Yang Z, Cotsarelis G (2003) Keratin 15 promoter targets putative epithelial stem cells in the hair follicle bulge. J Invest Dermatol 121(5):963–968
16. Snippert HJ, Haegebarth A, Kasper M, Jaks V, van Es JH, Barker N et al (2010) Lgr6 marks stem cells in the hair follicle that generate all cell lineages of the skin. Science 327(5971):1385–1389
17. Gurtner GC, Werner S, Barrandon Y, Longaker MT (2008) Wound repair and regeneration. Nature 453(7193):314–321
18. Driskell RR, Lichtenberger BM, Hoste E, Kretzschmar K, Simons BD, Charalambous M et al (2013) Distinct fibroblast lineages determine dermal architecture in skin development and repair. Nature 504(7479):277–281
19. Abbasi S, Sinha S, Labit E, Rosin NL, Yoon G, Rahmani W et al (2020) Distinct regulatory programs control the latent regenerative potential of dermal fibroblasts during wound healing. Cell Stem Cell 27(3):396-412.e6
20. Millar SE (2002) Molecular mechanisms regulating hair follicle development. J Invest Dermatol 118(2):216–225
21. Myung PS, Takeo M, Ito M, Atit RP (2013) Epithelial Wnt ligand secretion is required for adult hair follicle growth and regeneration. J Invest Dermatol 133(1):31–41
22. Gay D, Kwon O, Zhang Z, Spata M, Plikus MV, Holler PD et al (2013) Fgf9 from dermal γδ T cells induces hair follicle neogenesis after wounding. Nat Med 19(7):916–923
23. Chiang C, Swan RZ, Grachtchouk M, Bolinger M, Litingtung Y, Robertson EK et al (1999) Essential role for Sonic hedgehog during hair follicle morphogenesis. Dev Biol 205(1):1–9
24. St-Jacques B, Dassule HR, Karavanova I, Botchkarev VA, Li J, Danielian PS et al (1998) Sonic hedgehog signaling is essential for hair development. Curr Biol 8(19):1058–1068
25. Sennett R, Rendl M (2012) Mesenchymal-epithelial interactions during hair follicle morphogenesis and cycling. Semin Cell Dev Biol 23(8):917–927
26. Woo W-M, Zhen HH, Oro AE (2012) Shh maintains dermal papilla identity and hair morphogenesis via a Noggin-Shh regulatory loop. Genes Dev 26(11):1235–1246
27. Ouspenskaia T, Matos I, Mertz AF, Fiore VF, Fuchs E (2016) WNT-SHH antagonism specifies and expands stem cells prior to niche formation. Cell 164(1–2):156–169
28. Wang N, Wu Y, Zeng N, Wang H, Deng P, Xu Y, et al (2016) E2F1 hinders skin wound healing by repressing vascular endothelial growth factor (VEGF) expression, neovascularization, and macrophage recruitment. PLoS One 11(8):e0160411
29. Nelson AM, Loy DE, Lawson JA, Katseff AS, Fitzgerald GA, Garza LA (2013) Prostaglandin D2 inhibits wound-induced hair follicle neogenesis through the receptor, Gpr44. J Invest Dermatol 133(4):881–889
30. Fu J, Hsu W (2013) Epidermal Wnt controls hair follicle induction by orchestrating dynamic signaling crosstalk between the epidermis and dermis. J Invest Dermatol 133(4):890–898
31. Cohen J (1961) The transplantation of individual rat and guineapig whisker papillae. J Embryol Exp Morphol 9:117–127
32. Hughes MW, Jiang T-X, Plikus MV, Guerrero-Juarez CF, Lin C-H, Schafer C et al (2018) Msx2 supports epidermal competency during wound-induced hair follicle neogenesis. J Invest Dermatol 138(9):2041–2050

33. Gat U, DasGupta R, Degenstein L, Fuchs E (1998) De Novo hair follicle morphogenesis and hair tumors in mice expressing a truncated β-catenin in skin. Cell 95(5):605–614
34. Sun X, Are A, Annusver K, Sivan U, Jacob T, Dalessandri T, et al (2020) Coordinated hedgehog signaling induces new hair follicles in adult skin. Elife [Internet]. http://dx.doi.org/https://doi.org/10.7554/eLife.46756
35. Shah M, Foreman DM, Ferguson MW (1995) Neutralisation of TGF-beta 1 and TGF-beta 2 or exogenous addition of TGF-beta 3 to cutaneous rat wounds reduces scarring. J Cell Sci 108(Pt 3):985–1002
36. Lorenz HP, Lin RY, Longaker MT, Whitby DJ, Adzick NS (1995) The fetal fibroblast: The effector cell of scarless fetal skin repair. Plast Reconstr Surg 96(6):1251–1259
37. Duffield JS, Forbes SJ, Constandinou CM, Clay S, Partolina M, Vuthoori S et al (2005) Selective depletion of macrophages reveals distinct, opposing roles during liver injury and repair. J Clin Invest 115(1):56–65
38. Kasuya A, Ito T, Tokura Y (2018) M2 macrophages promote wound-induced hair neogenesis. J Dermatol Sci [Internet]. http://dx.doi.org/https://doi.org/10.1016/j.jdermsci.2018.05.004
39. Gay D, Ghinatti G, Guerrero-Juarez CF, Ferrer RA, Ferri F, Lim CH, et al (2020) Phagocytosis of Wnt inhibitor SFRP4 by late wound macrophages drives chronic Wnt activity for fibrotic skin healing. Sci Adv 6(12):eaay3704

Chapter 14
Hair Follicles in Wound Healing and Skin Remodelling

Magdalena Plotczyk and Francisco Jimenez

Abstract *Introduction*: Besides the primary role of hair follicles in producing hair shafts, these mini-organs have been shown to contribute to the healing and remodelling of the skin in the response to injury. There is an increasing number of clinical and experimental studies exploring the mechanisms by which hair follicles contribute to skin recovery. Mouse tracing experiments demonstrated that after skin injury, epithelial stem cells migrate out of the follicle to support wound re-epithelialization, while the mesenchymal cells from the hair follicle are mobilised to migrate into the wound bed and contribute to the repair of the dermis. Accordingly, there is a significant delay in wound healing in the absence of hair follicles, observed both in experimental studies in mice and in humans during clinical practice. In the healthy state, hair follicles induce changes to the surrounding skin, which occur in synchrony with the hair follicle cycle. Besides supporting wound healing and healthy skin remodelling, it remains to be investigated whether hair follicles could also remodel the fibrotic tissue formed after skin injuries. *Methods*: For the selection of literature covered in this chapter, we used Pubmed database with the keywords hair follicles in wound healing and skin remodelling, and hair follicles transplantation and wound healing. Publications related to the clinical use of hair follicle grafts for healing chronic cutaneous wounds are also discussed. *Results*: A large body of evidence supports the role of hair follicles in wound healing and skin remodelling. As demonstrated in several clinical studies, transplantation of autologous hair follicle grafts into chronic ulcers and difficult to heal wounds accelerate the healing of wounds. *Conclusions*: There is a great potential in exploring the role of hair follicles in wound healing and skin regeneration. Once we understand the exact mechanism by which hair follicles contribute to skin recovery, we can try to mimic this effect with therapeutic solutions. A potential approach involves deciphering the populations of hair follicle cells that contribute to the observed effect and injecting them directly into wounds or scar tissue to improve the clinical outcomes. Alternatively, elucidating the

---

M. Plotczyk
Department of Bioengineering, Imperial College London, London, UK

F. Jimenez (✉)
Mediteknia Dermatology and Hair Transplant Clinic, University Fernando Pessoa Canarias, Gran Canaria, Spain
e-mail: fjimenez@mediteknia.com

paracrine effect of transplanted follicles would open up new avenues for therapeutic discovery, to replicate the combination of required factors that facilitate healing or scar remodelling.

Keywords Hair follicles · Wound healing · Skin remodelling · Hair follicle cycle · Autologous hair transplantation

Key Points

- Hair follicles in the skin play an important role in stimulating healing. Hair follicle epithelial stem cells migrate and are known to contribute to re-epithelialization, while dermal remodelling is postulated to be regulated by the hair follicle dermal sheath.
- Clinical studies using hair follicles transplanted into wounds have been performed with successful outcomes in chronic ulcers and difficult-to-heal wounds and this type of intervention seems to be a promising novel therapeutic tool.
- Hair follicles induce changes in healthy skin during hair follicle cycles (skin remodelling). Their role in skin remodelling after injuries needs to be investigated and translated to clinical practice.

Summary

The role of hair follicles in human skin is not to simply produce hair shafts but to participate in many other functions including skin homeostasis and the wound healing response to cutaneous injuries.

In this chapter, we will first explore the role that human hair follicles play in wound healing and the clinical and experimental evidence that supports this connection. Then, we will summarize the latest translational clinical work that has been performed using autologous hair follicle grafts (hair transplantation) as a therapeutic tool to stimulate wound healing and discuss possible mechanisms through which hair transplantation stimulates the healing response. Finally, we will comment on the role of hair follicles in the remodelling of normal skin and fibrotic/scar tissue.

Hair Follicles and Wound Healing: Historical Studies

The first clinical studies suggesting an active role of hair follicles in the wound healing response date back to the work of Brown and McDowell in 1942 [1]. In their article, they suggest that during wound repair cells from the hair follicles dedifferentiate to cells that contribute to the healing of the wound. However, without doubt the most enlightening and convincing paper demonstrating that hair follicles are key contributors to the wound healing response after skin injuries in humans was written by Bishop, a neuroanatomist from Washington University in St. Louis [2]. In this paper, published in 1945, Bishop performed his experiments in the most objective way possible by self-inflicting cutaneous wounds on his own forearm at different depths and observing the clinical and histological healing process. Bishop demonstrated that the remaining hair follicles left in the wound bed played a pivotal role in wound healing and that their presence or absence influenced the outcome of the healing response. More specifically, he concluded that: (1) re-epithelialization of the wound starts not only from the marginal epithelium but also from the remaining hair follicles; (2) scar formation occurs when the wound is sufficiently deep to destroy the bases of hair follicles; and (3) when the skin is destroyed down to the reticular layer, the granulation tissue regenerates most readily from the connective tissue surrounding the hair follicles. This granulation tissue is necessary for migration of the epithelial cells and subsequent healing of the wound surface.

This pioneering work of Bishop supports the well-known observation of clinicians nowadays that a wound made in areas of high hair density (and with terminal hair follicles) heals faster than a wound of the same size in a less hairy area. For example, the healing time of a 0.2–0.3 mm depth wound made on the scalp averages 5–6 days against the 10–14 days that it takes to re-epithelialize a wound of the same size made on the thighs, buttocks or abdomen [3]. This faster healing time is the reason why the scalp seems to be a very efficient area to harvest split thickness skin grafts for burn wounds [4].

The contribution of hair follicles in wound healing described by Bishop were later confirmed by Ito et al. who showed that follicular stem cells played a key role in re-epithelialization of the wounded epithelium, as will be discussed later in more detail [5].

Hair Follicles as a Therapeutic Tool to Stimulate Wound Healing: Clinical Evidence

In view of this ability of hair follicles to promote wound healing, several clinical studies have been successfully undertaken to demonstrate the use of autologous hair follicle grafts harvested from the scalp in stimulating the healing of chronic ulcers or other difficult-to-heal wounds [6].

The first clinical cases showing that hair follicles could be used as a therapeutic tool to stimulate wound healing were reported in the plastic surgery literature. These reports described a few patients with deep and extensive burns successfully treated with a combination of an artificial dermis that provided an immediate coverage of the wound, followed by autologous hair follicle grafting that accelerated healing and re-epithelialization [7–9].

The methodology used for implantation of hair follicles in non-healing wounds is basically the same as is used in hair transplantation techniques for the treatment of baldness, where hair follicles are harvested from the patient's occipital scalp (donor area) and inserted into the skin of the balding area (recipient area) (Fig. 14.1). These autologous transplanted follicles (also called hair grafts or follicular unit grafts) have

Fig. 14.1 The process of hair transplantation into wounds. In the upper image, hair follicles harvested from the donor site (scalp) are inserted with fine tip forceps into slits created in the wound bed. The lower image represents follicular epithelial and mesenchymal cells proliferating and migrating out of the hair follicle into the interfollicular skin, increasing vascularization and re-epithelialization, and supporting the healing of wounds

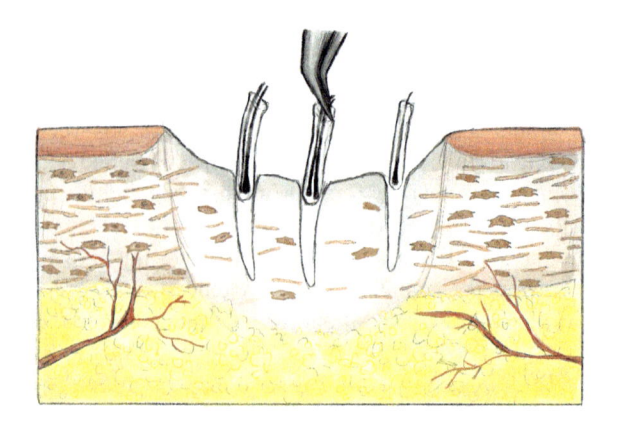

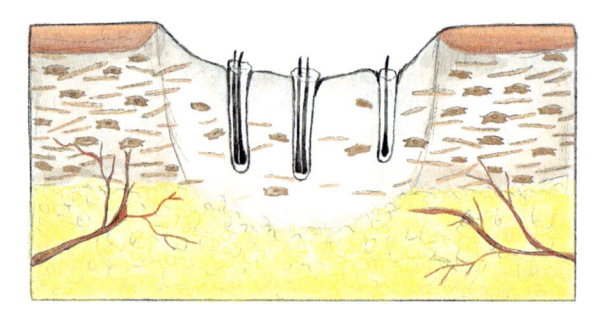

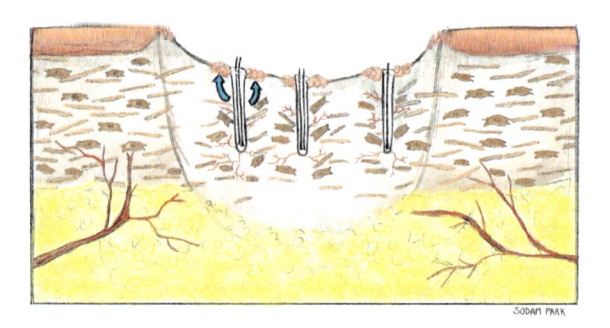

the attribute of "donor dominance", which means that they maintain their original characteristics after transplantation from a donor site to a new region [10]. They regenerate and start cycling, producing hair shafts approximately 2–4 months after their implantation. It is pertinent to say here that, in addition to terminal and vellus hair follicles, the follicular units (FUs) harvested with 1 mm punches may contain other cell types such as perifollicular dermal fibroblasts, dermal adipocytes, and eccrine epithelial cells [11] (Fig. 14.2). Although the contribution of all these extrafollicular cell types in wound healing has been well established [12–16], their relative

Fig. 14.2 Follicular unit harvested from the scalp during a hair transplantation procedure. This FU contain 3 hair follicles (1 terminal and 2 intermediate), perifollicular dermis, sebaceous gland, and adipose tissue. The presence of eccrine coil is not visible unless stained with specific dye such as methylene blue

proportion in the skin punches is minor in comparison with the number of follicular cells.

Hair follicle transplantation into wounds is performed under local anesthesia as an outpatient procedure. There are two surgical methods that can be used for hair follicle harvesting: the *follicular unit excision (FUE)* technique, which involves the excision of FUs one at a time using small circular punches, and the strip harvesting method also known as *follicular unit transplantation (FUT)*, which involves excising a strip from the mid-occipital scalp followed by its stereomicroscopic dissection into follicular units. Most published cases of hair transplantation in wounds have used the FUE technique because it is less invasive, and the donor wounds heal by second intention with no need for suturing. However, although in hair restoration surgery the widest punch used in FUE measures 1 mm in diameter, bigger punches of up to 2 mm have been used in transplantation for wound healing because the goal is not to provide a cosmetic natural hair coverage but to provide a source of epithelial and mesenchymal hair follicle cells that migrate into the wound site and stimulate the healing process. The hair follicle grafts are inserted into slits created in the wound bed, the size of which must be sufficient to accommodate the diameter of the hair follicle graft. For example, for the insertion of FU grafts, which normally measure

1 mm in diameter and 4–5 mm in depth, the slits can be made with a 20-gauge needle or a 1 mm surgical blade. The insertion can then be made with fine tip forceps or with hair implanters.

The first pilot study using hair transplant grafts to heal wounds included 10 patients with chronic leg ulcers of 10.5 years average duration and of venous pressure and mixed etiology [17]. Each ulcer had an experimental area that was transplanted with 2 mm punch scalp grafts and a non-transplanted control area. A significant reduction in ulcer size was noted in the experimental area compared with the control, and most patients presented an increase of granulation tissue, decrease of pain, and ulcer border reactivation. After this pilot study in 2012 [17], a further eight papers have been published reporting on the successful use of transplanted hair follicles to stimulate the healing of difficult-to-heal or complex wounds. These papers are summarized in Table 14.1. The following paragraphs describe some of the most important points of interest from these publications. Table 14.2 summarizes the clinical changes most commonly described after hair transplantation in wounds.

Martinez et al. published a randomized controlled trial of 12 patients with chronic leg ulcers comparing the healing capacity of hair follicle punch scalp grafts versus punch grafts harvested from non-hairy areas [18]. Each ulcer was divided longitudinally into two halves, one half receiving the hair follicle grafts and the other half the same number of grafts with no hair. At the 18-week endpoint a 75% ulcer reduction was observed in the area transplanted with hairy grafts compared with 34% in the area transplanted with non-hairy grafts, demonstrating that grafts containing hair follicles induced a significantly ($P = 0.002$) better healing response.

Another study published by Yang et al. compared the outcomes of 40 chronic wounds, 20 of which were treated with transplanted follicles and 20 with the more conventional therapy of split thickness skin grafts [19]. An interesting observation of this study was that wounds treated with the hair follicles achieved in general a better residual scar quality, which was more elastic and less contracted.

A 54-year-old intermediate recessive dystrophic epidermolysis bullosa patient with chronic ulcers was successfully treated with hair follicle transplantation [20]. This case is particularly interesting from a clinical point of view because the wounds of this congenital disease due to mutations in the COL7A1 gene are very recalcitrant and have a high risk of developing squamous cell carcinomas as result of poor wound healing. These patients suffer from skin fragility, blistering, and chronic non-healing ulcers from infancy and current treatments are mostly supportive. In this particular patient, a total of 360 follicular units were transplanted in nine sessions over 5 years, resulting in rapid healing of most of the treated ulcers. This novel indication of using hair follicles as a treatment of epidermolysis bullosa and other blistering diseases seems to be promising, especially since a recent publication has also demonstrated the key role of hair follicles stem cells in the healing of subepidermal blisters produced in mice [21].

It is also interesting to note that in several of the publications the authors described that the hairs transplanted in the wound grew in far less quantity than would be expected in a normal hair transplant procedure performed in androgenetic alopecia [19, 22], sometimes growing in different patterns in the central and peripheral

Table 14.1 Publications showing hair transplantation as a therapy for wound healing

Patients/type of wound treated	Procedure	Outcome	Reference
1 patient with full thickness burn on the scalp	Artificial skin (Integra) followed by autologous scalp hair follicle transplantation 12 days later	Complete re-epithelialization	Navsaria et al. [8]
2 patients with scalp defects	Artificial dermis in the first phase until granulation; then hair graft transplantation	Complete healing	Narushima et al. [9]
15 patients with third-degree burns on limbs and hands	Hair follicle containing dermal grafts from scalp	Complete re-epithelialization	Zakine et al. [7]
10 patients with chronic venous leg ulcers	Hair transplantation (2 mm punch)	At 18 weeks, 27% reduction in transplanted area compared with 6% reduction in the non-transplanted control area. Transplanted area increased in granulation tissue	Jimenez et al. [17]
14 patients with chronic wounds	Hair transplantation. No control group	Complete healing in all patients after 2 months	Liu et al. [22]
40 patients with traumatic or surgical wounds	20 patients treated with hair grafts and 20 treated with split thickness skin grafts	Better skin/scar quality and clinical outcome was found in the ulcers treated with hair follicle grafts	Yang et al. [19]
12 patients with chronic venous leg ulcers	2 mm punch hair transplant. Half of the ulcer transplanted with punch scalp hair grafts and the other half with punch skin grafts from non-hairy abdominal skin	At 18 week end point of the study, 75% ulcer area reduction in the side of transplanted hair versus 34% reduction in the side of non-hairy grafts	Martínez et al. [18]
1 patient with chronic leg ulcer	One third of the ulcer was transplanted with scalp hair grafts, one third with skin grafts from the back, and one third served as control	The area transplanted with scalp graft healed significantly better	Fox et al. [49]

(continued)

Table 14.1 (continued)

Patients/type of wound treated	Procedure	Outcome	Reference
15 patients with chronic leg ulcers	Hair follicular unit transplant	At 18 week end-point of the study, average ulcer area improved by 49% and volume by 72%. Two patients did not respond	Budamakuntla et al. [50]
2 patients with non-healing traumatic and iatrogenic ulcers	Hair follicular unit transplant	Ulcers completely healed	Alam et al. [23]
1 patient with dystrophic epidermolysis bullosa and non-healing ulcers	Hair follicular unit transplant	Most treated ulcers healed completely. 360 follicular units transplanted in 9 sessions over 5 years	Wong et al. [20]
10 patients with 14 chronic leg ulcers	Hair follicular unit transplant	Wound closure in 8–12 weeks	Saha et al [26]

Table 14.2 Clinical response after hair follicle transplantation in wounds

●	Overall faster and greater healing response (significant wound reduction)
●	Increased re-epithelialization from the borders (wound border reactivation) and from the transplanted follicles
●	Decrease in pain
●	Greater development of granulation tissue
●	Better skin/scar quality (more elastic and less contracted scar)

zones of the wounds [23]. This would seem to support the idea that the wound microenvironment dictates the fate of the transplanted hair follicles in the direction of wound healing and not in the direction of hair shaft production, a hypothesis suggested by Jahoda in 2001 [24, 25]. Interestingly, a recent study by Saha et al. [26] showed that in addition to inducing re-epithelialization of wounds the transplanted hair follicles grafts induces the reappearance of other important structures such as blood and lymphatic vessels, nerve fibres and sweat glands, potentially restoring crucial functions in the healed skin such as sweating and sensitive innervation.

Hair Follicles and Wound Healing: Experimental Studies

Years after the publication of Bishop's study, there exists an increasing body of evidence describing the role of hair follicles in wound healing. As mentioned above, experimental studies using transgenic mouse models demonstrated that bulge stem cells migrate out of the follicle in response to an injury to the skin, differentiate into epidermal progenitor cells, and contribute to the restoration of the epidermis [5, 27]. Lineage tracing in mouse skin confirmed that hair follicle stem cells contribute to the healing of wounds at the expense of hair development, causing a delay in hair follicle growth [21]. Hair follicles, rather than the interfollicular epidermis, were also shown to be the primary source of keratinocytes for the repair of blister wounds, which caused reduction in the hair follicle size. Accordingly, in the absence of hair follicles, there is a significant delay in re-epithelialization during the healing of murine wounds [28].

Besides the epidermal compartment of the hair follicle, the hair follicle mesenchyme also appears to participate in the wound healing response [25, 29, 30]. Lineage studies in mice demonstrated that alongside the skin-derived precursors (SKPs), hair follicle dermal stem cells, as well as dermal papilla and dermal sheath cells from follicles surrounding the site of injury are mobilised to migrate into the wound bed and support skin repair [31]. When injected into the site of skin injury, dermal stem cells repopulate full-thickness wounds, simultaneously rejuvenating hair follicle growth [31]. Similarly, dermal sheath cells were shown to have an ability to both reconstitute the hair follicle mesenchyme and generate interfollicular dermal fibroblasts after transplantation [29]. Importantly, the clinical potential of dermal cell populations lies in their ability to be expanded in vitro, whilst maintaining the capacity to give rise to various mesenchymal cell types [31]. Although there is strong evidence that follicular cells can migrate to support wound healing, lineage tracing and single-cell studies examining the long-term fate and progeny of these cells will be essential to understand the full potential of their contribution to the repair of wounds.

Importantly, it is not only the presence of hair follicles, but also the stage of the cycle that impact on the wound healing capacity of the skin. Mouse wounding studies demonstrated that skin containing hair follicles in an active anagen stage heal faster than skin with hair follicles in a resting telogen stage [32]. Alongside accelerated healing, the presence of anagen hair follicles was linked with faster re-epithelialization, increased angiogenesis, and deposition of extracellular matrix in the dermis [32].

A phenomenon of wound-induced hair neogenesis (WIHN) provides further evidence for the link between hair follicle growth and wound healing. In the mouse model of WIHN, large-size excisional skin wounds can heal by activating spontaneous regeneration of new hair follicles in the wound centre [33]. In the process, myofibroblasts transdifferentiate into adipocytes, reducing the formation of fibrotic tissue, with the conversion triggered by hair follicle-derived BMP (bone morphogenic protein) signals [13]. Similarly, in a coculture system, human scalp follicles induced BMP4-dependent trans differentiation of human keloid fibroblasts into adipocytes, suggesting a potential of hair follicles to benefit the wound healing response, as well as reducing the formation of scars [13].

Hair Follicles and Skin Remodelling: Future Benefits

During the hair follicle cycle, the healthy interfollicular skin undergoes constant physiological remodelling, which can be clearly observed in mouse skin where follicles grow synchronously with one another [34, 35]. When murine hair follicles are in anagen, the epidermis, dermis, and hypodermis (adipose tissue) are respectively 2.0-, 1.6-, and 1.7-fold thicker compared to when all follicles are in telogen [36]. Despite fluctuations in dermal thickness, the number of cells in the dermis remains constant during the cycle [37, 38]. This raises the interesting possibility that changes to dermal thickness can be linked to a redistribution of the extracellular matrix to accommodate the growth of follicles during anagen [39]. Skin vasculature also varies with the stage of the hair follicle cycle, with angiogenesis occurring around growing hair follicles (anagen) followed by degeneration of capillaries when hair follicles regress (catagen) [40–42].

While hair follicle cycling and remodelling of interfollicular skin occur at the same time, it is not clear whether the skin changes drive the hair follicle cycle or if the transition through hair follicle stages induces changes to the skin. In support of the former hypothesis, a functional analysis of mouse skin revealed that intradermal adipocytes are necessary and sufficient to drive follicular stem cell activation and induction of anagen, while adipocyte defects result in reduced hair growth [43]. Similarly, neurons surrounding the hair follicle were found to interact with bulge cells via release of paracrine factors, changing the lineage of bulge cells into epidermal stem cells [44]. These two studies raised the possibility that other cell types residing in the interfollicular skin may provide molecular input that could direct the hair follicle cycle. On the other hand, there is substantial evidence that cycling hair follicles induce the growth of vascularization and innervation in the interfollicular zone [40–42]. Among others, several in vitro and in vivo studies described the pro-angiogenic properties of the hair follicle, including the expression of VEGF (vascular endothelial growth factor) in various follicular compartments [45–48]. A possible explanation for the observed effect could be a bi-directional independence whereby cycling hair follicles influence remodelling of the interfollicular skin, while the interfollicular skin has the capacity to induce transition of the follicle through the cycle, depending on the strength of the induced effect and timing of the cycle stage.

Besides inducing changes to healthy skin during a natural hair cycle and contributing to the wound healing response, it remains to be investigated whether hair follicles could also remodel the fibrotic tissue formed after skin injuries. In fact, anecdotal unpublished observations by hair surgeons have noted the clinical aesthetic improvement of scalp scars (mainly burn scars and scars from split thickness skin grafts) as a result of hair transplantation (scientific poster presented by Dr. Richard Keller at ISHRS meeting in Montreal, 2008). Thus, it could be possible that anagen hair follicles transplanted into scars could remodel mature scars, similar to how they remodel healthy tissue, with an increase in the thickness of skin layers, vascularization, adipogenesis, and innervation of the fibrotic tissue. This effect could

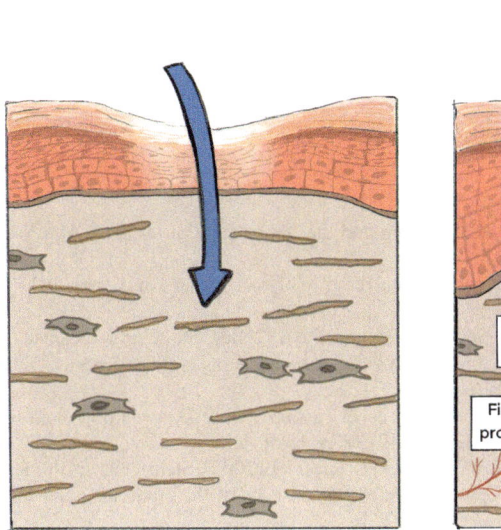

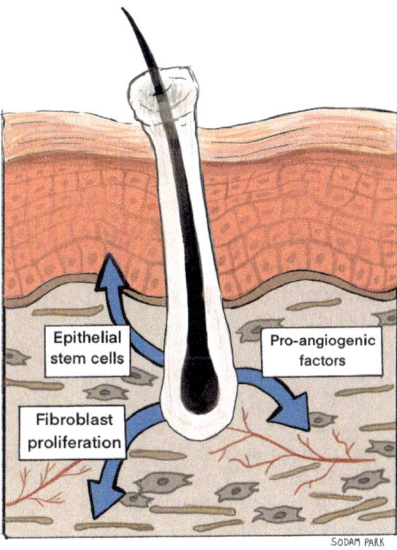

Fig. 14.3 Hair follicle transplantation into scars. We hypothesize that anagen hair follicles transplanted into scars could remodel mature scars, similar to how they remodel healthy interfollicular tissue

be facilitated by either migration of epidermal and dermal stem cells out of hair follicles to support skin remodelling, or the release of remodelling factors directly from the follicle (Fig. 14.3).

Concluding Remarks

Hair follicle transplantation for wound healing applications is a relatively recent and still underutilized minimally invasive surgical technique. It is a safe and relatively inexpensive procedure that does not require cell manipulation in the laboratory. The clinical indications, which currently have been limited to stimulate the healing of difficult-to-heal wounds or chronic ulcers, will become better defined as further clinical experience is gained with this form of therapy. However, there remain important questions to be answered, including the minimum number of hair follicles to be transplanted in the wound bed per cm^2 in order to reach an optimal healing response.

Once we understand the exact mechanism behind the role of hair follicles in wound healing and scar remodelling, we can try to mimic the effect of hair follicles on skin repair and regeneration with therapeutic solution. A potential approach involves deciphering the populations of hair follicle cells that contribute to the observed effect and injecting them directly into wounds or scar tissue in order to improve the clinical outcomes. Alternatively, elucidating the paracrine effect of transplanted follicles would open up new avenues for therapeutic discovery, to replicate the combination of required factors that facilitate healing or scar remodelling.

References

1. Brown J, McDowell F (1942) Epithelial healing and the transplantation of skin. Ann Surg 115(6):1166–1181
2. Bishop GH (1945) Regeneration after experimental removal of skin in man. Am J Anat (2):153–181
3. Mimoun M, Chaouat M, Picovski D, Serroussi D, Smarrito S (2006) The scalp is an advantageous donor site for thin-skin grafts: a report on 945 harvested samples. Plast Reconstr Surg 118(2):369–373
4. Weyandt G, Bauer B, Berens N, Hamm H, Broecker E (2009) Split-skin grafting from the scalp: the hidden advantage. Dermatol Surg 35(12):1873–1879
5. Ito M, Liu Y, Yang Z, Nguyen J, Liang F, Morris RJ et al (2005) Stem cells in the hair follicle bulge contribute to wound repair but not to homeostasis of the epidermis. Nat Med 11(12):1351–1354
6. Nuutila K (2019) Hair follicle transplantation for wound repair. Adv Wound Care 10(3):153–163
7. Zakine G, Mimoun M, Pham J, Chaouat M (2012) Reepithelialization from stem cells of hair follicles of dermal graft of the scalp in acute treatment of third-degree burns: first clinical and histologic study. Plast Reconstr Surg 130(1):42e–50e.
8. Navsaria H, Ojeh N, Moiemen N, Griffiths M, Frame J (2004) Reepithelialization of a full-thickness burn from stem cells of hair follicles micrografted into a tissue-engineered dermal template (Integra). Plast Reconstr Surg 113(3):978–981
9. Narushima M, Mihara M, Yamamoto Y, Iida T, Koshima I, Matsumoto D (2011) Hair transplantation for reconstruction of scalp defects using artificial dermis. Dermatol Surg 37(9):1348–1350
10. Orentreich N (1959) Autografts in alopecias and other selected dermatological conditions. Ann N Y Acad Sci 83(3):463–479
11. Poblet E, Jimenez F, Escario-Travesedo E, Hardman JA, Hernández-Hernández I, Agudo-Mena JL et al (2018) Eccrine sweat glands associate with the human hair follicle within a defined compartment of dermal white adipose tissue. Br J Dermatol 178(5):1163–1172
12. Shook BA, Wasko RR, Mano O, Rutenberg-Schoenberg M, Rudolph MC, Zirak B et al (2020) Dermal adipocyte lipolysis and myofibroblast conversion are required for efficient skin repair. Cell Stem Cell 26(6):880–895
13. Plikus MV, Guerrero-Juarez CF, Ito M, Li YR, Dedhia PH, Zheng Y et al (2017) Regeneration of fat cells from myofibroblasts during wound healing. Science (80)8792:1–12
14. Schmidt BA, Horsley V (2013) Intradermal adipocytes mediate fibroblast recruitment during skin wound healing. Development 140(7):1517–1527
15. Chan CKF, Longaker MT (2017) Fibroblasts become fat to reduce scarring: pathologic scarring could be avoided by manipulating fibroblast plasticity. Science 355(6326):693
16. Rittié L, Sachs D, Orringer J, Voorhees J, Fisher G (2013) Eccrine sweat glands are major contributors to reepithelialization of human wounds. Am J Pathol 182(1):163–171
17. Jimenez F, Garde C, Poblet E, Jimeno B, Ortiz J, Martínez ML et al (2012) A pilot clinical study of hair grafting in chronic leg ulcers. Wound Repair Regen 20(6):806–814
18. Martínez M-L, Escario E, Poblet E, Sanchez D, Buchon F-F, Izeta A et al (2016) Hair follicle-containing punch grafts accelerate chronic ulcer healing: a randomized controlled trial. J Am Acad Dermatol 75(5):1007–1014
19. Yang Z, Liu J, Zhu N, Qi F (2015) Comparison between hair follicles and split-thickness skin grafts in cutaneous wound repair. Int J Clin Exp Med 8(9):15822–15827

20. Wong TW, Yang CC, Hsu CK, Liu CH, Yu-Yun Lee J (2020) Transplantation of autologous single hair units heals chronic wounds in autosomal recessive dystrophic epidermolysis bullosa: a proof-of-concept study. J Tissue Viability 30(1):36–41
21. Fujimura Y, Watanabe M, Ohno K, Kobayashi Y, Takashima S, Nakamura H et al (2021) Hair follicle stem cell progeny heal blisters while pausing skin development. EMBO Rep 22(7):e50882
22. Liu J-Q, Zhao K-B, Feng Z-H, Qi F-Z (2015) Hair follicle units promote re-epithelialization in chronic cutaneous wounds: a clinical case series study. Exp Ther Med 10(1):25
23. Alam M, Cooley J, Plotczyk M, Martínez-Martín MS, Izeta A, Paus R et al (2019) Distinct patterns of hair graft survival after transplantation into 2 nonhealing ulcers. Dermatol Surg 45(4):557–565
24. Aamar E, Avigad Laron E, Asaad W, Harshuk-Shabso S, Enshell-Seijffers D (2021) Hair-follicle mesenchymal stem cell activity during homeostasis and wound healing. J Invest Dermatol 141(12):2797–2807
25. Jahoda CAB, Reynolds AJ (2001) Hair follicle dermal sheath cells: unsung participants in wound healing. Lancet 358:1445–1448
26. Saha D, Thannimangalath S, Budamakuntla L, Loganathan E, Jamora C (2021) Hair follicle grafting therapy promotes re-emergence of critical skin components in chronic nonhealing wounds. JID Innovations 1:100041
27. Heidari F, Yari A, Rasoolijazi H, Soleimani M, Dehpoor A, Sajedi N et al (2016) Bulge hair follicle stem cells accelerate cutaneous wound healing in rats. Wounds 28(4):132–141
28. Langton AK, Herrick SE, Headon DJ (2008) An extended epidermal response heals cutaneous wounds in the absence of a hair follicle stem cell contribution. J Invest Dermatol 128(5):1311–1318
29. Biernaskie J, Paris M, Morozova O, Fagan BM, Marra M, Pevny L et al (2009) SKPs derive from hair follicle precursors and exhibit properties of adult dermal stem cells. Cell Stem Cell 5(6):610–623
30. Gharzi A, Reynolds A, Jahoda C (2003) Plasticity of hair follicle dermal cells in wound healing and induction. Exp Dermatol 12(9):126–136
31. Agabalyan NA, Rosin NL, Rahmani W, Biernaskie J (2017) Hair follicle dermal stem cells and skin-derived precursor cells, exciting tools for endogenous and exogenous therapies. Exp Dermatol 26(6):505–509
32. Ansell DM, Kloepper JE, Thomason HA, Paus R, Hardman MJ (2011) Exploring the "hair growth-wound healing connection": anagen phase promotes wound re-epithelialization. J Invest Dermatol 131(2):518–528
33. Ito M, Yang Z, Andl T, Cui C, Kim N, Millar SE et al (2007) WNT-dependent de novo hair follicle regeneration in adult mouse skin after wounding. Nature 447(7142):316–320
34. Jahoda CAB, Christiano AM (2011) Niche crosstalk: intercellular signals at the hair follicle. Cell 146(5):678–681
35. Plikus MV, Chuong CM (2008) Complex hair cycle domain patterns and regenerative hair waves in living rodents. J Invest Dermatol 128(5):1071–1080
36. Hansen LS, Coggle JE, Wells J, Charles MW (1984) The influence of the hair cycle on the thickness of mouse skin. Anat Rec 210(4):569–573
37. Joost S, Annusver K, Jacob T, Sun X, Dalessandri T, Sivan U et al (2020) The molecular anatomy of mouse skin during hair growth and rest. Cell Stem Cell 26(3):441–457
38. Chase HB, Montagna W, Malone JD (1953) Changes in the skin in relation to the hair growth cycle. Anat Rec 116(1):75–81
39. Moffat GH (1968) The growth of hair follicles and its relation to the adjacent dermal structures. J Anat 102:527–540
40. Durward A, Rudall KM (1958) The vascularity and patterns of growth of hair follicles. Biol Hair Growth 189–218
41. Mecklenburg L, Tobin DJ, Müller-Röver S, Handjiski B, Wendt G, Peters EMJ et al (2000) Active hair growth (anagen) is associated with angiogenesis. J Invest Dermatol 114(5):909–916

42. Ellis RA, Moretti G (1959) Vascular patterns associated with catagen hair follicles in the human scalp. Ann N Y Acad Sci 2125
43. Festa E, Fretz J, Berry R, Schmidt B, Rodeheffer M, Horowitz M et al (2011) Adipocyte lineage cells contribute to the skin stem cell niche to drive hair cycling. Cell 146(5):761–771
44. Brownell I, Guevara E, Bai C, Loomis C, Joyner A (2011) Nerve-derived sonic hedgehog defines a niche for hair follicle stem cells capable of becoming epidermal stem cells. Cell Stem Cell 8(5):552–565
45. Yano K, Brown LF, Detmar M (2001) Control of hair growth and follicle size by VEGF-mediated angiogenesis. J Clin Invest 107(4):409–417
46. Kozlowska U, Blume-Peytavi U, Kodelja V, Sommer C, Goerdt S, Majewski S et al (1988) Expression of vascular endothelial growth factor (VEGF) in various compartments of the human hair follicle. Arch Dermatol Res 290(12):661–668
47. Lachgar S, Moukadiri H, Jonca F, Charveron M, Bouhaddioui N, Gall Y et al (1996) Vascular endothelial growth factor is an autocrine growth factor for hair dermal papilla cells. J Invest Dermatol 106(1):17–23
48. Bassino E, Gasparri F, Giannini V, Munaron L (2015) Paracrine crosstalk between human hair follicle dermal papilla cells and microvascular endothelial cells. Exp Dermatol 24(5):388–390
49. Fox JD, Baquerizo-Nole KL, Van Driessche F, Yim E, Nusbaum B, Jimenez F, Kirsner RS (2016) Optimizing skin grafting using hair-derived skin grafts: the healing response of hair follicle pluripotent stem cells. Wounds 28:109–111
50. Budamakuntla L, Loganathan E, Sarvajnamurthy S, Nataraj HO (2017) Follicular unit grafting in chronic nonhealing leg ulcers: a clinical study. J Cutan Aesthet Surg 10:200–206

Index

A
Aderans Research Institute, 31
Adipocyte derived stem cells, 82
Adipose cells (adipocytes), 44, 49
Adipose-derived exosomes, 211, 213, 229
Adipose tissue, dWAT, 258, 260
Alpha SMA, 67, 93, 96, 97, 99, 102
Anagen follicles, 4, 13, 14, 16
Androgenetic alopecia (AGA), 3, 4, 15–18, 26, 28, 29, 31–33, 63, 64, 211, 213, 219, 220, 222
Androgen receptors, 26, 43, 66, 67, 80
Angiopoietin-like protein 7 (ANGPTL7), 144, 145
Appendage formation, 7

B
Balding, 64, 80
Beta catenin, 209, 210, 212
Bioengineering, 255, 256, 259, 260, 264, 270
Biomaterials, 255, 260–262, 264, 267, 269, 270
Bioprinting, 264, 266–269
Blood vessels, 39, 40, 41, 46, 47
BMP pathway, 43
Bone morphogenetic protein (BMP), 68–72, 78, 83, 111, 113, 118, 121, 122, 241, 243, 248
Bulge, 39, 42–44, 46, 47
Bulge stem cells, 42, 46

C
Cell based therapy, 98, 100
Cell culture, 157, 159, 160, 162–165, 167, 169, 171, 173
Cell-free therapy, 224
Cell signaling, 167, 210, 248
Chronic ulcers, 291–293, 296, 301
Clinical trials, 205, 211, 213, 214, 219, 222, 224, 229
Conditioned medium, 219, 222, 224
Cultured DP cells, 67–69, 71, 72, 74, 78, 79

D
Dermal adipose, 39, 44
Dermal condensate, 4, 7, 61–63
Dermal fibroblasts, 92, 93, 95, 97, 98
Dermal inductivity, 59, 60, 68
Dermal papilla, 39, 41–43, 49, 59–68, 77–85, 92–99, 155, 157–159, 163–168, 170–175, 178, 179, 181–183
Dermal papilla (DP) cells, 256, 257, 259–265, 267–269
Dermal papillae, 6, 12, 15, 18
Dermal sheath, 4, 5, 7, 14, 15, 17, 39, 43, 49, 63–65, 81, 91–100, 102
Dermal sheath stem cells, 17
Dermal white adipose tissue (dWAT), 107–112, 114–124
Dihydrotestosterone (DHT), 26
DP inductivity, 158, 179, 181

E
Embryonic hair follicle, 277, 278, 281, 283, 285, 286

Epithelial-mesenchymal interactions, 125, 257, 260–262, 264, 269
Exosomes, 39, 40, 48, 49, 205–208, 210, 211, 213
Extracellular matrix (ECM), 255, 257, 260–262, 264, 267, 269
Extracellular vesicles, 39, 48, 205, 207, 208, 214
Ex vivo models, 155, 156, 162, 163
Ex vivo organ culture, 180, 181

F
FGF20, 61, 62
FGF7, 14
Fibroblast, 240–242, 278, 280–283, 285–288
Fibroblast growth factor (FGF), 68, 113, 118, 241
Fibrosis, 278, 281, 287
Finasteride, 28
5 alpha reductase, 41, 43, 64

H
Hair cycle, 26, 27, 32, 33, 39, 63, 80, 111–113, 116, 118, 123, 175, 181, 183, 300
Hair development, 41
Hair follicle anatomy, 108
Hair follicle cycle, 4, 16
Hair follicle dermal stem cells (HFDSCs), 62, 64, 65, 68, 81, 96, 97, 99, 102
Hair follicle development, 4, 6, 237, 239, 242, 244
Hair follicle development or hair follicle morphogenesis, 68
Hair follicle germ (HFG), 262, 265
Hair follicle mesenchyme, 3, 5, 7, 18
Hair follicle miniaturisation, 4
Hair follicle neogenesis, 13
Hair follicle regeneration, 28, 39, 155, 237, 240
Hair follicle rejuvenation, 28
Hair follicle stem cells (HFSCs), 39, 40, 42–48, 135, 141
Hair growth, 205, 206, 210, 211, 213
Hair miniaturization, 60, 63, 64, 68
Hair neogenesis, 258
Hair organoids, 237, 247
Hair regeneration, 209, 211
Hair transplantation, 65, 292–295, 300
Hair transplant surgery, 28, 29
Hedgehog, 110, 113, 118, 119

HF stem cell, 113, 118, 259, 260

I
Immune cells, 39–41, 45–47, 49
Induced dermal papilla substituting cells, 241, 242
Induced pluripotent stem cells (iPSCs), 237, 238, 241, 242, 248, 250, 251
Intercytex, 29–33
In vitro models, 170, 184
In vitro organ culture, 184
IPSC HF models, 160, 164, 173

L
Lactobacillus-derived EVs, 210
Laser ablation, 266, 268
Lymphatics, 39–41, 46
Lymphatic vessels, 135–146

M
Macrophage derived extracellular vesicles (MAC-EVs), 209
Macrophages, 280, 281, 288
Mast cells, 39, 45, 47, 48
Mechanic transduction, 277
Mesenchymal stem cell exosomes, 219
Mesenchymal stem cells derived extracellular vesicles, 170
Microfabrication, 255, 257, 264, 265
Micro moulding, 264, 265
Miniaturised follicle, 26, 33
Minoxidil, 28, 31
Myofibroblasts, 97, 98, 280, 282–286

N
Nerves, 39–41, 47

O
Organ culture, 163, 165, 166, 175, 179–181, 183, 184
Organoids, 60, 83, 85, 155, 158, 160, 162–164, 166, 167, 171–175, 181–183
Organotypic substitutes, 263

P
PDGF signaling, 96, 102
Placodes, 4, 7–11, 61

Index

Plant-derived extracellular vesicles (EVs), 205, 210
Platelet-derived growth factor (PDGF), 44, 111, 118, 119
Platelet-rich plasma (PRP), 219, 222, 223
PPAR gamma, 44
Proto-hairs, 161, 173

R
Regeneration, 277–279, 286–288
Replicel Life Sciences, 32
Reprogramming cells, 82

S
Scaffolds, 59, 60, 68, 77–80, 84
Secondary germ, 39, 42, 43
Secretome, 219, 220, 224, 225, 227
Shiseido, 32
Skin equivalents, 155, 161, 163, 165, 174, 175, 184
Skin remodeling, 291, 292
SKP (skin derived precursor) cells, 96, 97
Soft lithography, 265
Sonic Hedgehog (Shh), 277, 279, 285
Sox2, 13, 97
Spheroid cultures, 31, 43
Stem cell conditioned media, 205, 211, 219
Stem cell-derived EVs, 211
Stem cell therapy, 219, 220, 222
Stromal vascular fraction cells (SVFs), 219, 223, 224

T
T cells, 39, 46
Telogen follicles, 4, 14, 16
Terminal Hypertrichosis, 12
3D culture, 75, 77, 78, 80, 85, 162, 174, 175, 237, 240
3D dermal papilla culture, 77–80, 85
3D organoids, 246, 249
3D printed moulds, 79
3D spheroids, 159, 163, 171
Tissue engineering, 59, 60, 66, 68, 79, 83
Transient amplifying cells (TAC), 61–63

V
Vasculature, 270
Vibrissa (whisker) follicles, 6

W
Wnt, 67, 68, 70–77, 80, 113, 118–120, 123, 124, 241, 242, 244, 277, 279, 283–288
Wnt/β-catenin, 43–45, 49
Wnt signalling, 7, 9, 10, 13
Wound healing, 97, 98, 114, 120, 121, 124, 291–296, 298–301
Wound-induced hair neogenesis, 278–283, 285–288

Printed in Great Britain
by Amazon